Ecological and Cultural Ecosystem Services in Coastal Areas

Ecological and Cultural Ecosystem Services in Coastal Areas

Guest Editors

Mita Drius

Maria Laura Carranza

Robert Philipp Wagensommer

MDPI

Basel • Beijing • Wuhan • Barcelona • Belgrade • Novi Sad • Cluj • Manchester

Guest Editors

Mita Drius
Faculty of Education
Free University
of Bozen-Bolzano
Bressanone
Italy

Maria Laura Carranza
Envix-Lab, Department of
Bioscience and Territory
University of Molise
Pesche
Italy

Robert Philipp Wagensommer
Faculty of Education
Free University
of Bozen-Bolzano
Bressanone
Italy

Editorial Office
MDPI AG
Grosspeteranlage 5
4052 Basel, Switzerland

This is a reprint of the Special Issue, published open access by the journal *Land* (ISSN 2073-445X), freely accessible at: https://www.mdpi.com/journal/land/special_issues/L048DAU77J.

For citation purposes, cite each article independently as indicated on the article page online and as indicated below:

Lastname, A.A.; Lastname, B.B. Article Title. *Journal Name* **Year**, *Volume Number*, Page Range.

ISBN 978-3-7258-6039-5 (Hbk)
ISBN 978-3-7258-6040-1 (PDF)
https://doi.org/10.3390/books978-3-7258-6040-1

Contents

About the Editors . . . . **vii**

Mita Drius, Maria Laura Carranza and Robert Philipp Wagensommer
Ecological and Cultural Ecosystem Services in Coastal Areas
Reprinted from: *Land* **2025**, *14*, 1673, https://doi.org/10.3390/land14081673 . . . . **1**

Federica Compagnone, Marco Varricchione, Michele Innangi, Mirko Di Febbraro, Anna Loy, Angela Stanisci, et al.
Coastal Biodiversity Assessment Aided by Citizen Science Volunteers: A Look at the Italian Central Adriatic
Reprinted from: *Land* **2023**, *12*, 2023, https://doi.org/10.3390/land12112023 . . . . **5**

Saverio Sciandrello, Veronica Ranno and Valeria Tomaselli
The Role of Vegetation Monitoring in the Conservation of Coastal Habitats N2000: A Case Study of a Wetland Area in Southeast Sicily (Italy)
Reprinted from: *Land* **2024**, *13*, 62, https://doi.org/10.3390/land13010062 . . . . **25**

Kristina Veidemane, Agnese Reke, Anda Ruskule and Ivo Vinogradovs
Assessment of Coastal Cultural Ecosystem Services and Well-Being for Integrating Stakeholder Values into Coastal Planning
Reprinted from: *Land* **2024**, *13*, 362, https://doi.org/10.3390/land13030362 . . . . **55**

Yuqing Zhao, Zenglin Han, Changren Zhang, Yuqiao Wang, Jingqiu Zhong and Mengfan Gao
Coastal Cultural Ecosystem Services: A Bridge between the Natural Ecosystem and Social Ecosystem for Sustainable Development
Reprinted from: *Land* **2024**, *13*, 1352, https://doi.org/10.3390/land13091352 . . . . **89**

Bryanna Norlin, Andrew E. Scholl, Andrea L. Case and Timothy J. Assal
Mapping an Indicator Species of Sea-Level Rise along the Forest–Marsh Ecotone
Reprinted from: *Land* **2024**, *13*, 1551, https://doi.org/10.3390/land13101551 . . . . **111**

Gerald Schernewski, Mara Jekat, Frank Kösters, Thomas Neumann, Swantje Steffen and Miriam von Thenen
Ecosystem Services Supporting Environmental Impact Assessments (EIAs): Assessments of Navigation Waterways Deepening Based on Data, Experts, and a 3D Ecosystem Model
Reprinted from: *Land* **2024**, *13*, 1653, https://doi.org/10.3390/land13101653 . . . . **131**

Zeduo Zou, Xiaodie Yuan, Zhuo Zhang, Xingyan Li and Chunshan Zhou
Spatiotemporal Evolution and Influencing Factors of Coupling and Coordination between the Ecosystem Service Value and Economy in the Pearl River Delta Urban Agglomeration of China
Reprinted from: *Land* **2024**, *13*, 1670, https://doi.org/10.3390/land13101670 . . . . **156**

Hao Li, Yuxin Zhang, Xiaoli Wang, Peng Guo and Kai Liu
Monitoring and Evaluation of Coastal Ecological Carrying Capacity in the Context of Sustainable Development: A Case Study of Shandong Province
Reprinted from: *Land* **2024**, *13*, 1844, https://doi.org/10.3390/land13111844 . . . . **183**

Jian Liang, Hai-Rui Huang, Meng-Yuan Shu and Chae-Woo Ma
Assessing the Impact of Land-Based Anthropogenic Activities on the Macrobenthic Community in the Intertidal Zones of Anmyeon Island, South Korea
Reprinted from: *Land* **2025**, *14*, 62, https://doi.org/10.3390/land14010062 . . . . **202**

About the Editors

Mita Drius

Mita Drius works as a Researcher at the Free University of Bozen-Bolzano, Faculty of Education, in Italy. Her research focuses on landscape ecology, ecosystem services, green infrastructure, and ecology education. Since 2021 Mita Drius has been part of the MedECC (Mediterranean Experts on Climate and environmental Change) international working group.

Maria Laura Carranza

Maria Laura Carranza is a Full Professor in Ecology at the University of Molise in Italy. Her research focuses on the fields of plant ecology, landscape ecology and conservation biology. Her research interest includes quantitative ecology mainly applied to these disciplines.

Robert Philipp Wagensommer

Robert Philipp Wagensommer is an Associate Professor at the Free University of Bozen-Bolzano, Faculty of Education, in Italy. His research focuses on flora, systematics, taxonomy, Red Lists, and the conservation of threatened plant species, especially in the Mediterranean area. He also works in STEM education.

land

Editorial

Ecological and Cultural Ecosystem Services in Coastal Areas

Mita Drius [1,*], Maria Laura Carranza [2,3] and Robert Philipp Wagensommer [1]

1 Faculty of Education, Free University of Bozen-Bolzano, Viale Ratisbona 16, 39042 Bressanone, Italy; robertphilipp.wagensommer@unibz.it
2 EnviX-Lab, Department of Bioscience and Territory, University of Molise, 86090 Pesche, Italy; carranza@unimol.it
3 National Biodiversity Future Center (NBFC), Piazza Marina 61, 90133 Palermo, Italy
* Correspondence: mita.drius@unibz.it

Coastal areas provide various benefits through regulating, provisioning, and cultural ecosystem services, including carbon sequestration, protection from coastal erosion, recreation, and ecotourism [1]. They also support a unique biodiversity, which includes endangered species requiring conservation measures [2,3]. Achieving the balanced and sustainable management of coastal areas necessitates interdisciplinary approaches that integrate ecological, economic, and governance-related aspects. Additionally, incorporating cultural ecosystem service education is essential for favoring positive perceptions and values associated with coastal areas [4].

Science can play a crucial role in sustainable coastal management by bridging the gap between conceptual models and tangible applications [5]. The development of socio-ecological indicators for sustainable coastal tourism [6], the contribution of citizen science in coastal biodiversity monitoring [7], and the ecosystem services assessment for integrated coastal spatial planning [8] are valuable examples of how applied research can offer concrete solutions.

The Special Issue "Ecological and Cultural Ecosystem Services in Coastal Areas", published in the journal *Land*, includes nine contributions (eight research articles and one review) in line with the above-mentioned directions.

A set of three articles focused on plant studies, performed through surveys, cartographic analysis, and remote sensing. The study by Compagnone et al. (List of Contributions) implemented a citizen science approach supported by the iNaturalist project called Wild Coast CASCADE to monitor biodiversity in the central Italian Adriatic coast. Between 2020 and 2023, 3784 records covering 742 species were collected, with 81% meeting the "research grade criteria". The results underline the potential of citizen science for biodiversity conservation and emphasize the importance of public engagement in conservation efforts. Sciandrello et al. (List of Contributions) examined the state of the preservation of coastal wetlands in the Mediterranean area, specifically the "Saline di Priolo" Special Area of Conservation (SAC) in southeastern Sicily (Italy). Their aim was to provide in-depth knowledge on the natural vegetation through phytosociological surveys and on coastal landscape changes over time by means of a cartographic analysis. The authors concluded that, although they could identify three new plant associations, including two in wetlands and one in rocky coast environments, they recorded a significant reduction in wetland habitats over the last 70 years and a strong alteration of the ecological succession of psammophilous-hygrophilous vegetation. Norlin et al. (List of Contributions) determined the current distribution of *Chamaecyparis thyoides* (L.) Britton, Sterns & Poggenb. (Atlantic

White Cedar) in an area experiencing higher-than-average sea level rising (southern New Jersey, USA) by employing a combination of field investigations, aerial photo interpretation, and satellite data to develop a species distribution model. The model provided a transferable framework for mapping the Atlantic White Cedar and highlighted the importance of including remotely sensed data in species distribution models.

Two articles explored the evaluation of coastal ecosystem services. Veidemane et al. (List of Contributions) operationalized the ecosystem services approach for the integrated assessment and mapping of coastal cultural ecosystem services, to provide evidence for sustainable tourism development in the Southwestern Kurzeme coastal area (Latvia). Through an interdisciplinary approach, combining methods from landscape ecology and social sciences, they could identify the main perceived well-being benefits that people gain from coastal areas, highlighting different stakeholder perspectives. Schernewski et al. (List of Contributions) discussed the role of ecosystem services in supporting Environmental Impact Assessments (EIAs) for two navigation waterway-deepening projects: the Warnow Estuary in Germany and the Szczecin Lagoon in Poland. The authors conducted expert- and data-based ecosystem service assessments and performed 3D ecosystem model simulations to evaluate the environmental and societal consequences of waterway deepening. The results show that waterway deepening increased nutrient burial in sediments, but had minor overall impacts on the ecosystems. The expert-based assessments highlighted the relevance of cultural services and suggested that these assessments could complement EIAs by improving stakeholder communication and reducing conflicts.

Two articles employed the coupling and coordination models to assess coastal sustainable development. Zou et al. (List of Contributions) assessed the coupling and coordination between the coastal ecosystem and economy to achieve sustainable development in urban agglomerations, using the Pearl River Delta (China) as a case study. By means of remote sensing and statistical data, the researchers built a balanced development model depicting the ecological dynamics and the socio-economy in the Pearl River Delta region. Li et al. (List of Contributions) calculated the coastal ecological carrying capacity (CECC) of Shandong Province (China) and its significance for sustainable development. Using multi-source data and ecological analysis models (including coupling and coordination models), the authors also identified key factors influencing the CECC, including the per capita GDP, environmental protection investment, and per capita water resources. Concrete measures such as promoting water conservation, upgrading marine industries, and enhancing regional cooperation were offered to improve the CECC.

Liang et al. (List of Contributions) assessed the impact of land-based anthropogenic activities, such as land reclamation and industrialization, on the macrobenthic community in the intertidal zones of Anmyeon Island (South Korea). Remote sensing analyses and macrobenthic sample collection allowed a Remote Sensing Ecological Index, four traditional diversity indices, and two taxonomic distinctness indices to be applied. The results show that the western stations, characterized by sandy beaches, had higher species abundance compared to the eastern stations, which were influenced by shellfish farming and agricultural wastewater.

Finally, in their review, Zhao et al. (List of Contributions) first explored the concept of cultural ecosystem services in coastal areas, emphasizing their role in connecting natural ecosystems with human society and promoting sustainable development. Then, they identified trends in coastal cultural ecosystem services research, noting an increase in publications since 2009, with a significant rise after 2018. They also highlighted the importance of integrating multi-source data, interdisciplinary development, and incorporating cultural ecosystem services into policy making.

The Special Issue "Ecological and Cultural Ecosystem Services in Coastal Areas" offers nine valuable papers that, although different from each other, share the common goal of offering approaches for balancing coastal protection and management through tools and methodologies supported by the ecosystem services framework. The papers are authored by researchers from around the world, highlighting the international scope of contemporary coastal research. They showcase research conducted in China, Germany, Italy, Latvia, Poland, South Korea, and the USA. This Special Issue promotes interdisciplinary research that integrates methods from plant ecology, landscape ecology, and the social sciences. It seeks to inspire researchers to find optimal solutions for mitigating the risks associated with unsustainable coastal land use and to encourage sustainable landscape management in the context of climate change.

Author Contributions: Conceptualization, M.D.; writing—original draft preparation, M.D.; writing—review and editing, M.D., M.L.C. and R.P.W. All authors have read and agreed to the published version of the manuscript.

Acknowledgments: We want to thank all the authors that contributed to this Special Issue, the reviewers for their valuable work, and the academic editors of the journal, who helped us greatly during the editorial process.

Conflicts of Interest: The authors declare no conflicts of interest.

List of Contributions:

1. Compagnone, F.; Varricchione, M.; Innangi, M.; Di Febbraro, M.; Loy, A.; Stanisci, A.; De Francesco, M.C.; Matteucci, G.; Carranza, M.L. Coastal Biodiversity Assessment Aided by Citizen Science Volunteers: A Look at the Italian Central Adriatic. *Land* **2023**, *12*, 2023. https://doi.org/10.3390/land12112023.
2. Sciandrello, S.; Ranno, V.; Tomaselli, V. The Role of Vegetation Monitoring in the Conservation of Coastal Habitats N2000: A Case Study of a Wetland Area in Southeast Sicily (Italy). *Land* **2024**, *13*, 62. https://doi.org/10.3390/land13010062.
3. Norlin, B.; Scholl, A.E.; Case, A.L.; Assal, T.J. Mapping an Indicator Species of Sea-Level Rise along the Forest–Marsh Ecotone. *Land* **2024**, *13*, 1551. https://doi.org/10.3390/land13101551.
4. Veidemane, K.; Reke, A.; Ruskule, A.; Vinogradovs, I. Assessment of Coastal Cultural Ecosystem Services and Well-Being for Integrating Stakeholder Values into Coastal Planning. *Land* **2024**, *13*, 362. https://doi.org/10.3390/land13030362.
5. Schernewski, G.; Jekat, M.; Kösters, F.; Neumann, T.; Steffen, S.; Von Thenen, M. Ecosystem Services Supporting Environmental Impact Assessments (EIAs): Assessments of Navigation Waterways Deepening Based on Data, Experts, and a 3D Ecosystem Model. *Land* **2024**, *13*, 1653. https://doi.org/10.3390/land13101653.
6. Zou, Z.; Yuan, X.; Zhang, Z.; Li, X.; Zhou, C. Spatiotemporal Evolution and Influencing Factors of Coupling and Coordination between the Ecosystem Service Value and Economy in the Pearl River Delta Urban Agglomeration of China. *Land* **2024**, *13*, 1670. https://doi.org/10.3390/land13101670.
7. Li, H.; Zhang, Y.; Wang, X.; Guo, P.; Liu, K. Monitoring and Evaluation of Coastal Ecological Carrying Capacity in the Context of Sustainable Development: A Case Study of Shandong Province. *Land* **2024**, *13*, 1844. https://doi.org/10.3390/land13111844.
8. Liang, J.; Huang, H.-R.; Shu, M.-Y.; Ma, C.-W. Assessing the Impact of Land-Based Anthropogenic Activities on the Macrobenthic Community in the Intertidal Zones of Anmyeon Island, South Korea. *Land* **2025**, *14*, 62. https://doi.org/10.3390/land14010062.
9. Zhao, Y.; Han, Z.; Zhang, C.; Wang, Y.; Zhong, J.; Gao, M. Coastal Cultural Ecosystem Services: A Bridge between the Natural Ecosystem and Social Ecosystem for Sustainable Development. *Land* **2024**, *13*, 1352. https://doi.org/10.3390/land13091352.

References

1. Drius, M.; Jones, L.; Marzialetti, F.; De Francesco, M.C.; Stanisci, A.; Carranza, M.L. Not just a sandy beach. The multi-service value of Mediterranean coastal dunes. *Sci. Total Environ.* **2019**, *668*, 1139–1155. [CrossRef] [PubMed]
2. Sperandii, M.G.; Barták, V.; Carboni, M.; Acosta, A.T.R. Getting the measure of the biodiversity crisis in Mediterranean coastal habitats. *J. Ecol.* **2021**, *109*, 1224–1235. [CrossRef]
3. Turco, A.; Wagensommer, R.P.; Medagli, P.; D'Emerico, S.; Ippolito, F.; Scordella, G.; Albano, A. *Centaurea pumilio* (Asteraceae): Conservation status, threats and population size of a critically endangered species in Italy. *Plants* **2025**, *14*, 1074. [CrossRef] [PubMed]
4. Hutchison, L.; Montagna, P.; Yoskowitz, P.; Scholz, D.; Tunnell, J. Stakeholder Perceptions of Coastal Habitat Ecosystem Services. *Estuaries Coasts* **2015**, *38*, 67–80. [CrossRef]
5. Tintoré, J.; Medina, R.; Gómez-Pujol, L.; Orfila, A.; Vizoso, G. Integrated and interdisciplinary scientific approach to coastal management. *Ocean Coast. Manag.* **2009**, *52*, 493–505. [CrossRef]
6. Drius, M.; Pugnetti, A.; Bongiorni, L. Disentangling Trade-Offs Between the State of Coastal Ecosystems with Human Well-Being and Activities as a Strategy Addressing Sustainable Tourism. In *Human-Nature Interactions*; Misiune, I., Depellegrin, D., Egarter Vigl, L., Eds.; Springer International Publishing: Cham, Switzerland, 2022; pp. 25–36. [CrossRef]
7. Hart; Blenkinsopp, C. Using Citizen Science to Collect Coastal Monitoring Data. *J. Coast. Res.* **2020**, *95*, 824. [CrossRef]
8. Sousa, L.P.; Alves, F.L. A model to integrate ecosystem services into spatial planning: Ria de Aveiro coastal lagoon study. *Ocean Coast. Manag.* **2020**, *195*, 105280. [CrossRef]

land

Article

Coastal Biodiversity Assessment Aided by Citizen Science Volunteers: A Look at the Italian Central Adriatic

Federica Compagnone [1,†], Marco Varricchione [1,†], Michele Innangi [1,*], Mirko Di Febbraro [1], Anna Loy [1,2], Angela Stanisci [2,3], Maria Carla de Francesco [3], Giorgio Matteucci [2,4] and Maria Laura Carranza [2,3]

1 EnvixLab, Department of Biosciences and Territory, University of Molise, Contrada Fonte Lappone, I-86090 Pesche, IS, Italy; f.compagnone@studenti.unimol.it (F.C.); marco.varricchione@unimol.it (M.V.); mirko.difebbraro@unimol.it (M.D.F.); a.loy@unimol.it (A.L.)
2 National Biodiversity Future Center (NBFC), I-90133 Palermo, PA, Italy; stanisci@unimol.it (A.S.); giorgio.matteucci@ibe.cnr.it (G.M.); carranza@unimol.it (M.L.C.)
3 EnvixLab, Department of Biosciences and Territory, University of Molise, Via Duca degli Abruzzi, I-86039 Termoli, CB, Italy; maria.defrancesco@unimol.it
4 Institute of BioEconomy, National Research Council of Italy, v. Madonna del Piano 10, I-80056 Sesto Fiorentino, FI, Italy
* Correspondence: michele.innangi@unimol.it; Tel.: +39-0874-404-185
† These authors contributed equally to this work.

Abstract: Coastal ecosystems, encompassing land and marine environments and hosting substantial biodiversity, are among the most threatened worldwide. The European Habitats Directive prioritises coastal habitats and species, requiring legislative, direct protection, monitoring, and informational measures. Accurate habitat and species monitoring is crucial to conservation efforts, yet biodiversity research in complex, ever-changing environments like coastal areas is difficult. Citizen Science may bridge biodiversity assessment and eco-friendly monitoring by incorporating non-scientists into the data collection for scientists and stakeholders. A Citizen Science approach supported by a dedicated iNaturalist project (called Wild Coast CASCADE) was implemented to obtain a complete monitoring framework that includes observations of many taxa in terrestrial, aquatic, and transitional dynamic coastal environments in the Central Italian Adriatic coast. We explored data gathered focusing on the IUCN Red List species, the species and habitats of European conservation concern, and the non-native species. Between 2020 and 2023, we collected 3784 records covering 742 species, with 81% meeting the "research grade criteria", and these were retained for subsequent research. Citizen Science volunteers have collected 291 georeferenced animal records from the global IUCN Red List, 51 plant species from 14 species that are indicators of the presence of habitats of European Conservation Concern, and 44 non-native plants and animals. Our results provide evidence that citizen research projects can effectively assist in monitoring coastal–marine habitats and species. They also underline the potential of Citizen Science for biodiversity conservation and emphasize the importance of public engagement in conservation efforts.

Keywords: cross taxa monitoring; iNaturalist; Habitats Directive; European regulation on alien species; IUCN Red List; Long-Term Ecological Research sites

1. Introduction

Coastal ecosystems are characterised by dynamic transitional zones between terrestrial and marine environments, exhibiting significant levels of specialised biodiversity [1,2]. Not only do they possess significant quantities of unique flora and fauna, but they also offer a multitude of benefits to society and ecosystem services [3,4]. Simultaneously, coastal areas are ranked as one of the most at risk in the world [5,6], with the Mediterranean region being particularly vulnerable [7,8]. To mitigate the ongoing deterioration of these imperilled habitats, they are subject to conservation restrictions that have been imposed at both the global and regional level. The Convention on Biological Diversity [9] underscores the importance of allocating focused attention and financial resources to the preservation of biodiversity in coastal areas. The Member States of the European Union have collectively embraced the Council Directive 92/43/EEC, more often known as the Habitats Directive (HD). This legislation represents an important step forward in the development of a European approach to nature conservation. The HD designates numerous coastal ecosystems as habitats that require conservation efforts. These habitats are specifically mentioned in Annex I of the directive. Additionally, the HD identifies certain coastal species of flora and fauna that are of conservation importance. These species are included in Annexes II, IV, and V of the Habitats Directive, deserving either direct (the species) or indirect (their habitat) protection. Art. 17 of the directive impose all member states to monitor and report every six years on the distribution and conservation status of all species and habitats listed in Annexes I, II, IV and V. Furthermore, it establishes a comprehensive network of biologically significant areas referred to as the Natura 2000 network, aiming to preserve these ecosystems and species in a state of "good conservation status". Each Natura 2000 site is obligated to develop and implement the essential measures and actions required to guarantee the enduring preservation of the designated species and habitats as stipulated by HD. These measures must encompass regulatory, direct protection, monitoring, and information-based strategies. The importance of precise monitoring of habitats and species is widely acknowledged as a crucial element of successful conservation strategies [10,11]. However, conducting periodic biodiversity assessments in complex and ever-changing environments such as coastal transitions can be problematic. This difficulty stems from a variety of factors, including the high costs involved [12,13], poor accessibility to specific sampling locations [14], and the intricate nature of organising field campaigns that require frequent and regular intervals [15].

Citizen Science (CS) is a potential solution for bridging the divide between biodiversity assessment and the implementation of efficient and economical monitoring practises [16]. Citizen Science encompasses the participation of non-professional citizens in many scientific activities, including data collection and analysis, technology development, species occurrence assessment, and the dissemination of these endeavours [16]. The continuous expansion of online platforms and the introduction of novel digital technologies are facilitating a significant increase in the engagement of CS activities [16,17]. Volunteer participation in environmental data collection fosters their connection to research subjects, while also raising public awareness about pressing issues such as biodiversity loss, climate change, environmental contamination, and the spread of invasive species [18]. Furthermore, CS efforts have shown favourable improvements in citizen behaviour, encouraging a more environmentally concerned and scientifically engaged citizenry [19]. At the same time, there is empirical support for the impact of data acquired by CS endeavours in the surveillance of distinct ecological systems, such as forests or savannas [20–22]. CS may also contribute to increasing public awareness in long-term monitoring areas [23,24]. Certain studies have specifically focused on investigating the application of CS in the monitoring of fauna, e.g., [25,26], or flora [27]. Additionally, there are studies that have explored the use of CS for monitoring specific domains, such as water or land [28,29]. CS initiatives are frequently aimed at specific tasks, such as implementing environmental education for youth [30–32], studying the presence of rare or less attractive species [33,34], or even monitoring biological invasions [35,36]. Yet, the studies that simultaneously analyse organisms belonging to distinct kingdoms are still scarce [37–39]. Moreover, the use of CS to support the implementation of a comprehensive observation framework, which includes observations of many taxa in various terrestrial, aquatic, and transitional dynamic systems in coastal regions, has not been attempted thus far, and it is necessary to enhance our comprehension of how CS can function in such a context.

iNaturalist [40] is a social networking platform that was founded in 2008 to facilitate the collaboration of individuals from various scientific backgrounds, including professionals and laypeople, to collectively document and disseminate discoveries related to biodiversity [40]. This platform is well recognised as an efficient networking service for the purpose of observing biodiversity and preserving associated data within the realm of CS activities. iNaturalist relies on a substantial user base of over six million registered participants, serving as a hub for scientists, naturalists, biologists, and citizens alike who are interested in advancing online CS endeavours and facilitating the gathering of biodiversity data [40]. The collection consists of more than 160 million verifiable observations, with a continuous growth rate, and each observation is accompanied by details, including the geographic location and date of observation. The dataset excludes observations of organisms that are held in captivity or those that have been cultivated. Notably, around 100 million of these observations have achieved the designation of "research grade", indicating their credibility and suitability for scientific applications (as of 10 October 2023). The iNaturalist programme has been recognised as a key tool for collecting biodiversity data in multiple nations globally [41]. Moreover, it has gained significant popularity as a social platform for consolidating nature-related data [12]. According to [42], the platform has the capacity to significantly contribute to the implementation of conservation strategies by providing useful data to many stakeholders, including research initiatives, land managers, environmental organisations, and the wider public.

Within this framework, the current study aims to examine the capacity of CS initiatives facilitated by the iNaturalist platform in monitoring biodiversity in coastal–marine regions, with a specific focus on the Central Adriatic Italian coast as a designated experimental site. For this purpose, we established a specialised iNaturalist project known as the "Wild Coast CASCADE" (WCC) and subsequently investigated critical questions pertaining to the preservation of biodiversity within these regions by means of CS. Subsequently, we focused our attention on those data pertaining to taxonomic groups that are of conservation significance. These groups include (a) species that are listed in the IUCN red book and are also present in the annexes of HD, (b) plant species that serve as indicators for the occurrence of habitats that are of conservation concern in Europe (HD), and (c) non-native species. Focusing on these three species groups can meet different needs in conservation biology. Firstly, the updated distribution of vulnerable species might potentially enhance devoted monitoring and conservation efforts. Secondly, investigating the occurrence of plant species that serve as indicators of European habitats may allow us to propose specific management priorities for the analysed coasts. Finally, the inclusion of current and accurate data regarding the occurrence of alien organisms can enhance the effectiveness of the early detection systems for invasive species.

2. Materials and Methods

2.1. Study Area

Biodiversity data were gathered from the coastal areas of the Abruzzo and Molise regions in Central Italy, along the Italian Central Adriatic coast. The study area, spanning a total distance of roughly 166 km and encompassing a landward extent of 300 m and a seaward extent of 500 m, exhibits a Mediterranean climate. This stretch of coast is distinguished by expanses of sandy or gravelly/pebbly beaches that are occasionally interrupted by rocky promontories and interspersed with the mouths of rivers. Previous studies evidenced the high ecological significance of the analysed coasts [43,44]. The southern sector is characterised by seashores mostly composed of gravel and pebbles (Abruzzo), while the southern tract (Molise) is predominantly characterised by low dunes. As far as bathymetry is concerned, the entire study area is characterised by very shallow depths, with the depths sloping gently from the coast towards the open sea [45].

The study area comprises a total of eleven areas under differ types of protection, including 1 marine protected area, 3 regional reserve sites, and 7 Natura 2000 sites (Figure 1), all requiring biodiversity monitoring activities, as mandated national and international regulations. In addition, the area includes two sites of the Long-Term Ecological Research Network (LTER), specifically referred to as "Foce Saccione-Bonifica Ramitelli" (IT20-003-T) and "Foce Trigno-Marina di Petacciato" (IT20-002-T; [46]; https://deims.org/6d7ffd99-40e1-4f0d-ad26-6904581dbe9b (accessed on 1 August 2023)).

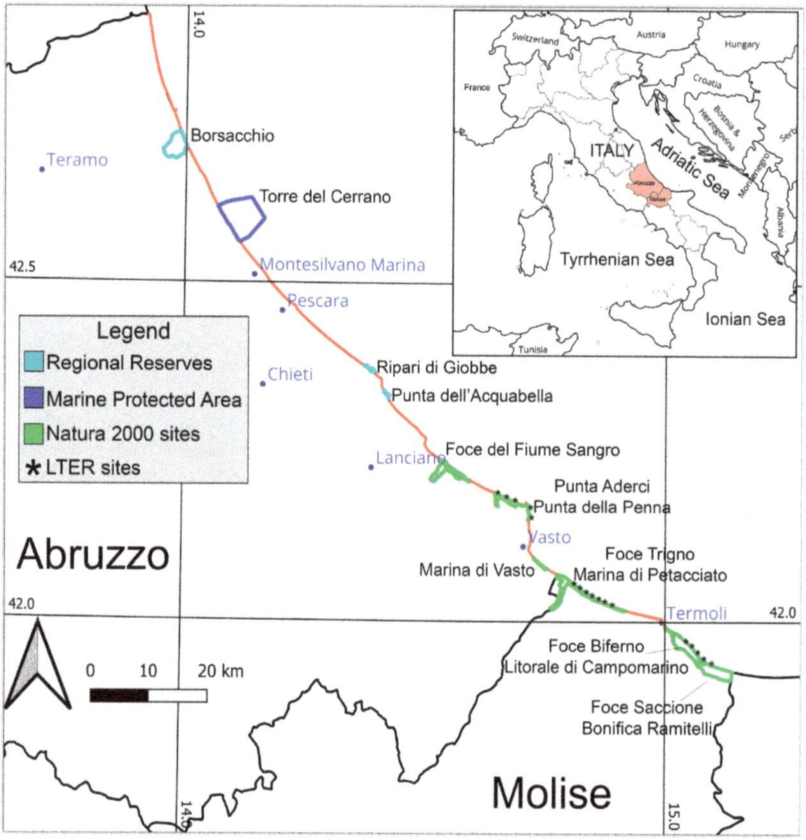

Figure 1. Central Adriatic coast of the Abruzzo and Molise Region (Italy). Protected areas are reported in different colours, with the Natura 2000 sites in green (IT7140107-Lecceta litoranea di Torino di Sangro e foce del Fiume Sangro; IT7140108-Punta Aderci—Punta della Penna; IT7140109-Marina di Vasto; IT7228221-Foce Trigno—Marina di Petacciato; IT7222216-Foce Biferno—Litorale di Campomarino; and IT7222217-Foce Saccione—Bonifica Ramitelli), regional reserves in cyan (Riserva naturale del Borsacchio; Ripari di Giobbe-Foce Fiume Moro; Punta dell'Acquabella), and the Marine Protected Area in blue (Torre del Cerrano, annexing on emerged dunes a Natura 2000 site IT7120215). Asterisks indicate the LTER sites (IT20-002-T in the north and IT20-003-T in the south). The name of the main cities (with more than 30,000 residents) are reported in blue text as the analysed seashore (Abruzzo and Molise Regions) is indicated with a red line.

2.2. Data Collection

Data were collected through CS initiatives, which included active engagement in public outreach and educational activities that began in August 2020 (for details, see Figure 2). We sent invitations to a diverse group of people, requesting their cooperation and participation in the process of collecting data on the presence of wild species in both natural and semi-natural sites. From 2020 to 2022, a total of 12 CS initiatives along the Central Adriatic coast were coordinated and implemented in collaboration with local NGO's, such as "Ambiente Basso Molise" (http://ambientebassomolise.blogspot.com/ (accessed on 1 August 2023)) and the Abruzzo Institution for Protected Areas (IAAP, https://www.iaap.it/

(accessed on 1 August 2023)), along with protected area management bodies such as the Marine Protected Area "Torre del Cerrano" and the Special Area of Conservation Punta Aderci–Punta della Penna (IT7140108). Our initiative was partially supported by the INTERREG Italy–Croatia CASCADE project (https://www.italy-croatia.eu/web/cascade (accessed on 1 August 2023)) and the LIFE17 NAT/IT/000565 CALLIOPE project (https://lifecalliope.eu/ (accessed on 1 August 2023)).

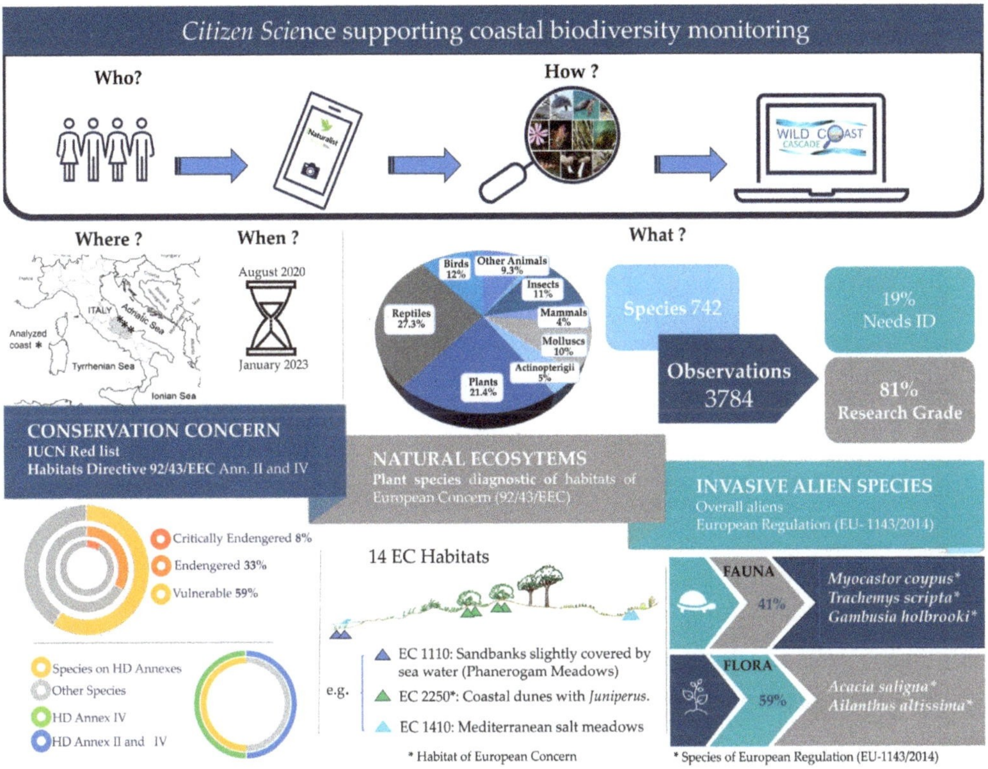

Figure 2. Flowchart depicting a synthetic overview of the adopted approach.

2.3. Data Storage and Organization

The WCC records were obtained from a diverse range of sources, including organised CS events and the independent leisure activities of individuals. These records were collected from the beginning of August 2020 to the end of January 2023 and involved the participation of hundreds of observers (Figure 2). Data were systematically collected within a dedicated iNaturalist project known as the "Wild Coast CASCADE" (WCC) (https://www.inaturalist.org/projects/wild-coast-cascade (accessed on 1 August 2023)). Each individual observation submitted to the WCC project consisted of one or more field photographs, precise geographic coordinates, and a proposed taxonomic identification provided by the iNaturalist criteria (for the full explanation of how iNaturalist works, please refer to https://www.inaturalist.org/pages/help#general3 (accessed on 1 August 2023)). In addition to the iNaturalist platform, we established a specialised Facebook webpage that offered people with low computer experience the chance to submit observations that the managers of the WCC will then upload to iNaturalist on their behalf.

The comprehensive evaluation of the quality of each observation involved an initial classification of the newly submitted entry as "Needs ID", indicating the requirement for verification and potential correction by members of the iNaturalist community who possess expertise in the relevant taxonomic categories.

After achieving a consensus among a majority of specialists (i.e., more than two-thirds) who verified the identification of the organism at the species level, the observation was subsequently elevated to the category of "research grade". This categorization enhances the suitability of the data for scientific utilisation (Figure 2). The data collected consisted of a wide range of living organisms, representing different taxonomic levels. The aforementioned observations were categorised according to the 13 taxonomic groups acknowledged by iNaturalist, namely Actinopterygii, amphibians, arachnids, birds, chromists, fungi and lichens, insects, mammals, molluscs, plants, protozoa, reptiles, and other animals [40].

2.4. Data Extraction and Analysis

We first classified data on iNaturalist taxonomic groups and evaluated the reliability of the observations collected, keeping for further analysis the records that had reached the "research grade".

The number of threatened species (Figure 2) collected within the WCC project was assessed by relying on the global [47] and Italian [48] IUCN Red Lists. Additionally, the conservation concern level of these species in Europe was determined by referring to the annexes of the HD. The IUCN Red List provides a comprehensive and up-to-date assessment of the conservation status and risk of extinction for various biological species. These species are classified into nine categories, and, among these, we focused on critically endangered (CR), endangered (EN), vulnerable (VU), and near threatened (NT). The annexes of the HD consist of a list of species that member states of the European Union have agreed to safeguard and regulate through a range of measures. These measures include the designation of Areas of Special Conservation Interest (Annex II), the establishment of stringent protection rules (Annex IV), and the development of national management plans (Annex V).

As for the analysis on plant species that represent "diagnostic taxa" (Figure 2) according to the HD, we followed the nomenclature recommended in the updated checklist of "Flora d'Italia" [49]. These species are of outmost importance in influencing the composition and operation of these habitats. Because of their importance to the general functionality and survival of the ecosystem, they are also known as "focal species" [50]. The identification of diagnostic species was based on the "Italian Vegetation Prodrome" [51] and the European Union's Habitat Interpretation Manual [52].

Subsequently, we shifted our attention towards an analysis of alien and invasive species (IAS; Figure 2) that are included in the European regulation known as "on the prevention and management of the introduction and spread of invasive alien species" (Regulation (EU) No. 1143/2014, hereafter referred to as the IAS Regulation). The IAS Regulation delineates a catalogue of invasive alien species categorised as being of "Union concern". These species pose a threat to biodiversity and ecosystem services, necessitating coordinated efforts at the European Union level [53]. The determination of the origin of non-native species was also established by Galasso et al. [54] and Loy et al. [55]. In our analysis of plant species, we incorporated the research conducted by Galasso et al. [54] and its subsequent revisions [56,57]. Regarding the fauna, we conducted a thorough examination of the data provided by the Global Biodiversity Information Facility (GBIF) platform (https://www.gbif.org (accessed on 1 August 2023)). Additionally, we conducted a quick analysis of the plant species that are poisonous to humans based on information from the iNaturalist platform and consulting some of the relevant literature (e.g., [58]).

3. Results and Discussion

As of 31 January 2023, the WCC project has documented a total of 3,784 observations by 250 distinct users pertaining to 742 species across diverse ecosystems (terrestrial and aquatic; Table 1). The research has acquired a large number of observations collected systematically through time and across diverse coastal ecosystems (terrestrial and aquatic).

These observations provide compelling evidence of the efficacy of the dedicated CS activities, as well as the valuable contributions made by people during their leisure activities. During the summer season, the coastal regions under study attract a significant number of tourists. However, during the winter and fall seasons, these sites are frequented by schools and local people [4]. Scaling with the study area extent, our collection of data showed figures that are comparable to previous studies conducted in various regions, such as California, Scandinavia, South America, and North Asia [59]. Our findings also align with the research conducted by international institutions, including natural history museums [30].

Table 1. Number of observations that are "research grade" and species per iNaturalist taxa, along with the respective percentages.

iNaturalist Macro-Categories	Number of Observations	% of Observations	Number of Species	% of Species
Actinopterygii	168	5%	53	7%
Amphibians	10	<1%	3	<1%
Arachnids	50	1%	18	3%
Birds	434	12%	82	11%
Chromists	8	<1%	2	<1%
Fungi and lichens	11	<1%	2	<1%
Insects	415	11%	161	22%
Mammals	162	4%	8	1%
Molluscs	358	10%	79	11%
Other animals	350	9%	94	13%
Plants	784	21%	230	31%
Reptiles	1034	27%	10	1%
Total	3784		742	

As for the number of records per taxonomic group, reptiles account for 27% of total records, making it the most observed group, followed by plants (21%), birds (12%), and insects (11%). The less represented categories (with less than 1%), include amphibians, fungi and lichens, and chromists (Table 1). A substantial percentage of species is also present among the category of other animals, i.e., 13%. On iNaturalist, this macro-category collects a series of taxonomic categories, such as Cnidaria, Arthropoda, and Chordata, without any distinction and in which we have identified species of considerable importance.

The high number of reptile observations is mainly attributed to records of *Caretta caretta*. In recent years, this species has been recolonising and returning to historical nesting sites along the coasts of Spain, France, and Western Italy [60], with the WCC project providing evidence of this ongoing process also in the Adriatic.

Plants was the most diverse taxonomic category, with 236 species recorded (31%), followed by insects (163 species; 22%). Birds are another taxonomic group that has a significant number of reports and species (82 species, 11%). In this case, it is highly probable that this pattern is associated with the sea–land ecotone, which provides a diverse range of trophic niches [61].

Conversely, the lowest represented taxonomic groups, each comprising less than 1% of the total, were fungi and lichens (two taxa), chromists (two taxa), and amphibians (three taxa) (Table 1). Such a distribution of reports is congruent with earlier iNaturalist studies that have identified plants and insects as the most often observed groups [59]. Plants and insects in the studied area showed a great species diversity, which is indicative of the extensive ecological mosaic influenced by a pronounced environmental sea–land ecotone. This ecotone, found within relatively narrow coastal strips, facilitates substantial turnover rates of plant species [62] and arthropods [63].

The "research grade" status achievement, indicating species that have undergone thorough review, was observed in 81% of the cases (Figure 3). The "research grade" was reached according to iNaturalist criteria of at least 2/3 of users agreeing on an identification, and, in our case, most of the observations (81%) reached "research grade", with 1 or 2 users in agreement with the submitting user. Less than 1% of the "research grade" observations showed any remaining disagreement among users on the determination of the specimen. This highlights the significant potential of CS activities in involving a high number of well-trained citizens who are able to improve collected data reliability, thus confirming the usefulness of the iNaturalist platform in the context of future research and conservation endeavours, despite some research highlighting limitations and biases in the iNaturalist approach [64–66], including errors or undersampling in less known and charismatic taxa, while well-known taxa or charismatic species might be oversampled. Accordingly, it is worth mentioning that both amphibians and reptiles have attained a "research grade" status of 100% for all their recorded observations (Figure 3). The higher quality of these observations can be primarily attributed to the number of records related to a few easily recognizable species. For example, the abundance of reptile observations at the "research grade" is mainly due to the loggerhead turtle (*Caretta caretta*) and the Italian wall lizard (*Podarcis siculus*) records, while for amphibians, it includes several observations of the Balearic green toad (*Bufotes balearicus*) and the marsh frog (*Pelophylax ridibundus*).

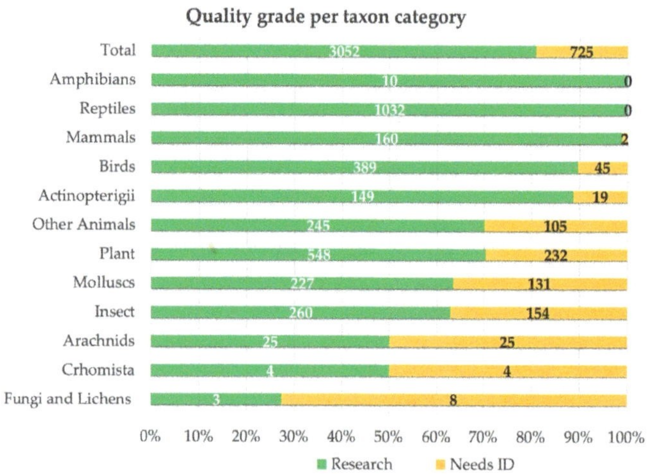

Figure 3. Number and percentage of observations at research level and on need of ID for each iNaturalist macro-category.

Among the other macro-categories, most taxonomic groups reached the "research grade" for at least the 50% of the records. Only fungi and lichens had less than 30% of their observations at the "research grade" (Figure 3). As observed by Barbato et al. [34] in regard to marine environments, most of the reports that did not reach the "research grade" (19% requiring identification) are not readily identifiable and classifiable by photographic evidence. The 30% of plant records that have not achieved the research status can be ascribed to the phenology and seasonal variability of species features (e.g., flowers and fruits), which enable identification only within a short time frame [26].

3.1. Species of Conservation Concern

The WCC has documented a total of 291 observations (Table 2) of species that fall within the global IUCN Red List [47]. These observations pertain to 17 species that have been classified as critically endangered (CR—5.9%), endangered (EN—17.7%), vulnerable (VU—47.0%), or near threatened (NT—29.4%). Among the most threatened taxa, we documented the occurrence of one critically endangered species (Fan Mussel—*Pinna nobilis*) and three endangered (EN) species (namely *Testudo hermanni*, *Cladocora caespitosa*, and *Chelonia mydas*).

The critically endangered Fan Mussel (*Pinna nobilis* Linnaeus, 1758), the largest long-lived endemic bivalve species in the Mediterranean Sea [67], has faced a significant decline since 2016, primarily attributed to infections caused by pathogens. This infection likely heightened the species' susceptibility to changes in water temperature, resulting in sporadic occurrences around the Adriatic seashore [68]. The reported occurrence of *Pinna nobilis* within the WCC project can help enhance scientific research and guide future monitoring efforts. The occurrence areas of *Pinna nobilis* can serve as pilot sites for implementing reinforcement actions aimed at boosting the population of the species [69]. Moreover, they can guide habitat restoration plans and facilitate the implementation of protection actions required by the HD for species listed on Annex IV.

The Hermann's tortoise (*Testudo hermanni* Gmelin, 1789) is an endangered terrestrial Testudinidae that is native to the northern coastal region of the Mediterranean and occurs in Italy with a fragmented distribution [70]. The Hermann's tortoise exhibits a preference for well-preserved coastal dune mosaics throughout the Adriatic coast that consist of many habitats that are of conservation concern [71]. Being highly susceptible to habitat deterioration [72], its occurrence underscores the high conservation status of certain coastal tracts in the Central Adriatic.

Updated data on the presence of Annex II species outside the Natura 2000 network can also help us organise monitoring campaigns targeted at expanding the network. The occurrence data collected within protected areas can aid in the implementation of targeted conservation measures for species listed on Annex IV of the HD.

The WCC also documented the presence of the endangered green sea turtle (*Chelonia mydas* Linnaeus, 1758), which has a preferential distribution on tropical and subtropical waters and a reduced presence in temperate regions [73]. The green turtle population in the Mediterranean Sea is highly susceptible to the effects of climate change and human activities, which have a detrimental impact on all stages of its life cycle, from eggs to adult individuals [74]. The significance of our observations is heightened by recent research that highlights the strategic importance of the Adriatic basin for the conservation of green turtle populations [60]. Furthermore, the new data on green sea turtle's occurrence in the Adriatic coastal area provide valuable insights for planning additional monitoring campaigns and developing specific protection measures, as mandated by the European Union for species listed on Annex IV (HD).

Table 2. Status of threatened species registered on the WCC iNaturalist according to the global and the Italian IUCN Red Lists [48,75] along with their inclusion in HD Annexes II and IV. The number of observations and the iNaturalist taxonomic category are also reported. CR, critically endangered; EN, endangered; VU, vulnerable; NT, near threatened; LC, least concern; NE, not evaluated.

Species	GLOBAL IUCN	Italian IUCN	HD Ann.	No. of Obs.	iNaturalist Category
Pinna nobilis	CR	CR	IV	3	Mollusks
Cladocora caespitosa	EN	LC		9	Other animals
Chelonia mydas	EN	NE	IV	3	Reptiles
Testudo hermanni	EN	EN	II; IV	8	Reptiles
Caretta caretta	VU	EN	II; IV	239	Reptiles
Passer italiae	VU	VU		8	Birds
Streptopelia turtur	VU	LC		5	Birds
Palinurus elephas	VU	NE		2	Other animals
Balaenoptera physalus	VU	EN	IV	1	Mammals
Epinephelus marginatus	VU	EN		2	Fishes
Aythya ferina	VU	VU		1	Birds
Cerambyx cerdo	VU	LC	II; IV	1	Insects
Haematopus ostralegus	NT	VU		3	Birds
Calidris ferruginea	NT	NE		2	Birds
Aythya nyroca	NT	EN		1	Birds
Raja asterias	NT	NE		2	Other animals
Sciaena umbra	NT	VU		1	Fishes

The Mediterranean Pillow Coral (*Cladocora caespitosa* Linnaeus, 1767) is an endangered long-lived reef-building species indigenous to the Mediterranean Sea, where it forms scattered colonies, beds, or vast banks [76,77]. *Cladocora caespitosa* is declining in the Adriatic Sea due to its great sensitivity to pollution [78], invasive algae [79], eutrophication [80], trawling, dredging [81], and climate change. *Cladocora caespitosa* was discovered in the setting of infra-littoral algal biocenosis, providing additional evidence of the presence of the Habitat of European Concern 1170 "Reefs" on several coastal tracts where the habitat had not previously been documented. Moreover, the WCC's new knowledge may enable future scientific campaigns to adequately record and map the Reef EU habitat, updating its distribution range in the Central Adriatic area and extending the execution of necessary conservation actions [77].

The loggerhead sea turtle (*Caretta caretta* Linnaeus, 1758) has a broad distribution in the Mediterranean Sea [11]. The Italian breeding population of *Caretta caretta* plays a crucial role in preserving the genetic diversity of the species [82]. The species is classified in the national Red List as endangered [48], as it suffered a serious decline in Italy that were caused by multiple threats, such as urban expansion, human trampling, and pollution [83]. The recorded occurrences and nesting sites for *Caretta caretta* in the WWC database fill the information gap concerning the Central Adriatic coast, where, until about ten years ago, there was a lack of recorded nesting sites [84]. Furthermore, the current distribution data can be instrumental in guiding scientific research and the development of suitable conservation measures for Italian populations, as advocated for species listed on Annexes II and IV of the Habitats Directive (42/93/EEC).

The investigation of the WCC database, with a focus on endangered species, demonstrates how CS can be used to monitor various types of wildlife in need of conservation. This indicates an essential issue that necessitates additional research [85]. The abundance of georeferenced recordings of species listed on the IUCN Red List and in the HD annexes allows us to generate a comprehensive picture of the priority areas of the Central Adriatic coast that require concerted conservation efforts. Our findings highlight the potential of CS for biodiversity monitoring, a task that is frequently performed for a single taxon. The biodiversity data acquired in these broad areas can play a critical role in the development of integrated coastal management strategies. These strategies should successfully include the conservation needs of each taxon or group of taxa within the geographical framework offered by protected area networks such as the Natura 2000 network and regional/national protected areas.

3.2. EU Habitats

In WCC, we recorded 51 plant species which are diagnostic of 14 habitats of European Conservation Concern (Table 3). Our results reveal the presence of the typical ecosystems of dune zonation from the halo-nitrophilous formations on the drift-line to the fore dune woody formations (e.g., 1210: Annual vegetation of drift lines; 2110: Embryonic shifting dunes; 2120: Shifting dunes with *Ammophila arenaria*; 2130 *: Fixed coastal dunes with herbaceous vegetation; 2230: Malcolmietalia dune grasslands; 2250 *: Coastal dunes with *Juniperus* spp.; 2260: Cisto-Lavanduletalia dune sclerophyllous scrubs; and 2270 *: Wooded dunes with *Pinus pinea* and/or *Pinus pinaster*), some cliff formations (1240: Vegetated sea cliffs of the Mediterranean coasts with endemic *Limonium* spp.), a rich and heterogeneous salt marsh mosaic (1410: Mediterranean salt meadows *Juncetalia maritimi*; 1420: Mediterranean and thermo-Atlantic halophilous scrubs *Sarcocornietea fruticosi*; 3170*: Mediterranean temporary ponds; and 6420: Mediterranean tall humid herb grasslands of the *Molinio-Holoschoenion*), and few river (3250: Constantly flowing Mediterranean rivers with *Glaucium flavum*) and marine habitats (1110: Sandbanks slightly covered by seawater).

The data acquired validates earlier research indicating the great diversity and well-preserved dune mosaic characterising the Molise coast [86], which, in accordance with HD, is in large part included in the Natura 2000 network [4]. It also points to the necessity of restoration efforts along the Abruzzo coastal dunes, where natural ecosystems are fragmented and poorly covered by European protection areas [87]. Conservation actions aimed at reinforcing populations of dune-building perennial grasses (e.g., *Ammophila arenaria* and *Elymus farctus*) and reducing trampling pressure, for example, may play a critical role in dune consolidation, ensuring both biodiversity values and ecological functions [88].

Table 3. List of the recorded diagnostic plant species of habitats of European Conservation Concern, along with the respective habitat code and name [89]. * Priority habitats.

Species	Habitat Code and Name
Cymodocea nodosa (Ucria) Asch.	1110: Sandbanks which are slightly covered by seawater all the time
Cakile maritima Scop. subsp. *Maritima* *Salsola kali* L. *Salsola soda* L.	1210: Annual vegetation of drift lines
Crithmum maritimum L.	1240: Vegetated Sea cliffs of the Mediterranean coasts with endemic *Limonium* spp.
Calystegia soldanella (L.) R. Br. *Elymus farctus* (Viv.) Runemark ex Melderis *Eryngium maritimum* L. *Euphorbia paralias* L. *Euphorbia peplis* L. *Lotus creticus* L. *Medicago marina* L. *Polygonum maritimum* L.	2110: Embryonic shifting dunes
Achillea maritima (L.) Ehrend. et Y.P. Guo *Ammophila arenaria* LK. *Anthemis maritima* L. *Cyperus capitatus* Vand. *Echinophora spinosa* L. *Eryngium maritimum* L. *Lotus creticus* L. *Pancratium maritimum* L. *Sixalix atropurpurea* (L.) Greuter and Burdet	2120: Shifting dunes along the shoreline with *Ammophila arenaria* (white dunes)
Glaucium flavum Crantz	3250: Constantly flowing Mediterranean rivers with *Glaucium flavum*
Euphorbia terracina L. *Verbascum niveum* subsp. *garganicum* (Ten.) Murb.	2130 *: Fixed coastal dunes with herbaceous vegetation (grey dunes)
Artemisia campestris L. *Erodium laciniatum* (Cav.) Willd. *Lagurus ovatus* L. *Malcolmia ramosissima* (Desf.) Thell. *Matthiola incana* (L.) R. Br. *Ononis variegata* L. *Silene colorata* Poir. *Verbascum niveum* subsp. *garganicum* (Ten.) Murb.	2230: *Malcolmietalia* dune grasslands
Asparagus acutifolius L. *Juniperus macrocarpa* Sm. *Myrtus communis* L. *Phillyrea angustifolia* L. *Pistacia lentiscus* L. *Rhamnus alaternus* L. *Smilax aspera* L.	2250 *: Coastal dunes with *Juniperus* spp.
Asparagus acutifolius L. *Cistus creticus* L. *Cistus salviifolius* L. *Phillyrea angustifolia* L. *Pistacia lentiscus* L. *Rhamnus alaternus* L. *Rosmarinus officinalis* L. *Smilax aspera* L.	2260: *Cisto-Lavanduletalia* dune sclerophyllous scrubs
Phillyrea angustifolia L. *Pinus halepensis* Mill. *Pinus pinea* L. *Pistacia lentiscus* L.	2270 *: Wooded dunes with *Pinus pinea* and/or *Pinus pinaster*

Table 3. *Cont.*

Species	Habitat Code and Name
Atriplex prostrata Boucher ex DC. *Blackstonia perfoliata* (L.) Huds. *Juncus acutus* L. *Limbarda crithmoides* (L.) Dumort.	1410: Mediterranean salt meadows (*Juncetalia maritimi*)
Sarcocornia fruticosa (L.) A.J. Scott	1420: Mediterranean and thermo-Atlantic halophilous scrubs (*Sarcocornietea fruticosi*)
Schoenus nigricans L. *Scirpoides holoschoenus* (L.) Soják *Tripidium ravennae* (L.) H. Scholz	6420: Mediterranean tall humid herb grasslands of the Molinio-Holoschoenion

The coastal dunes with *Juniperus* spp. form a priority habitat characterised by a distinctive *Juniperus* scrub which is mainly threatened by urban expansion, tourist pressure [90], and invasive alien plants [91]. These dune juniper thickets, which reach the northern limit of their geographical range in Molise [86], are included in a Natura 2000 site, where they provide suitable habitats for several species of fauna of conservation concern, such as the Hermann tortoise [71].

The recorded salt meadows (*Juncetalia maritimi*) are characteristic of well-preserved back-dune depressions that are periodically flooded by brackish water. Unfortunately, as with most salt meadows, they are impinged by land reclamation, changes in water regimes, urbanization, invasive species, and coastal erosion [92]. The observed habitat, which represents one of the few remaining coastal wetlands in the Mediterranean [93], is encompassed within a Natura 2000 site. These meadows may play a pivotal role in biodiversity conservation, serving as essential habitats for both resident and migratory avian species, as well as various aquatic organisms, like fish and arthropods [94].

We also have detected the presence of *Cymodocea nodosa* (Ucria) Ascherson formations on the Central Adriatic coastal seabed. *Cymodocea* meadows may play a crucial role in supporting fish populations by offering refuge to schools, facilitating the establishment of trophic niches, and serving as nursery sites [95]. Additionally, these meadows have a significant function in safeguarding coastal areas from various environmental impacts [96]. The areas of occurrence, all identified outside the Natura 2000 network, warrant dedicated field campaigns to verify and map the presence of *Cymodocea nodosa* meadows.

The record of species that are diagnostic of habitats of European concern has enabled their identification on the Central Adriatic coast. Despite the partial degradation of coastal areas, these habitats remain valuable components, serving as important seed banks for the recolonization of degraded areas and providing the natural support for their recovery. Our findings highlight the potential of CS for vegetation monitoring, a task that, in line with the Habitat Directive, is typically scheduled every 5 years. However, this time interval may be inadequate for studying highly dynamic ecosystems like dunes [97]. Citizen Science could also help indirectly to monitor the anthropogenic pressures that most affect habitat and species of conservation concern, providing essential information for their proper management.

3.3. Alien Species

We registered 109 records of 44 non-native species (Tables 4 and 5). We specifically reported the presence of 25 alien plants (57%) and 19 species of non-native fauna (43%). It is worth noting that six of the alien plants belong to the taxonomic family *Asteraceae*, which is distinguished by its mostly anemophilous dispersal mechanism, making them particularly effective in the processes of expansion and colonization. The substantial number of *Asteraceae* species aligns well with the overall composition of alien flora in Italy [54] and Europe [98]. Instead, the number of *Poaceae* records appears relatively low in comparison to Italy's alien flora. This is likely due to the difficulty in identifying these species, which poses a challenge for both regular citizens in the field and professionals who rely solely on uploaded photos.

In terms of life forms, we found a significantly higher number of perennial plants, including herbs, shrubs, and trees, than the national average [99]. This is most likely due to their extensive and long-lasting vegetative structures, which allow them to be identified all year. In contrast, we have documented fewer annuals than the national list of non-native plants [54], which can be attributed to the limited season observational window of annuals.

Two of the alien plants, namely *Acacia saligna* and *Ailanthus altissima*, have been identified as invasive species of European Union concern [53]. For these two species, the Italian government is required to undertake monitoring, management, and control measures. *Acacia saligna* is among the most invasive taxa in Italy and Europe, and in the Central Adriatic coast, it mostly colonises back dune habitats, impinging several habitats of conservation concern (e.g., coastal dunes with *Juniperus* spp., dunes with sclerophyllous vegetation of Cisto-Lavanduletalia, and wooded dunes with *Pinus pinea* and/or *Pinus pinaster*). This information is supported by the studies conducted by Marzialetti et al. [100]. *Ailanthus altissima*, commonly known as the Tree of Heaven, is a highly invasive plant that is present on every continent except Antarctica. Its primary distribution in its invasive range includes urban areas, from where it has expanded into neighbouring natural and semi-natural environments [101]. *Ailanthus altissima* provides significant negative impacts on biodiversity by altering the vegetation structure, degrading soil properties, and threatening ecosystem stability [102]. The WCC has identified the existence of several invasive species of European Union concern within the Adriatic coast's network of protected areas. These occurrence data can be used to lead future dedicated field investigations targeted at detecting and mapping alien plants, which are required for the implementation of control or eradication methods outlined in European legislation.

Table 4. List of non-native species, along with the number of reports, taxonomic family, and life form: "Tree" (phanerophytes); "Shrub" (nano-phanerophytes and woody climbers); "Dwarf" (chamaephytes); "Herb" (hemicryptophytes and geophytes); "Annual" (therophytes) (Sensu [103]) and geographic origin. Bold font highlights invasive alien species of European Union concern [53]. Superscript letters: NI, invasive neophytes; NC, casual neophytes; NN, naturalised neophytes; (Sensu [54]); TX, toxic.

Species	No.	Family	Life Form	Origin
Xanthium orientale subsp. *italicum* (Moretti) Greuter^{NI}	11	*Asteraceae*	Annual	North America
***Acacia saligna* (Labill.) H.L. Wendl.**^{NI}	9	*Fabaceae*	Tree	Western Australia
Oxalis pes-caprae L.^{NI}	7	*Oxalidaceae*	Herb	South Africa
Carpobrotus acinaciformis (L.) L. Bolus^{NI}	5	*Aizoaceae*	Dwarf	South Africa
Carpobrotus edulis (L.) N.E. Br.^{NI}	5	*Aizoaceae*	Dwarf	South Africa
Erigeron bonariensis L.^{NI}	3	*Asteraceae*	Annual	South America
Opuntia stricta (Haw.) Haw.^{NI}	3	*Cactaceae*	Shrub	Central America
Robinia pseudoacacia L.^{NI}	3	*Fabaceae*	Tree	North America
Symphyotrichum squamatum (Spreng.) G.L. Nesom^{NI}	3	*Asteraceae*	Annual	South America
***Ailanthus altissima* (Mill.) Swingle**^{NI}	2	*Simaroubaceae*	Tree	China
Datura stramonium L.$^{NI\text{-}TX}$	2	*Solanaceae*	Annual	Central/North Am.
Euphorbia maculata L.^{NI}	2	*Euphorbiaceae*	Annual	North America
Oxalis articulata Savigny^{NI}	2	*Oxalidaceae*	Herb	South America
Agave americana L.^{NI}	1	*Asparagaceae*	Shrub	Central/North Am.
Amorpha fruticosa L.^{NI}	1	*Fabaceae*	Shrub	North America
Erigeron canadensis L.^{NI}	1	*Asteraceae*	Annual	North America
Helianthus annuus L.^{NC}	1	*Asteraceae*	Annual	North America
Mirabilis jalapa L.^{NI}	1	*Nyctaginaceae*	Herb	South America
Nicotiana glauca Graham$^{NI\text{-}TX}$	1	*Solanaceae*	Shrub	South America
Oenothera glazioviana Micheli^{NI}	1	*Onagraceae*	Herb	North America
Opuntia ficus-indica (L.) Mill.^{NI}	1	*Cactaceae*	Shrub	Central America
Paspalum dilatatum Poir.^{NI}	1	*Poaceae*	Herb	South America
Pittosporum tobira (Thunb.) W.T. Aiton^{NN}	1	*Pittosporaceae*	Shrub	Eastern Asia
Senecio inaequidens DC$^{NI\text{-}TX}$	1	*Asteraceae*	Herb	South Africa
Yucca gloriosa L.^{NI}	1	*Asparagaceae*	Shrub	North America

It is also worth noting the existence of some toxic alien plants (e.g., *Datura stramonium*, *Senecio inaequidens*, and *Nicotiana glauca*) [58] that pose an additional threat to human health. Following the CS approach, special campaigns for their identification might be organised, raising people's knowledge of the risks connected with their ingestion, usage, or exposure.

The WCC recorded 40 observations of 19 alien species of fauna (Table 5). *Myocastor coypus* was the most recorded species, followed by *Trachemys scripta*. It is noteworthy that these two species, in addition to *Gambusia holbrooki*, are widely recognised as being highly invasive in Europe, thus warranting their inclusion in the European regulation [53].

Table 5. List of non-native species, along with the relative number of reports (40 for fauna), taxonomic category (taxon), and geographic origin (origin). Bold font highlights invasive alien species of European Union concern [53].

Species	No.	iNaturalist Taxon	Origin
Myocastor coypus (Molina, 1782)	13	Mammals	South America
Trachemys scripta (Schoepff, 1792)	4	Reptiles	North America
Hystrix cristata (Linnaeus, 1758)	2	Mammals	Africa
Magallana gigas (Thunberg, 1793)	2	Molluscs	Asia
Paysandisia archon (Burmeister, 1880)	2	Insects	South America
Rapana venosa (Valenciennes, 1846)	2	Molluscs	Asia
Sceliphron caementarium (Drury, 1773)	2	Insects	North America
Threskiornis aethiopicus (Latham, 1790)	2	Birds	Africa; Asia
Anadara transversa (Say, 1822)	1	Molluscs	North America
Blatta orientalis (Linnaeus, 1758)	1	Insects	Asia; Africa
Cairina moschata (Linnaeus, 1758)	1	Birds	South America
Complex *Pelophylax ridibundus* (Pallas, 1771)	1	Amphibians	Europe; Asia
Gambusia holbrooki (Girard, 1859)	1	Actinopterygii	North America
Harmonia axyridis (Pallas, 1773)	1	Insects	Asia
Isodontia mexicana (de Saussure, 1867)	1	Insects	North America
Leptoglossus occidentalis (Heidemann, 1910)	1	Insects	North America
Rhynchophorus ferrugineus (Olivier, 1790)	1	Insects	Asia
Steatoda nobilis (Thorell, 1875)	1	Arachnids	Macaronesia
Trichopoda pictipennis (Fabricius, 1781)	1	Insects	South America

The coypu (*Myocastor coypus*) is an aquatic rodent that is native to South America and was imported to Italy for fur farming [104]. The escaped individuals tend to establish themselves preferentially along riverbanks and in wetlands, leading to detrimental effects on biological diversity, ecosystems, crops, and irrigation systems. Given the severity of this invasion and the high ecological value of the studied coasts, our data underscore the urgent need to implement control and eradication measures. A management and restriction strategy for the coypu should be implemented following a thorough assessment of its current distribution along the Central Adriatic coast [104].

The pond slider (*Trachemis scripta*) is a tortoise native to North America that was introduced through the pet trade. Mostly sold as small hatchlings, they are often released by owners into natural systems during their adulthood. The pond slider can establish breeding populations and competes with native species for food resources and basking sites, also exerting strong predation pressure on the biodiversity of invaded water bodies [105]. Considering this, a deeper field evaluation aiming at better defining and comprehending the ongoing invasion became critical [106]. Such updated data could be utilised to identify the proper control mechanisms claimed by the European Union and the Italian management plan [106].

4. Conclusions

Citizen Science data from the WCC project showed the promise of this strategy, which directly engages citizens in biodiversity monitoring. Citizen research projects that are supported by the iNaturalist platform have shown that they can monitor habitats and species in complex and ever-changing landscapes, like coastal–marine transitions. Existing and archived data, as well as prospective data from the next phase of the project, may give biodiversity study and management opportunities. The WCC is a pioneering programme that collects multi-taxa and multi-domain data, which may be useful for integrated biodiversity monitoring, especially in long-term observation sites like the LTER network [23] in the examined area.

The WCC submitted new and amended spatial data on various IUCN red book or HD annex species. The simultaneous analysis of the spatial distribution and consistency of several conservation-important species across the coastal gradient can help determine comprehensive habitat restoration strategies and implement specific protective measures claimed by international agreements and policies like the Convention on Biological Diversity [9] and the HD. The high number of species on the WCC, indicating the presence of multiple European habitats of conservation concern (HD), showed a wide range of ecosystems along the Central Adriatic coast, some of which are new to the area. In line with previous research by Drius et al. [4], the WCC's coastal transition mosaic is still largely intact.

The citizen research strategy using iNaturalist has also been helpful in catching alien species across large areas at a cheap expense. Our findings suggested that CS can help implement targeted monitoring and control actions against invasive alien species, as required by European law [53]. The WCC may help identify foreign plant and animal species, thus enabling early detection and eradication. Additionally, it may help monitor invasive taxa's regional expansion and population dynamics to ensure effective control. Lozano et al. [107] and Di Febbraro et al. [108] emphasise that the data derived from CS can improve predictions of the spread of invasive alien species, especially when used in nationwide projects.

Author Contributions: Conceptualisation, M.L.C., F.C., M.V. and M.D.F.; methodology, F.C., M.V. and M.D.F.; data collection: F.C., M.V., A.S. and M.C.d.F.; software, M.D.F.; formal analysis, F.C. and M.V.; investigation, F.C., M.V., M.I., M.D.F., A.L., A.S., G.M. and M.L.C.; resources, M.L.C. and A.S.; data curation, F.C. and M.V.; writing—original draft preparation, F.C., M.V., M.I. and M.L.C.; writing—review and editing, F.C., M.V., M.I., M.D.F., A.L., A.S., G.M. and M.L.C.; visualisation, F.C., M.V. and M.I.; supervision, M.L.C., A.S., A.L., M.D.F. and M.I.; project administration, M.L.C. and A.S. All authors have read and agreed to the published version of the manuscript.

Funding: The research was partially supported by the INTERREG V-A IT-HR CBC Program (CoAStal and marine Waters Integrated Monitoring Systems for Ecosystems Protection and Management—CASCADE—ID: 10255941) and by LIFE17 NAT/IT CALLIOPE ("Coastal dune hAbitats, subLittoraL sandbanks, marIne reefs: cOnservation, Protection, and thrEats mitigation) grant number 000565.

Data Availability Statement: Data are available on iNaturalist platform (https://www.inaturalist.org/projects/wild-coast-cascade (accessed on 1 August 2023).

Acknowledgments: Thanks to the INPS (National Institute of Social Security) for a PhD scholarship and the LTER network (Long-Term Ecological Research). Thanks to the Italian Ministry of University and Research (MIUR) for supporting a Researcher contract on Green axes. We are thankful to the Basso Molise Association, the Marine Protected Area Torre del Cerrano, the Abruzzo Institute for Protected Areas and Legambiente for their valuable support in involving students, local people, and tourists in Citizen Science activities. Thank you to Denis Carlucci for his collaboration and support. We are also grateful to the reviewers and to the editor for helping us improve this manuscript. Finally, we would like to sincerely thank all the citizens and iNaturalist members who contributed to the success of the project by uploading their observations and/or contributing to the recognition of taxa.

Conflicts of Interest: The authors declare no conflict of interest.

References

1. Walker, S.; Wilson, J.B.; Steel, J.B.; Rapson, G.L.; Smith, B.; King, W.M.; Cottam, Y.H. Properties of ecotones: Evidence from five ecotones objectively determined from a coastal vegetation gradient. *J. Veg. Sci.* **2003**, *14*, 579–590. [CrossRef]
2. Martínez, M.L.; Psuty, N.P.; Lubke, R.A. A Perspective on Coastal Dunes. In *Coastal Dunes*; Martínez, M.L., Psuty, N.P., Eds.; Ecology and Conservation; Springer: Berlin/Heidelberg, Germany, 2004; pp. 3–10. [CrossRef]

3. Carpenter, S.R. *Ecosystems and Human Well-Being: Scenarios*; Findings of the Scenarios Working Group; Island Press: Washington, DC, USA, 2005; Volume 2.
4. Drius, M.; Jones, L.; Marzialetti, F.; De Francesco, M.C.; Stanisci, A.; Carranza, M.L. Not just a sandy beach. The multi-service value of Mediterranean coastal dunes. *Sci. Total. Environ.* **2019**, *668*, 1139–1155. [CrossRef]
5. Schlacher, T.A.; Dugan, J.; Schoeman, D.S.; Lastra, M.; Jones, A.; Scapini, F.; McLachlan, A.; Defeo, O. Sandy beaches at the brink. *Divers. Distrib.* **2007**, *13*, 556–560. [CrossRef]
6. Defeo, O.; McLachlan, A.; Schoeman, D.S.; Schlacher, T.A.; Dugan, J.; Jones, A.; Lastra, M.; Scapini, F. Threats to sandy beach ecosystems: A review. *Estuar. Coast. Shelf Sci.* **2009**, *81*, 1–12. [CrossRef]
7. Hesp, P.A.; Martínez, M.L. Disturbance Processes and Dynamics in Coastal Dunes. In *Disturbance Ecology: The Process and the Response*; Johnson, E.A., Miyanishi, K., Eds.; Elsevier: Amsterdam, The Netherlands, 2007; pp. 215–247. [CrossRef]
8. Malavasi, M.; Bartak, V.; Carranza, M.L.; Simova, P.; Acosta, A.T.R. Landscape pattern and plant biodiversity in Mediterranean coastal dune ecosystems: Do habitat loss and fragmentation really matter? *J. Biogeogr.* **2018**, *45*, 1367–1377. [CrossRef]
9. CBD 1992. Convention on Biological Diversity. Available online: http://www.cbd.int/doc/legal/cbd-en.pdf (accessed on 1 August 2023).
10. Gigante, D.; Acosta, A.T.R.; Agrillo, E.; Armiraglio, S.; Assini, S.; Attorre, F.; Bagella, S.; Buffa, G.; Casella, L.; Giancola, C.; et al. Habitat conservation in Italy: The state of the art in the light of the first European Red List of Terrestrial and Freshwater Habitats. *Rend. Lincei. Sci. Fis. E Nat.* **2018**, *29*, 251–265. [CrossRef]
11. La Mesa, G.; Paglialonga, A.; Tunesi, L. *Manuali per il Monitoraggio di Specie e Habitat di Interesse Comunitario (Direttiva 92/43/CEE e Direttiva 09/147/CE) in Italia: Ambiente Marino*; ISPRA, Serie Manuali e Linee Guida: Rome, Italy, 2019.
12. Goldsmith, G.R. The field guide, rebooted Identifying species in the field? There's an app for that. *Science* **2015**, *349*, 594. [CrossRef]
13. Ellwood, E.R.; Crimmins, T.M.; Miller-Rushing, A.J. Citizen science and conservation: Recommendations for a rapidly moving field. *Biol. Conserv.* **2017**, *208*, 1–4. [CrossRef]
14. Maccherini, S.; Bacaro, G.; Tordoni, E.; Bertacchi, A.; Castagnini, P.; Foggi, B.; Gennai, M.; Mugnai, M.; Sarmati, S.; Angiolini, C. Enough Is Enough? Searching for the Optimal Sample Size to Monitor European Habitats: A Case Study from Coastal Sand Dunes. *Diversity* **2020**, *12*, 138. [CrossRef]
15. Marzialetti, F.; Giulio, S.; Malavasi, M.; Sperandii, M.G.; Acosta, A.T.R.; Carranza, M.L. Capturing Coastal Dune Natural Vegetation Types Using a Phenology-Based Mapping Approach: The Potential of Sentinel-2. *Remote Sens.* **2019**, *11*, 1506. [CrossRef]
16. Silvertown, J. A new dawn for citizen science. *Trends Ecol. Evol.* **2009**, *24*, 467–471. [CrossRef] [PubMed]
17. Lemmens, R.; Antoniou, V.; Hummer, P.; Potsiou, C. Citizen Science in the Digital World of Apps. In *The Science of Citizen Science*; Vohland, K., Land-Zandstra, A., Ceccaroni, L., Lemmens, R., Perelló, J., Ponti, M., Samson, R., Wagenknecht, K., Eds.; Springer: Cham, Switzerland, 2021. [CrossRef]
18. Peter, M.; Diekötter, T.; Kremer, K. Participant Outcomes of Biodiversity Citizen Science Projects: A Systematic Literature Review. *Sustainability* **2019**, *11*, 2780. [CrossRef]
19. Peter, M.; Diekötter, T.; Höffler, T.; Kremer, K. Biodiversity citizen science: Outcomes for the participating citizens. *People Nat.* **2021**, *3*, 294–311. [CrossRef]
20. Conrad, C.C.; Hilchey, K.G. A review of citizen science and community-based environmental monitoring: Issues and opportunities. *Environ. Monit. Assess.* **2010**, *176*, 273–291. [CrossRef] [PubMed]
21. Steger, C.; Butt, B.; Hooten, M.B. Safari Science: Assessing the reliability of citizen science data for wildlife surveys. *J. Appl. Ecol.* **2017**, *54*, 2053–2062. [CrossRef]
22. de Groot, M.; Pocock, M.J.O.; Bonte, J.; Fernandez-Conradi, P.; Valdés-Correcher, E. Citizen Science and Monitoring Forest Pests: A Beneficial Alliance? *Curr. For. Rep.* **2022**, *9*, 15–32. [CrossRef]
23. Bergami, C.; Campanaro, A.; Davis, C.; L'astorina, A.; Pugnetti, A.; Oggioni, A. Environmental citizen science practices in the ILTER community: Remarks from a case study at global scale. *Front. Environ. Sci.* **2023**, *11*, 1130020. [CrossRef]
24. L'astorina, A.; Davis, C.; Pugnetti, A.; Campanaro, A.; Oggioni, A.; Bergami, C. Scientists' attitudes about citizen science at Long-Term Ecological Research (LTER) sites. *Front. Environ. Sci.* **2023**, *11*, 1130022. [CrossRef]
25. Mwango'mbe, M.G.; Spilsbury, J.; Trott, S.; Nyunja, J.; Wambiji, N.; Collins, T.; Gomes, I.; Pérez-Jorge, S. Cetacean Research and Citizen Science in Kenya. *Front. Mar. Sci.* **2021**, *8*, 642399. [CrossRef]
26. Rosa, R.M.; Cavallari, D.C.; Salvador, R.B. iNaturalist as a tool in the study of tropical molluscs. *PLoS ONE* **2022**, *17*, e0268048. [CrossRef]
27. Seregin, A.P.; Bochkov, D.A.; Shner, J.V.; Garin, E.V.; Pospelov, I.N.; Prokhorov, V.E.; Golyakov, P.V.; Mayorov, S.R.; Svirin, S.A.; Khimin, A.N.; et al. "Flora of Russia" on iNaturalist: A dataset. *Biodivers. Data J.* **2020**, *8*, e59249. [CrossRef] [PubMed]
28. dos Santos, P.V.R.; Bessa, E. Dolphin conservation can profit from tourism and Citizen science. *Environ. Dev.* **2019**, *32*, 100467. [CrossRef]
29. La Sorte, F.A.; Somveille, M. Survey completeness of a global citizen-science database of bird occurrence. *Ecography* **2020**, *43*, 34–43. [CrossRef]
30. Aristeidou, M.; Herodotou, C.; Ballard, H.L.; Young, A.N.; Miller, A.E.; Higgins, L.; Johnson, R.F. Exploring the participation of young citizen scientists in scientific research: The case of iNaturalist. *PLoS ONE* **2021**, *16*, e0245682. [CrossRef]

31. Echeverria, A.; Ariz, I.; Moreno, J.; Peralta, J.; Gonzalez, E.M. Learning Plant Biodiversity in Nature: The Use of the Citizen–Science Platform iNaturalist as a Collaborative Tool in Secondary Education. *Sustainability* **2021**, *13*, 735. [CrossRef]
32. Forti, L.R. Students as citizen scientists: Project-based learning through the iNaturalist platform could provide useful biodiversity data. *Biodiversity* **2023**, *24*, 76–78. [CrossRef]
33. Wilson, J.S.; Pan, A.D.; General, D.E.M.; Koch, J.B. More eyes on the prize: An observation of a very rare, threatened species of Philippine Bumble bee, *Bombus irisanensis*, on iNaturalist and the importance of citizen science in conservation biology. *J. Insect Conserv.* **2020**, *24*, 727–729. [CrossRef]
34. Barbato, D.; Benocci, A.; Guasconi, M.; Manganelli, G. Light and shade of citizen science for less charismatic invertebrate groups: Quality assessment of iNaturalist nonmarine mollusc observations in central Italy. *J. Molluscan Stud.* **2021**, *87*, eyab033. [CrossRef]
35. Hiller, T.; Haelewaters, D. A case of silent invasion: Citizen science confirms the presence of *Harmonia axyridis* (Coleoptera, Coccinellidae) in Central America. *PLoS ONE* **2019**, *14*, e0220082. [CrossRef]
36. Dimson, M.; Fortini, L.B.; Tingley, M.W.; Gillespie, T.W. Citizen science can complement professional invasive plant surveys and improve estimates of suitable habitat. *Divers. Distrib.* **2023**, *29*, 1141–1156. [CrossRef]
37. Callaghan, C.T.; Ozeroff, I.; Hitchcock, C.; Chandler, M. Capitalizing on opportunistic citizen science data to monitor urban biodiversity: A multi-taxa framework. *Biol. Conserv.* **2020**, *251*, 108753. [CrossRef]
38. Mesaglio, T.; Callaghan, C.T. An overview of the history, current contributions and future outlook of iNaturalist in Australia. *Wildl. Res.* **2021**, *48*, 289–303. [CrossRef]
39. Gorta, S.B.Z.; Callaghan, C.T.; Samonte, F.; Ooi, M.K.J.; Mesaglio, T.; Laffan, S.W.; Cornwell, W.K. Multi-taxon biodiversity responses to the 2019–2020 Australian megafires. *Glob. Chang. Biol.* **2023**. *online ahead of print*. [CrossRef] [PubMed]
40. iNaturalist. Available online: http://www.inaturalist.org (accessed on 30 September 2023).
41. Pimm, S.L.; Jenkins, C.N.; Abell, R.; Brooks, T.M.; Gittleman, J.L.; Joppa, L.N.; Raven, P.H.; Roberts, C.M.; Sexton, J.O. The biodiversity of species and their rates of extinction, distribution, and protection. *Science* **2014**, *344*, 1246752. [CrossRef] [PubMed]
42. Bowser, A.; Wiggins, A.; Shanley, L.; Preece, J.; Henderson, S. Sharing data while protecting privacy in citizen science. *Interactions* **2014**, *21*, 70–73. [CrossRef]
43. Miccadei, E.; Mascioli, F.; Piacentini, T.; Ricci, F. Geomorphological Features of Coastal Dunes along the Central Adriatic Coast (Abruzzo, Italy). *J. Coast. Res.* **2011**, *277*, 1122–1136. [CrossRef]
44. Drius, M.; Carranza, M.L.; Stanisci, A.; Jones, L. The role of Italian coastal dunes as carbon sinks and diversity sources. A multi-service perspective. *Appl. Geogr.* **2016**, *75*, 127–136. [CrossRef]
45. Di Paola, G.; Amodio, A.M.; Dilauro, G.; Rodriguez, G.; Rosskopf, C.M. Shoreline Evolution and Erosion Vulnerability Assessment along the Central Adriatic Coast with the Contribution of UAV Beach Monitoring. *Geosciences* **2022**, *12*, 353. [CrossRef]
46. Capotondi, L.; Ravaioli, M.; Acosta, A.; Chiarini, F.; Lami, A.; Stanisci, A.; Tarozzi, L.; Mazzocchi, M.G. (Eds.) *La Rete Italiana per la Ricerca Ecologica di Lungo Termine. Lo Studio della Biodiversità e dei Cambiamenti*; CNR Edizioni: Roma, Italy, 2011; ISBN 978-88-8080-214-3.
47. IUCN. Red List of Threatened Species. Available online: http://www.iucnredlist.org (accessed on 30 September 2023).
48. Rondinini, C.; Battistoni, A.; Teofili, C. *Lista Rossa IUCN dei Vertebrati Italiani 2022*; Comitato Italiano IUCN e Ministero dell'Ambiente e della Sicurezza Energetica: Roma, Italy, 2022.
49. Pignatti, S.; Guarino, R.; La Rosa, M. *Flora d'Italia*; Edagricole: Bologna, Italy, 2017.
50. Santoro, R.; Carboni, M.; Carranza, M.L.; Acosta, A.T.R. Focal species diversity patterns can provide diagnostic information on plant invasions. *J. Nat. Conserv.* **2012**, *20*, 85–91. [CrossRef]
51. Biondi, E.; Blasi, C.; Allegrezza, M.; Anzellotti, I.; Azzella, M.M.; Carli, E.; Casavecchia, S.; Copiz, R.; Del Vico, E.; Facioni, L.; et al. Plant communities of Italy: The Vegetation Prodrome. *Plant Biosyst. Int. J. Deal. Asp. Plant Biol.* **2014**, *148*, 728–814. [CrossRef]
52. European Commission DG Enviroment. Interpretation Manual of European Union Habitats. [Eur 28.Nature ENV B.3]. 2013. Available online: https://eunis.eea.europa.eu/references/2435 (accessed on 30 September 2023).
53. EUR-Lex Access to European Union Law. Available online: http://data.europa.eu/eli/reg_impl/2022/1203/oj (accessed on 30 September 2023).
54. Galasso, G.; Conti, F.; Peruzzi, L.; Ardenghi, N.M.G.; Banfi, E.; Celesti-Grapow, L.; Albano, A.; Alessandrini, A.; Bacchetta, G.; Ballelli, S.; et al. An updated checklist of the vascular flora alien to Italy. *Plant Biosyst. Int. J. Deal. Asp. Plant Biol.* **2018**, *152*, 556–592. [CrossRef]
55. Loy, A.; Aloise, G.; Ancillotto, L.; Angelici, F.M.; Bertolino, S.; Capizzi, D.; Castiglia, R.; Colangelo, P.; Contoli, L.; Cozzi, B.; et al. Mammals of Italy: An annotated checklist. *Hystrix It. J. Mamm.* **2019**, *30*, 87–106. [CrossRef]
56. Bartolucci, F.; Galasso, G.; Peruzzi, L.; Conti, F. Report 2020 on plant biodiversity in Italy: Native and alien vascular flora. *Nat. Hist. Sci.* **2021**, *8*, 41–54. [CrossRef]
57. Bartolucci, F.; Galasso, G.; Peruzzi, L.; Conti, F. Report 2021 on plant biodiversity in Italy: Native and alien vascular flora. *Nat. Hist. Sci.* **2022**, *10*, 41–50. [CrossRef]
58. Mazza, G.; Tricarico, E. *Invasive Species and Human Health*; CABI: Oxfordshire, UK, 2018.
59. Aristeidou, M.; Herodotou, C.; Ballard, H.L.; Higgins, L.; Johnson, R.F.; Miller, A.E.; Young, A.N.; Robinson, L.D. How Do Young Community and Citizen Science Volunteers Support Scientific Research on Biodiversity? The Case of iNaturalist. *Diversity* **2021**, *13*, 318. [CrossRef] [PubMed]

60. Casale, P.; Broderick, A.C.; Camiñas, J.A.; Cardona, L.; Carreras, C.; Demetropoulos, A.; Fuller, W.J.; Godley, B.J.; Hochscheid, S.; Kaska, Y.; et al. Mediterranean Sea turtles: Current knowledge and priorities for conservation and research. *Endanger. Species Res.* **2018**, *36*, 229–267. [CrossRef]
61. Gregory, R.D.; Eaton, M.A.; Burfield, I.J.; Grice, P.V.; Howard, C.; Klvaňová, A.; Noble, D.; Šilarová, E.; Staneva, A.; Stephens, P.A.; et al. Drivers of the changing abundance of European birds at two spatial scales. *Philos. Trans. R. Soc. B Biol. Sci.* **2023**, *378*, 20220198. [CrossRef]
62. Acosta, A.; Blasi, C.; Carranza, M.L.; Ricotta, C.; Stanisci, A. Quantifying ecological mosaic connectivity and hemeroby with a new topoecological index. *Phytocoenologia* **2003**, *33*, 623–631. [CrossRef]
63. Degli, E.I.; Defeo, O.; Scapini, F. Arthropodofauna richness and abundance across beach–dune systems with contrasting morphodynamics. *Reg. Stud. Mar. Sci.* **2021**, *44*, 101722. [CrossRef]
64. Hochmair, H.H.; Scheffrahn, R.H.; Basille, M.; Boone, M. Evaluating the data quality of iNaturalist termite records. *PLoS ONE* **2020**, *15*, e0226534. [CrossRef]
65. Koo, K.-S.; Oh, J.-M.; Park, S.-J.; Im, J.-Y. Accessing the Accuracy of Citizen Science Data Based on iNaturalist Data. *Diversity* **2022**, *14*, 316. [CrossRef]
66. Campbell, C.J.; Barve, V.; Belitz, M.W.; Doby, R.J.; White, E.; Seltzer, C.; Di Cecco, G.; Hurlbert, A.H.; Guralnick, R. Identifying the identifiers: How iNaturalist facilitates collaborative, research-relevant data generation and why it matters for biodiversity science. *BioScience* **2023**, *73*, 533–541. [CrossRef]
67. *Pinna nobilis*. The IUCN Red List of Threatened Species 2019. Available online: https://www.iucnredlist.org/species/160075998/160081499 (accessed on 11 August 2023).
68. Lattos, A.; Papadopoulos, D.K.; Giantsis, I.A.; Feidantsis, K.; Georgoulis, I.; Karagiannis, D.; Carella, F.; Michaelidis, B. Investigation of the highly endangered Pinna nobilis' mass mortalities: Seasonal and temperature patterns of health status, antioxidant and heat stress responses. *Mar. Environ. Res.* **2023**, *188*, 105977. [CrossRef]
69. Nebot-Colomer, E.; Álvarez, E.; Belando, M.D.; Deudero, S.; Catanese, G.; Bernardeau-Esteller, J.; García-Muñoz, R.; Ramos-Segura, A.; Ruiz, J.M.; Vázquez-Luis, M. Living under threat: Will one of the last *Pinna nobilis* populations be able to survive? *Aquat. Conserv. Mar. Freshw. Ecosyst.* **2022**, *32*, 1–13. [CrossRef]
70. Corti, C.; Bassu, L.; Biaggini, M.; Bressi, N.; Capula, M.; Di Cerbo, A.R.; Vanni, S. Updated distribution of *Testudo hermanni* hermanni in Italy. In Proceedings of the International Workshop on the Management and Restoration of Hermann's Tortoise Habitats and Populations, Soptom, Gonfaron, France, 18–20 September 2013.
71. Berardo, F.; Carranza, M.L.; Frate, L.; Stanisci, A.; Loy, A. Seasonal habitat preference by the flagship species *Testudo hermanni*: Implications for the conservation of coastal dunes. *Comptes Rendus Biol.* **2015**, *338*, 343–350. [CrossRef]
72. Cheylan, M.; Corti, C.; Carpaneto, G.M.; Mazzotti, S.; Zuffi, M.A.L. *Testudo hermanni* Gmelin, 1789. In *Fauna d'Italia*; Capula, M., Corti, C., Luiselli, L., Razzetti, E., Sindaco, R., Eds.; Reptilia, Edizioni Calderini: Bologna, Italy, 2011; Volume XLV, pp. 188–199.
73. Seminoff, J.A.; Allen, C.D.; Balazs, G.H.; Dutton, P.H.; Eguchi, T.; Haas, H.L.; Hargrove, S.A.; Jensen, M.P.; Klemm, D.L.; Lauritsen, A.M.; et al. *Green Turtle (Chelonia mydas) Status Review under the U.S. Endangered Species Act*; NOAA Technical Memorandum, NOAA-NMFS-SWFSC: La Jolla, CA, USA, 2015; pp. 539–571.
74. *Chelonia mydas*. The IUCN Red List of Threatened Species 2004. Available online: https://www.iucnredlist.org/species/pdf/11037468/attachmenten (accessed on 11 August 2023).
75. IUCN Liste Rosse Italiane. Available online: https://www.iucn.it/liste-rosse-italiane.php (accessed on 30 September 2023).
76. Casado de Amezua, P.; Kersting, D.; Linares, C.L.; Bo, M.; Caroselli, E.; Garrabou, J.; Cerrano, C.; Ozalp, B.; Terrón-Sigler, A.; Betti, F. *Cladocora caespitosa*. In *The IUCN Red List of Threatened Species*; International Union for Conservation of Nature: Gland, Switzerland, 2015. [CrossRef]
77. Roveta, C.; Coppari, M.; Calcinai, B.; Di Camillo, C.G.; Marrocco, T.; Mantas, T.P.; Puce, S.; Torsani, F.; Valisano, L.; Cerrano, C. What's the key for success? Translocation, growth and thermal stress mitigation in the Mediterranean coral *Cladocora caespitosa* (Linnaeus, 1767). *Front. Mar. Sci.* **2023**, *10*, 1199048. [CrossRef]
78. El Kateb, A.; Stalder, C.; Neururer, C.; Pisapia, C.; Spezzaferri, S. Correlation between pollution and decline of Scleractinian *Cladocora caespitosa* (Linnaeus, 1758) in the Gulf of Gabes. *Heliyon* **2016**, *2*, e00195. [CrossRef] [PubMed]
79. Kersting, D.K.; Cebrian, E.; Casado, C.; Teixidó, N.; Garrabou, J.; Linares, C. Experimental evidence of the synergistic effects of warming and invasive algae on a temperate reef-builder coral. *Sci. Rep.* **2015**, *5*, 18635. [CrossRef]
80. Kružić, P.; Požar-Domac, A. Impact of tuna farming on the banks of the coral *Cladocora caespitosa* in the Adriatic Sea. *Coral Reefs* **2007**, *26*, 665. [CrossRef]
81. *Cladocora caespitosa*. The IUCN Red List of Threatened Species 2015. Available online: https://www.iucnredlist.org/species/133142/165739749 (accessed on 11 August 2023).
82. Garofalo, L.; Mastrogiacomo, A.; Casale, P.; Carlini, R.; Eleni, C.; Freggi, D.; Gelli, D.; Knittweis, L.; Mifsud, C.; Mingozzi, T.; et al. Genetic characterization of central Mediterranean stocks of the loggerhead turtle (*Caretta caretta*) using mitochondrial and nuclear markers, and conservation implications. *Aquat. Conserv. Mar. Freshw. Ecosyst.* **2013**, *23*, 868–884. [CrossRef]
83. Giacoma, C.; Balletto, E.; Bentivegna, F.; Guarino, F.M.; Hochscheid, S.; Maio, N.; Mingozzi, A.T.; Piovano & Scaravelli, D. Caretta caretta (Linnaeus, 1758). In *Fauna d'Italia*; Corti, C., Capula, M., Luiselli, L., Razzetti, E., Sindaco, R., Eds.; Reptilia: Edizioni Calderini: Bologna, Italy, 2011; Volume XLV.

84. Di Renzo, L.; Mascilongo, G.; Berti, M.; Bogdanović, T.; Listeš, E.; Brkljača, M.; Notarstefano, V.; Gioacchini, G.; Giorgini, E.; Olivieri, V.; et al. Potential Impact of Microplastics and Additives on the Health Status of Loggerhead Turtles (Caretta caretta) Stranded along the Central Adriatic Coast. *Water Air Soil Pollut.* **2021**, *232*, 98. [CrossRef]
85. Fontaine, A.; Simard, A.; Brunet, N.; Elliott, K.H. Scientific contributions of citizen science applied to rare or threatened animals. *Conserv. Biol.* **2022**, *36*, e13976. [CrossRef] [PubMed]
86. Stanisci, A.; Acosta, A.T.R.; Carranza, M.L.; de Chiro, M.; Del Vecchio, S.; Di Martino, L.; Frattaroli, A.R.; Fusco, S.; Izzi, C.F.; Pirone, G.; et al. EU habitats monitoring along the coastal dunes of the LTER sites of Abruzzo and Molise (Italy). *Plant Sociol.* **2014**, *51*, 51–56. [CrossRef]
87. de Francesco, M.C.; Cerrano, C.; Pica, D.; D'Onofrio, D.; Stanisci, S. Characterization of Teatina Coast Marine Habitats (Central Adriatic Sea) toward an Integrated Coastal Management. *Oceanogr. Fish. Open Access J.* **2017**, *5*, 7–10. [CrossRef]
88. de Francesco, M.C.; Tozzi, F.P.; Buffa, G.; Fantinato, E.; Innangi, M.; Stanisci, A. Identifying Critical Thresholds in the Impacts of Invasive Alien Plants and Dune Paths on Native Coastal Dune Vegetation. *Land* **2022**, *12*, 135. [CrossRef]
89. Biondi, E.; Blasi, C.; Burrascano, S.; Casavecchia, S.; Copiz, R.; Del Vico, E.; Galdenzi, D.; Gigante, D.; Lasen, C.; Spampinato, G.; et al. *Manuale Italiano di Interpretazione degli Habitat della Direttiva 92/43/CEE*; Ministero dell'Ambiente e della Tutela del Territorio e del Mare, Direzione per la Protezione della Natura: Rome, Italy, 2009; pp. 1–16.
90. Carranza, M.L.; Drius, M.; Malavasi, M.; Frate, L.; Stanisci, A.; Acosta, A.T.R. Assessing land take and its effects on dune carbon pools. An insight into the Mediterranean coastline. *Ecol. Indic.* **2018**, *85*, 951–955. [CrossRef]
91. Arianoutsou, M.; Leone, V.; Moya, D.; Lovreglio, R.; Delipetrou, P.; de las Heras, J. Management of threatened, high conservation value, forest hotspots under changing fire regimes. In *Post-Fire Management and Restoration of Southern European Forests, Managing Forest Ecosystems*; Moreira, F., Arianoutsou, M., Corona, P., de las Heras, J., Eds.; Springer: Amsterdam, The Netherlands, 2012; pp. 257–291.
92. Adam, P. Salt marsh restoration. In *Coastal Wetlands*, 2nd ed.; Perillo, G.M.E., Wolanski, E., Cahoon, D.R., Hopkinson, C.S., Eds.; Elsevier: Amsterdam, The Netherlands, 2019; pp. 817–861. [CrossRef]
93. Tozzi, F.P.; Varricchione, M.; de Francesco, M.C.; Carranza, M.L.; Stanisci, A. Vegetation Dynamics on a Restored salt Marsh Mosaic: A Re-Visitation Study in a Coastal Wetland in Central Italy. *Wetlands* **2022**, *42*, 101. [CrossRef]
94. Perennou, C.; Gaget, E.; Galewski, T.; Geijzendorffer, I.; Guelmami, A. Evolution of wetlands in Mediterranean region. In *Water Resources in the Mediterranean Region*; Zribi, M., Brocca, L., Tramblay, Y., Molle, F., Eds.; Elsevier: Amsterdam, The Netherlands, 2018; pp. 297–320.
95. Casas, E.; Martín-García, L.; Otero-Ferrer, F.; Tuya, F.; Haroun, R.; Arbelo, M. Economic mapping and assessment of Cymodocea nodosa meadows as nursery grounds for commercially important fish species. A case study in the Canary Islands. *One Ecosyst.* **2021**, *6*, e70919. [CrossRef]
96. Da Ros, Z.; Corinaldesi, C.; Dell'Anno, A.; Gambi, C.; Torsani, F.; Danovaro, R. Restoration of *Cymodocea nodosa* seagrass meadows: Efficiency and ecological implications. *Restor. Ecol.* **2021**, *29*, e13313. [CrossRef]
97. Angelini, P.; Casella, L.; Grignetti, A.; Genovesi, P. *Manuali per il Monitoraggio di Specie e Habitat di Interesse Comunitario (Direttiva 92/43/CEE) in Italia: Habitat*; ISPRA Serie Manuali e Linee Guida: Rome, Italy, 2016.
98. Fristoe, T.S.; Chytrý, M.; Dawson, W.; Essl, F.; Heleno, R.; Kreft, H.; Maurel, N.; Pergl, J.; Pyšek, P.; Seebens, H.; et al. Dimensions of invasiveness: Links between local abundance, geographic range size, and habitat breadth in Europe's alien and native floras. *Proc. Natl. Acad. Sci. USA* **2021**, *118*, e2021173118. [CrossRef]
99. Lazzaro, L.; Bolpagni, R.; Buffa, G.; Gentili, R.; Lonati, M.; Stinca, A.; Acosta, A.T.R.; Adorni, M.; Aleffi, M.; Allegrezza, M.; et al. Impact of invasive alien plants on native plant communities and Natura 2000 habitats: State of the art, gap analysis and perspectives in Italy. *J. Environ. Manag.* **2020**, *274*, 111140. [CrossRef] [PubMed]
100. Marzialetti, F.; Frate, L.; De Simone, W.; Frattaroli, A.R.; Acosta, A.T.R.; Carranza, M.L. Unmanned Aerial Vehicle (UAV)-Based Mapping of *Acacia saligna* Invasion in the Mediterranean Coast. *Remote Sens.* **2021**, *13*, 3361. [CrossRef]
101. Espenschied-Reilly, A.L.; Runkle, J.R. Distribution and changes in abundance of *Ailanthus altissima* (Miller) Swingle in a Southwest Ohio Woodlot. *Ohio J. Sci.* **2008**, *108*, 16–22.
102. Castro-Díez, P.; Fierro-Brunnenmeister, N.; González-Muñoz, N.; Gallardo, A. Effects of exotic and native tree leaf litter on soil properties of two contrasting sites in the Iberian Peninsula. *Plant Soil* **2011**, *350*, 179–191. [CrossRef]
103. Guarino, R.; Chytrý, M.; Attorre, F.; Landucci, F.; Marcenò, C. Alien plant invasions in Mediterranean habitats: An assessment for Sicily. *Biol. Invasions* **2021**, *23*, 3091–3107. [CrossRef]
104. Cocchi, R.; Bertolino, S. *Piano di Gestione Nazionale della Nutria (Myocastor coypus)*; Manuali ISPRA: Rome, Italy, 2020.
105. Standfuss, B.; Lipovšek, G.; Fritz, U.; Vamberger, M. Threat or fiction: Is the pond slider (*Trachemys scripta*) really invasive in Central Europe? A case study from Slovenia. *Conserv. Genet.* **2016**, *17*, 557–563. [CrossRef]
106. Macchi, S.; Scali, S.; Bisi, F.; Martinoli, A.; Alonzi, A.; Carnevali, L. *Piano Nazionale per la Gestione della Testuggine Palustre Americana (Trachemys scripta)*; Manuali ISPRA: Rome, Italy, 2020.

107. Lozano, V.; Di Febbraro, M.; Brundu, G.; Carranza, M.L.; Alessandrini, A.; Ardenghi, N.M.G.; Barni, E.; Bedini, G.; Celesti-Grapow, L.; Cianfaglione, K.; et al. Plant invasion risk inside and outside protected areas: Propagule pressure, abiotic and biotic factors definitively matter. *Sci. Total. Environ.* **2023**, *877*, 162993. [CrossRef]
108. Di Febbraro, M.; Bosso, L.; Fasola, M.; Santicchia, F.; Aloise, G.; Lioy, S.; Tricarico, E.; Ruggieri, L.; Bovero, S.; Mori, E.; et al. Different facets of the same niche: Integrating citizen science and scientific survey data to predict biological invasion risk under multiple global change drivers. *Glob. Chang. Biol.* **2023**, *29*, 5509–5523. [CrossRef]

land

MDPI

Article

The Role of Vegetation Monitoring in the Conservation of Coastal Habitats N2000: A Case Study of a Wetland Area in Southeast Sicily (Italy)

Saverio Sciandrello [1,*], Veronica Ranno [1] and Valeria Tomaselli [2]

[1] Department of Biological, Geological and Environmental Sciences, University of Catania, Via A. Longo 19, 95125 Catania, Italy; veronica.ranno@phd.unict.it

[2] Department of Biosciences Biotechnologies and Environment, University of Bari "Aldo Moro", Via Orabona 4, 70125 Bari, Italy; valeria.tomaselli@uniba.it

* Correspondence: s.sciandrello@unict.it

Abstract: The coastal wetlands are among the most vulnerable and threatened environments of the Mediterranean area. Targeted actions for their conservation require in-depth knowledge of current and past natural vegetation. In this paper, we surveyed the vegetation composition and the spatio-temporal changes of a coastal wetland area in southeastern Sicily ("Saline di Priolo" SAC). Based on 128 phytosociological surveys and several plant collections, a total of 304 taxa, 28 plant communities, and 16 habitats have been identified. Furthermore, three new plant associations were described, including two in wetland and one in rocky coast environments. For the classification of plant communities and habitats, a hierarchical clustering was performed by using Euclidean coefficient and beta-flexible algorithm. The life form spectrum of the current flora highlights the dominance of therophytes and hemicryptophytes. The Mediterranean species are largely prevailing with 123 taxa. The cartographic analysis performed with ArcGis 10.3 shows a radical reduction in the wetland habitats in the last 70 years, and a strong alteration of the ecological succession of the psammophilous-hygrophilous vegetation. Moreover, landscape configuration of the coastal vegetation and habitat types was well highlighted by a set of specific landscape metrics. In particular, our outcomes identify three habitats (2110, 2210, and 5220* EU code) with bad conservation status, among which we identified one of priority conservation (*Zyziphus* arborescent matorral) that requires urgent restoration measures.

Keywords: conservation; diachronic analysis; landscape fragmentation; plant diversity

1. Introduction

Mediterranean wetlands and their adjacent areas are extremely important for biodiversity and human wellbeing, but their integrity has often been compromised [1]. Fortunately, in recent times, opinion movements and policy actions aimed at wetland recovery and protection have arisen [2]. Wetlands are essential ecosystems for preserving animal and plant diversity in the Mediterranean region. These biotopes host highly specialized plant communities linked to the substrate type, water quality, flooding time, and bioclimate, in which rare species grow with specific ecophysiological adaptations [3]. These habitats are disappearing due to human pressure, especially when these regions are subject to land reclamation [4]. This is the case of southern Italy, Sicily, and nearby islands, where numerous coastal wetlands have been subjected to severe degradation in recent decades [5,6]. In Sicily, coastal wetlands are located mainly in the west, between Trapani and Mazara del Vallo and in the southeastern sectors of the island, and along the coastal strip between Siracusa and Catania. The knowledge about the coastal saltmarsh vegetation in Sicily was investigated by several phytosociological studies [5,7–12]. In southeastern Sicily, the Special Area of Conservation (SAC) "Saline di Priolo" (ITA090013), despite several environmental changes occurring in the last 70 years, is distinguished by a significant diversity of habitats and plant species. This area includes the Natural Reserve "Saline di Priolo" and comprises different environments, including beaches, cliffs, brackish marshes, grasslands, temporary ponds, etc. The SAC "Saline di Priolo", in addition to an extensive system of coastal wetlands and saltmarshes, also includes the "Magnisi" peninsula, and the archaeological sites known as Thapsos and "Biggemi" lagoon. This area is under considerable anthropogenic pressure from nearby industrial activity; in fact, immediately next to the protected area is the Syracuse petrochemical hub. It is a vast industrialized area (the largest Italian petrochemical hub, and the second in Europe) created in the 1950s which extends over an area of 40 km^2 on the east coast of Sicily, including the territory of the municipalities of Augusta, Priolo Gargallo, and Melilli up to the gates of Syracuse, and exerting a dramatic environmental pressure on the surrounding territories [13]. As regards the flora and vegetation of the Priolo wetland, only a few significant investigations have been carried out [9,12,14–16]. This study is aimed at providing a comprehensive and updated list of the flora and habitats present in the area, also including species of conservational interest and alien species, with new insights for future management activities of the territory. A diachronic approach, using mapping and monitoring of the flora, vegetation, and habitat of the Nature Reserve and SAC "Saline di Priolo" over the last 70 years, has been carried out, as part of the "MEDISWET" project of the MAVA foundation (Action Activity 7.1.18). Moreover, quantifying the landscape's spatial structure provides a better understanding of the ongoing impact on ecological processes. In this framework, landscape metrics (LMs) are an essential tool to analyze the spatial arrangement of the landscape structure over time [17,18]. The application of LM is particularly suitable to coastal areas because these landscapes are prone to rapid transformations [19].

2. Materials and Methods

2.1. Study Area

The study area is located on the coastal side of the Hyblean plateau (southeastern Sicily), which is one of the northernmost promontories of the African plate. The Hyblaean plateau (or Hyblaean–Maltese Plateau) is a crust of the continental type isostatically raised and well-defined on its edges, different from the other ones in Sicily, and extending south to the Maltese Islands, from which it is separated by a continental shelf [20]. From a floristical point of view, the Hyblean territory hosts a lot of peculiar endemic and rare species, such as *Urtica rupestris*, *Zelkova sicula*, *Trachelium lanceolatum*, *Anthemis pignattiorum*, *Limonium pachynense*, *Romulea variicolor*, *Elatine macropoda*, and *Solenopsis laurentia* subsp. *hyblaea* [21,22]. Moreover, it is an area extremely rich in terms of vegetation, hosting a great variety of vegetation types, as well as numerous habitat types according to the Annex I of the 92/43 EEC Directive (or Habitat Directive), including several priority habitat types such as 3170* "Mediterranean Temporary Ponds". The "Saline di Priolo" SAC extends over 252.5 ha and its perimeter falls within the Priolo Gargallo municipal area. Average annual rainfall is around 500 mm and average annual temperatures are around 18 °C [23]. From a bioclimatic point of view, the area under investigation falls within the thermo-Mediterranean zone, with a dry ombrotype [24].

2.2. Phytosociological Data and Taxonomy

The field surveys were carried out mostly from spring 2021 to autumn 2022, according to the Braun–Blanquet phytosociological approach [25]; these relevés were integrated with other data previously collected in 2008, thus achieving a dataset consisting in a total of 128 relevés. Specimens were identified following the second edition of the Flora d'Italia [26–29], while the nomenclature of vascular plants follows Portal of the Flora of Italy and subsequent updates [30,31], and the regional distribution "rarity" follows Giardina et al. [32].

2.3. Statistical Analysis and Classification

For vegetation classification and identification of plant communities, a hierarchical clustering was performed in the program PC-ORD 6 [33], by using Euclidean coefficient and beta-flexible algorithm, according to [34]. Syntaxonomic classification follows Biondi et al. [35] and Mucina et al. [36]. Natural vegetational types were classified into habitat types according to the Italian interpretation manual of habitats [37].

2.4. Landscape and Diachronic Analysis

To evaluate the conservation status of the vegetation and habitat types throughout the last 70 years, a diachronic analysis using aerial orthophotos taken in 1955 and 2015 has been carried out, with the production of vegetation and habitat maps achieved with ArcGis 10.3 software. Finally, in order to analyze and evaluate vegetation mosaics and changes in spatial arrangement of the landscape structure over the considered timespan, a set of landscape metrics (LMs) was selected and applied to the two habitat maps (1955 and 2015). Considering the differences in spatial and thematic resolution between the two years of observation, implying a certain level of uncertainty in identifying the single plant associations in the 1955 orthophotos, the landscape metrics were applied to the habitat maps, which have a thematic resolution less detailed than the phytosociological vegetation map. In selecting the most appropriate LM for the fragmentation assessment in coastal Mediterranean environments, we referred to Cushman et al. [38] and to Tomaselli et al. [19]; a limited set of metrics was selected, specifically CA (Class Area), NumP (Number of Patches), MPS (Mean Patch Size), MPAR (Mean Perimeter–Area Ratio), MSI (Mean Shape Index), and AWMSI (Area-Weighted Mean Shape Index), and the analysis was carried out by using the Patch-Analyst extension of ArcGis.

3. Results and Discussions

3.1. The Vascular Flora: Trait Analysis

A total of 304 taxa were recorded, of which there were 196 for the "Saline di Priolo" and 178 for the Magnisi peninsula (Floristic Appendix A). Overall, 59 families were recorded, with the most represented being *Asteraceae* with 15% (45 sp.), followed by *Poaceae* with 13% (39 sp.) and *Fabaceae* with 12% (38 sp.). The life forms detected show the typical Mediterranean pattern. In fact, the therophytes, with 123 species, represent 40% of the whole flora, followed by the hemicryptophytes with 31% (94 sp.). The percentage of geophytes is also considerable, with 13% (39 sp.), followed by phanerophytes (16 sp.) with 6%, nanophanerophytes (6 sp.) with 2%, and hydrophytes (4 sp.) with 1%. The chorological analysis shows the clear prevalence of the *Mediterranean* chorotype, which represents 42% of the taxa (129 sp.), whereas the *Euro-Mediterranean* (62 sp.) and *Cosmopolitan* (33 sp.) make up, respectively, 20% and 11% of the flora. The *Mediterranean-Turanian* and *Paleotemperate* chorotypes follow (each with 14 sp.) with 15%, while the endemic species (6 sp.) constitute the 2%, which is still relevant considering the limited size of the study area.

3.2. Taxa of Relevant Interest

Field surveys highlighted some rare or endemic species, such as *Limonium syracusanum, Ziziphus lotus, Poterium spinosum, Bulliarda vaillantii, Damasonium bourgaei* (*Magnisi peninsula*), *Teucrium scordium* subsp. *scordioides, Limonium narbonense, Euphorbia hirsuta*, and *Cressa cretica* ("Saline di Priolo") (Figure 1).

The most interesting ones are briefly commented on in the following paragraphs.

Limonium syracusanum Brullo (Plumbaginaceae).

Suffruticose chamaephyte endemic to the Hyblean Ionian coast, between Augusta and Capo Passero (south-eastern Sicily) [39]. The species characterizes the halophilous phytocoenoses of the rocky coasts of the *Crithmo-Limonietea* class, together with *Crithmum maritimum* and *Arthrocaulon meridionale*. It is included in the red list of Italian flora [40] with the status of least concern (LC). The species has been observed along the reefs of the north and northeastern sector of the Magnisi peninsula, near the lighthouse.

Figure 1. Photo plate illustration of some species of the "Saline di Priolo" SAC: (**A**) *Limonium syracusanum;* (**B**) *Poterium spinosum;* (**C**) *Ziziphus lotus;* (**D**) *Damasonium bourgaei;* (**E**) *Euphorbia hirsuta;* (**F**) *Bulliarda vaillantii;* (**G**) *Convolvulus soldanella;* (**H**) *Limonium narbonense.*

Ziziphus lotus (L.) Lam. (Rhamnaceae).

Xerophilous deciduous shrub species with a southern Mediterranean–Saharan range. Reported in the red list of Italian flora [40] with the status of least concern (LC). In Sicily, the species is rather rare and, in addition to the studied site, it was reported only for the western sector in M. Pellegrino and Mazara del Vallo [32]. Recently, La Mantia and Scuderi [16] gave a more detailed distribution of the species, confirming the location close to

"Saline di Priolo". At the regional level, Conti et al. [41] reported this species as vulnerable, while Orsenigo et al. [42] listed it as near threatened (NT). In the Magnisi Peninsula, the species was observed for the first time by Zodda [43]; at present, it is represented by a few individuals, limited to a small area on the limestone plateau.

Poterium spinosum L. (Rosaceae).

Thorny chamaephyte with East Mediterranean distribution. In Italy, the species is present in Apulia, Basilicata, Calabria, Sicily, and Sardinia. In Sicily, the species is localized exclusively in the Hyblaean area, preferring both carbonate and volcanic substrates [44]. Gargano et al. [45] classified the species as endangered at the national level (EN). Recently, Orsenigo et al. [42] confirmed the status as endangered (EN) for the Italian territory. In the study area, the species was sporadically observed along the coast, in the southern part of the peninsula, and in the garrigues dominated by *Thymbra capitata* and *Micromeria graeca*.

Bulliarda vaillantii (Wild.) DC. (Crassulaceae).

Small therophyte with Mediterranean–tropical distribution. It is an amphibious species typical of Mediterranean temporary ponds (Habitat 3170*: Mediterranean temporary ponds). It is a characteristic species of the annual amphibious communities referred to as *Lythro hyssopifoliae-Crassuletum vaillantii* of the *Isoeto-Nanojuncetea* class. In Italy it is present in Sicily, Apulia, Basilicata, Lazio, Tuscany, Liguria, and Sardinia. In the study area, the species is localized in the small temporary ponds in the central part of the Magnisi peninsula.

Damasonium bourgaei Coss. (Alismataceae).

Rooting hydrophyte with an Atlantic–Mediterranean distribution. Annual species typical of Mediterranean temporary ponds, growing on carbonate and volcanic substrate. It characterizes the annual amphibious communities of the *Isoeto-Nanojuncetea* class (Habitat 3170*: Mediterranean temporary ponds). In Italy it is present in Apulia, Basilicata, Sardinia, and Sicily. *D. bourgaei* was reported for many coastal sites of Sicily, but at present, it can be considered quite rare. In the Magnisi peninsula, the species is rather localized.

Teucrium scordium L. subsp. *scordioides* (Schreb.) Arcang. (Lamiaceae).

Scapose hemicryptophyte with a Euro-Caucasian to NW African distribution range. It typically grows in wet meadows, where it forms hygrophilous communities together with *Juncus subulatus*, *Lotus corniculatus* subsp. *preslii*, *Phyla nodiflora*, and *Potentilla reptans*. In Sicily, it is threatened by the reduction in its natural habitat [32]. In the study area "Saline di Priolo", it is very localized.

Limonium narbonense Mill. (Plumbaginaceae).

Rosulated hemicryptophyte with Euro-Mediterranean distribution. Perennial halophilous species typical of the coastal saltmarshes. This species, together with other halophytes, such as *Arthrocaulon meridionale*, *Salicornia perennis* subsp. *alpini*, and *Halimione portulacoides*, characterizes the perennial halophilous vegetation belonging to *Salicornietea fruticosa* class. In the "Saline di Priolo", *L. narbonense* is quite common.

Euphorbia hirsuta L. (Euphorbiaceae).

Rhizomatous geophyte with Mediterranean–Macaronesian distribution. The species mainly grows in wet meadows. In the Saline di Priolo, it thrives together with *Teucrium scordium* subsp. *scordioides*, *Lotus corniculatus* subsp. *preslii*, *Phyla nodiflora*, and *Potentilla reptans*. In the Saline di Priolo, the species is very localized, on the edges of small wet areas in contact with *Tamarix africana*. In the last 50 years, the species has undergone a strong reduction in its range along the Sicilian coast; in fact, today there are few areas where the species is conserved [32].

Cressa cretica L. (Convolvulaceae).

It is a thermo-cosmopolitan halophilous species growing in sandy or muddy saline habitats. In the study area, the species is rare due to reduction in or alteration of its natural habitat (1310 "*Salicornia* and other annuals colonizing mud and sand"). *Cressa cretica*, together with other annual succulent plants with a summer cycle, characterizes the *Cressetum creticae*, halo-subnitrophilous and termophilous vegetation colonizing abandoned fields after farming on clayey and salty soils [5]. According to Oresenigo et al. [42], it is classified as "endangered" (EN) in Italy.

3.3. Plant Community Description

A total of 28 plant communities (seven annuals and twenty-one perennials) belonging to 18 phytosociological classes has been identified by cluster analysis (Appendix B).

3.3.1. Coastal Dune Vegetation (Figure 2; Table S1)

The psammophilous vegetation immediately adjacent the aphytoic belt is characterized by annual herbaceous communities ascribable to the association *Salsolo-Cakiletum maritimae* (habitat 1210 "Annual vegetation of drift lines"). On the embryonic dunes, it is possible to distinguish two different associations. The first one is the perennial herbaceous vegetation of the *Cypero capitati-Agropyretum juncei*, with the dominance of *Thinopyrum junceum* (habitat 2120 "Embryonic shifting dunes"); the second one is the *Sileno coloratae-Ononidetum variegatae*, annual vegetation characterized in particular by *Ononis variegata* and *Silene niceensis* (habitat 2230 "Malcolmietalia dune grasslands"). In the backdune, where the substrate is more stable and richer in organic matter, a chamaephytic vegetation occurs, known as the *Centaureo sphaerocephalae-Ononidetum ramosissimae*. In Sicily, these psammophilous communities are quite widely represented [11,46] despite their bad conservation status.

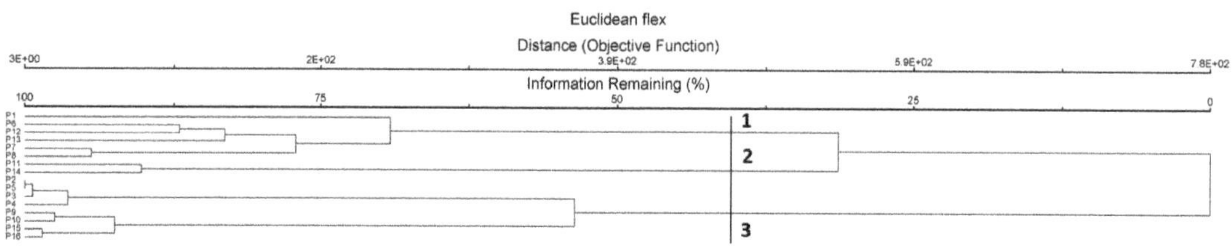

Figure 2. Cluster analysis of coastal dune communities. Plant communities: 1. *Cypero capitati-Agropyretum juncei*; 2. *Centaureo sphaerocephalae-Ononidetum ramosissimae*; 3. *Sileno coloratae-Ononidetum variegatae*.

3.3.2. Rocky Coast Vegetation (Figures 3 and 4; Table S2)

The coastal rocky outcrops of the Manghisi peninsula are colonized by halophytic communities belonging to the classes *Salicornietea fruticosae*, *Crithmo-Staticetea*, and *Saginetea maritimae*. The first belt, closest to the sea and subjected to salt spray, as well as the brackish rocky pools periodically inundated behind the cliffs, are home to a vegetation dominated by *Arthrocaulon meridionale*, a succulent chenopod shrub usually linked to seasonally inundated saltmarshes, and by *Limonium virgatum* and *Limbarda crithmoides*. This vegetation is attributable to the *Limonio virgati-Arthrocnemetum macrostachyi*, an association belonging to the class *Salicornietea fruticosae*, described by Biondi et al. [47] from southern Apulia and already reported for Sicily by Minissale and Sciandrello [11]. Immediately inwards follows the *Limonietum syracusani* (habitat 1240 "Vegetated sea cliffs of the Mediterranean coasts with endemic *Limonium* spp."), a perennial association of the *Crithmo-Staticetea* and dominated by *Limonium syracusanum* and *Crithmum maritimum*, with a few other halophilous/subalophilous species. On corroded surfaces, more or less flat, with thin soil, subjected to disturbance, a prostrate perennial vegetation dominated by *Lotus cytisoides* and *Frankenia hirsuta* develops, taking the place of the *Limonietum syracusani*. The subhalophilous ephemeral communities, characterized by small therophytic, xerophilous plants with a short spring cycle growing in the small limestone rocky pools covered by a thin layer of sandy–loamy soil rich in salts and nitrates and periodically inundated by salt water, are referred to *Parapholido incurvae-Spergularietum marinae*, a new association described here (Holotypus: rel.2, Table 1), characterized by the dominance of *Parapholis incurva* and *Spergularia marina* (=*S. salina*) and belonging to the *Sileno sedoidis-Catapodion balearici* alliance (class *Saginetea maritimae*, order *Saginetalia maritimae*). This annual spring vegetation is in catenal contact with the perennial communities belonging to the *Crithmo-Staticetea* class.

Table 1. *Parapholido incurvae-Spergularietum marinae* ass. nova hoc loco—sampling plots of main features of plant community investigated. According to the Braun-Blanquet method in each relevé the complete list of vascular plant species was recorded and for each species the cover value (percentage of soil surface) was assessed (+: <1% cover; 1: 1–5% cover; 2: 5–25% cover; 3: 25–50% cover; 4: 50–75% cover; 5: >75% cover). Asterisk (*) indicates the holotypus of the new association.

Relevé number	1	2 *	3	4	5	6	7	8	9	10	11	
Original relevé number	17	18	19	20	21	22	23	24	25	26	27	
Number Cluster	4	4	4	4	4	4	4	4	4	4	4	
Surface (mq)	2	2	3	3	2	2	2	2	2	2	2	
Coverage (%)	60	70	80	60	100	80	70	85	85	85	85	
Altitude (m a.s.l.)	2	3	2	3	2	2	8	6	6	6	6	
Floristic richness	6	8	10	4	11	14	4	7	6	7	7	presence
Char. Ass.												
Parapholis incurva (L.) C.E. Hubb.	2	2	3	3	3	2	2	4	3	4	3	11
Char. Sileno sedoidis-Catapodion balearici de Foucault and Bioret 2010 corr. Tomaselli et al. 2020												
Catapodium balearicum (Willk.) H. Scholz	+	1	.	.	.	.	.	.	.	.	.	2
Silene sedoides Poir.	.	+	.	.	.	.	.	.	.	.	+	2
Char. Saginetea maritimae Westhoff, Van Leeuwen and Adriani 1962, Saginetalia maritimae Westhoff, Van Leeuwen & Adriani 1962												
Spergularia marina (L.) Besser	3	4	+	1	+	3	1	1	+	1	2	11
Beta vulgaris subsp. *maritima* (L.) Arcang.	+	+	1	.	+	1	.	1	+	1	.	8
Medicago littoralis Rohde ex Loisel.	+	+	+	.	3	1	.	.	.	+	+	7
Plantago coronopus L.	+	1	1	1	.	.	.	.	.	.	.	4
Mesembryanthemum nodiflorum L.	.	.	1	.	.	.	.	+	.	.	.	2
Matthiola tricuspidata (L.) W.T. Aiton	.	.	.	.	.	+	.	.	.	.	.	1
Other species												
Lotus cytisoides L.	.	.	+	1	+	.	+	+	1	+	.	7
Frankenia hirsuta L.	.	.	+	.	+	.	2	1	1	+	1	7
Trifolium resupinatum L.	.	1	.	.	+	+	.	+	+	+	+	7
Anthemis arvensis L.	.	.	1	.	1	+	.	.	.	.	.	3
Silene colorata Poir.	.	.	.	.	+	+	.	.	.	.	.	2
Glaucium flavum Crantz	.	.	.	.	+	+	.	.	.	.	.	2
Crithmum maritimum L.	.	.	.	.	.	+	.	.	.	.	.	1
Plantago macrorhiza Poir.	.	.	.	.	.	+	.	.	.	.	.	1
Trifolium nigrescens Viv.	.	.	.	.	.	+	.	.	.	.	.	1
Medicago truncatula Gaertn.	.	.	+	.	.	.	.	.	.	.	.	1
Lotus ornithopodioides L.	.	.	.	.	.	+	.	.	.	.	.	1
Reichardia picroides (L.) Roth	.	.	.	.	+	.	.	.	.	.	.	1
Sonchus asper (L.) Hill	.	.	.	.	.	+	.	.	.	.	.	1
Tamarix africana Poir.	.	.	.	.	.	.	.	.	.	.	+	1

3.3.3. Coastal Wetland Vegetation (Figure 5; Table S3)

Helophytic and Herbaceous Perennial Communities of Fresh and Brackish Waters

The wetland areas close to the industrial area, subjected to long periods of submersion, are covered by perennial vegetation dominated by helophytes, known as the class *Phragmito-Magnocaricetea*. In particular, according to the flooding period and water depth, the following zonation has been observed: in deeper waters (60–80 cm), the *Typhetum domingensis*; in shallow waters (50–60 cm), the *Eleocharido-Alismetum lanceolati*, characterized by the dominance of *Eleocharis palustris* and *Alisma lanceolatum*; in correspondence of small ponds in proximity of the sea, with 40–50 cm water depth, drying in summer, with clayey-sandy soils, the *Bolboschoeno-Alismetum lanceolati* occurs, a new association described here (Holotypus: rel. 6, Table 2), characterized by the dominance of *Alisma lanceolatum* and *Bolboschoenus maritimus*, and which can be referred to the *Glycerio-Sparganion neglecti* alliance (*Nasturtio-Glycerietalia fluitantis* order). The edges of the "Saline di Priolo" are covered by monophytic vegetation dominated by *Phragmites australis*, belonging to the *Phragmitetum communis*. This latter association is widespread in Sicily along both the coastal strip and the middle and final stretch of watercourses, where there is stagnant water with a certain degree of eutrophication. Often, when favored by waters rich in nitrates, it replaces the natural halophilic vegetation of the *Salicornietea fruticosae* class [5].

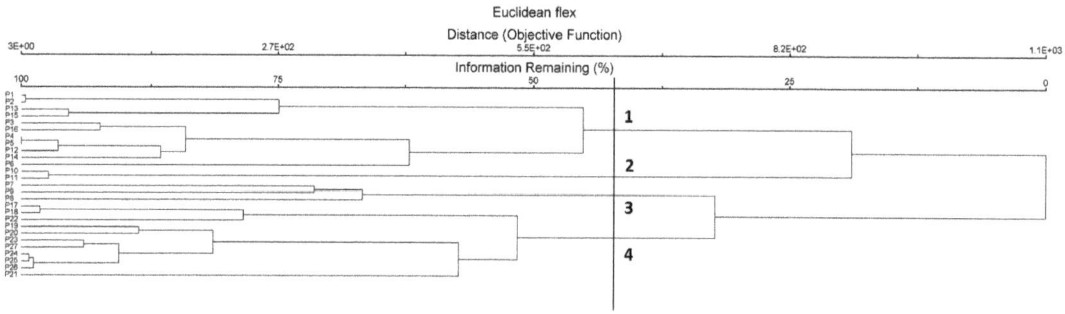

Figure 3. Cluster analysis of rocky coast communities. Plant communities: 1. *Limonietum syracusani*; 2. *Limonio-Arthrocnemetum macrostachyi*; 3. *Lotus cytisoides* and *Frankenia hirsuta* comm.; 4. *Parapholido incurvae-Spergularietum marinae* ass. nova.

The edges of depressed areas, with clayey–sandy soils periodically flooded, develop a perennial herbaceous vegetation enriched by floristic elements typical of the *Molinio-Arrhenatheretea* class, such as *Lythrum junceum*, *Potentilla reptans*, *Juncus articulatus*, *Oenanthe globosa*, *Trifolium fragiferum*, *Euphorbia hirsuta*, *Teucrium scordium*, *Kickxia commutata*, *Phyla nodiflora*, etc. Due to its ecological and floristic features, a new association with the name *Euphorbio hirsutae-Lotetum preslii* is described here (Holotypus: rel. 2, Table 3), characterized by the dominance of *Lotus corniculatus* subsp. *preslii* and *Euphorbia hirta*, which is referred to the *Paspalo-Agrostion semiverticillati* alliance (*Paspalo-Heleochloetalia*).

The large wetland area of the "Saline di Priolo", permanently flooded by more or less deep salt and brackish waters, is characterized by the hydrophytic submerged communities of *Enteromorpho intestinalidis-Ruppietum maritimae*; this vegetation is characterized by the dominance of *Ruppia maritima*, which occasionally grows with green algae as *Enteromorpha intestinalis*. In the study area, these communities are quite frequent, favored by waters rich in nitrates from the surrounding agricultural fields, and are usually in catenal contact with the bank vegetation, generally represented by annual halophilous communities of the *Thero-Suaedetea* class (habitat 1310 "Annual pioneer vegetation in Salicornia and other species of muddy and sandy areas") and by perennial halophilous communities of the *Salicornietea fruticosae* class (habitat 1420 "Mediterranean and thermo-Atlantic grasslands and fruit groves (*Sarcocornetea fruticosi*)"). These latter communities are characterized by the presence of perennial Amaranthaceae species, and differ in species composition and cover in relation to the flooding period, thus characterizing a typical zonation from the innermost to the outer part of the brackish basin. Usually, in the inner part area of the saltmarsh, subjected to long periods of submersion, the *Junco subulati-Sarcocornietum alpini* and the *Arthrocaulo meridionalis-Juncetum subulati* are found; in the more peripheral part of the saltmarsh, rarely subjected to submersion, the *Agropyro scirpei-Inuletum crithmoidis* and the *Halimiono-Suaedetum verae* communities occur. These latter are halo-nitrophilic communities that, at present, have a very scattered distribution in the study area. The innermost parts of the saltmarshes, drying up in the summer–autumn period, are populated by succulent annual species of Amarantaceae with a summer cycle, such as *Salicornia perennans* (=*Salicornia patula*) and *Suaeda maritima*, ascribable to the *Suaedo-Salicornietum patulae* association. This vegetation is typical of salty soils rich in organic matter. Usually, it alternates with *Ruppia maritima* communities over the growing season, with *Ruppietum maritimae* hydrophytic communities growing during the flooding period, and the *Suaedo-Salicornietum patulae* in the summer/autumn season, on the dried-up substrata. On soils rich in sandy–silty components and subjected to short periods of submersion, the presence of *Juncus acutus* communities is significant. This pythocoenosis comprises hygro-halophilous hemicryptophytes and geophytes, such as *Scirpoides holoschoenus*, *Juncus subulatus*, and *Carex extensa*, which allow us to refer, despite the absence of *Juncus maritimus*, to the *Juncetum maritimo-acuti*. Small marginal areas rarely subject to flooding, on sandy soils, are covered by an herbaceous dense vegetation dominated by *Elytrygia atherica*, a species that forms dense and paucispecific populations. In correspondence with depressed areas characterized by a periodic supply of sandy–silty material, a sub-halophilous arboreal vegetation characterized by *Tamarix africana* and *T. gallica* develops. In some stretches, this woody vegetation is enriched with halophilous species typical of the class *Salicornietea fruticosae*. In particular, the presence of *Limbarda crithmoides* allows us to ascribe this phytocoenosis to the *Inulo crithmoidis-Tamaricetum africanae*. This association, included in the *Tamaricion africanae* (*Nerio-Tamaricetea*), has already been reported for the saltmarshes of southeastern Sicily [5]. Moreover, in the "Saline di Priolo", close to the industrial area, grows a patch of woody vegetation dominated by *Ulmus minor*, probably of anthropogenic origin, in contact with the *Juncus acutus* community.

Temporary Ponds

On the carbonatic substrata of the Magnisi peninsula, in correspondence with small carbonate rocky pools with silty soil periodically flooded, this amphibious ephemeral vegetation occurs during the winter–spring period. From a phytosociological point of view, this vegetation type falls within the *Isoëto-Nanojuncetea* class, referable to the habitat 3170* "Mediterranean temporary ponds". The pools with shallow waters and thin soil, subjected to early drying in spring, are covered by amphibious vegetation dominated by *Bulliarda vaillantii* and *Lythrum hyssopifolia* and referrable to the *Lythro hyssopifoliae-Crassuletum vaillantii* association. In the nearby area of "Capo Murro di Porco", the association *Pulicario grecae-Damasonietum bourgaei* has been described, which is very similar to the *Lythro hyssopifoliae-Crassuletum vaillantii* community detected in the Magnisi peninsula due to the presence of *Damasonium bourgaei*.

Figure 4. Photo plate illustrating different habitat types. (**A**,**B**) Psammophylous vegetation of the coast dunes (Saline di Priolo); (**C**,**D**) rocky coast vegetation with *Limonium syracusanum* and *Crithmum maritimum*. (Penisola di Magnisi); (**E**) rocky coast with *Artemisia arborescens* vegetation mixed with *Hyparrhenia hirta* dry grassaland; (**F**) Garrigues with *Tymbra capitata* (Penisola di Magnisi); (**G**,**H**) Subhalophilous ephemeral communities with *Spergularia marina* and *Parapholis incurva* (Penisola di Magnisi).

Table 2. *Bolboschoeno-Alismetum lanceolati* ass. nova hoc loco—sampling plots of main features of plant community investigated. According to the Braun-Blanquet method in each relevé the complete list of vascular plant species was recorded and for each species the cover value (percentage of soil surface) was assessed (+: <1% cover; 1: 1–5% cover; 2: 5–25% cover; 3: 25–50% cover; 4: 50–75% cover; 5: >75% cover). Asterisk (*) indicates the holotypus of the new association.

Relevé number	1	2	3	4	5	6 *	7	8	
Original relevé number	22	23	24	26	27	30	31	32	
Number Cluster	4	4	4	4	4	4	4	4	
Surface (mq)	16	16	16	16	16	16	16	16	
Coverage (%)	80	75	85	90	100	90	90	90	
Altitude (m a.s.l.)	5	5	5	5	3	2	2	2	
Floristic richness	8	17	11	9	10	12	11	7	presence
Char. Ass.									
Bolboschoenus maritimus (L.) Palla	4	1	2	1	4	2	4	2	8
Char. Glycerio-Sparganion neglecti Br.-Bl. and Sissing in Boer 1942									
Alisma lanceolatum With.	2	4	4	4	3	4	4	4	8
Char. Phragmito-Magnocaricetea Klika in Klika and Novák 1941									
Phragmites australis (Cav.) Trin. ex Steud.	+	.	.	1	.	+	1	1	5
Veronica anagallis-aquatica L.	+	1	+	.	.	+	.	+	5
Eleocharis palustris (L.) Roem. and Schult.	.	+	.	.	1	1	.	.	3
Typha domingensis (Pers.) Steud.	.	.	.	.	.	.	1	1	2
Carex otrubae Podp.	.	+	.	.	.	.	.	.	1
Transgr. Molinio-Arrhenatheretea R.Tx.1937									
Potentilla reptans L.	.	+	+	+	+	1	+	.	6
Lotus corniculatus L. subsp. *preslii* (Ten.) P.Fourn.	.	1	+	+	.	1	1	.	5
Lythrum junceum Banks and Sol.	+	1	1	+	.	.	.	.	4
Phyla nodiflora (L.) Greene	.	.	.	.	.	2	1	1	3
Juncus articulatus L.	.	.	.	.	.	2	+	1	3
Trifolium fragiferum L.	.	+	+	.	.	.	.	.	2
Other species									8
Symphyotrichum squamatum (Spreng.) G.L. Nesom	.	+	+	+	+	+	+	.	6
Mentha pulegium L.	+	1	+	+	+	.	.	.	5
Ranunculus trilobus Desf.	.	+	1	+	+	.	.	.	4
Tamarix africana Poir.	1	.	.	.	1	.	1	.	3
Plantago media L.	.	+	.	.	.	+	+	.	3
Holcus lanatus L.	.	1	+	.	.	.	.	.	2
Juncus hybridus Brot.	.	+	.	.	.	.	.	.	1
Isolepis cernua (Vahl) Roem. and Schult.	.	.	.	.	.	+	.	.	1
Juncus acutus L.	.	.	.	.	+	.	.	.	1
Carex extensa Gooden.	.	+	.	.	.	.	.	.	1
Vitex agnus-castus L.	.	.	.	.	+	.	.	.	1
Rumex pulcher L.	.	+	.	.	.	.	.	.	1
Chara sp.	1	.	.	.	.	.	.	.	1

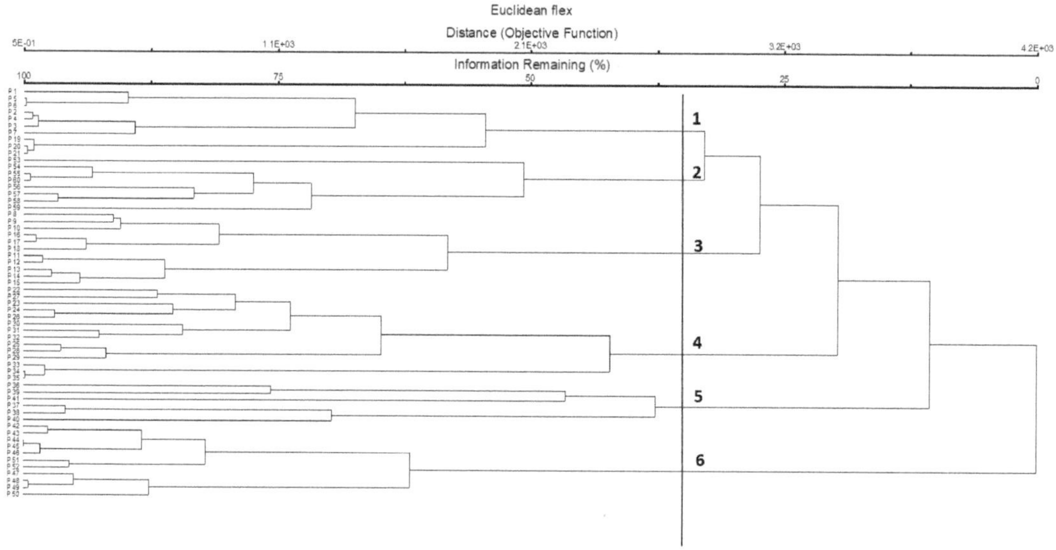

Figure 5. Cluster analysis of coastal wetland vegetation. Plant communities: 1. *Junco subulati-Sarcocornietum alpini/Arthrocaulo meridionalis-Juncetum subulati/Suaedo-Salicornietum patulae;* 2. *Lythro hyssopifoliae-Crassuletum vaillantii;* 3. *Agropyro scirpei-Inuletum crithmoidis/Halimiono-Suaedetum verae;* 4. *Phragmition/Scirpion compacti/Glycerio-sparganion;* 5. *Euphorbio hirsutae-Lotetum preslii/Juncus acutus* comm.; 6. *Limbardo crithmoidis-Tamaricetum africanae.*

Table 3. *Euphorbio hirsutae-Lotetum preslii* ass. nova hoc loco—sampling plots of main features of plant community investigated. According to the Braun-Blanquet method in each relevé the complete list of vascular plant species was recorded and for each species the cover value (percentage of soil surface) was assessed (+: <1% cover; 1: 1–5% cover; 2: 5–25% cover; 3: 25–50% cover; 4: 50–75% cover; 5: >75% cover). Asterisk (*) indicates the holotypus of the new association.

Relevé number	1	2 *	
Original relevé number	40	41	
Number Cluster	5	5	
Surface (mq)	16	16	
Coverage (%)	100	90	
Altitude (m a.s.l.)	2	6	
Floristic richness	18	20	presence
Char. Ass.			
Lotus corniculatus L. subsp. *preslii* (Ten.) P.Fourn.	3	3	2
Euphorbia hirsuta L.	2	2	2
Char. Paspalo-Agrostion semiverticillati Br.-Bl. in Br.-Bl. Roussine and Negre 1952 and Paspalo-Heleochloetalia Br.-Bl. ex Rivas Goday 1956			
Symphyotrichum squamatum (Spreng.) G. L. Nesom	1	+	2
Char. Molinio-Arrhenatheretea R.Tx.1937			
Phyla nodiflora (L.) Greene	1	3	2
Lythrum junceum Banks & Sol.	3	1	2
Potentilla reptans L.	+	2	2
Juncus articulatus L.	1	+	2
Scirpoides holoschoenus (L.) Soják	1	+	2
Teucrium scordium L.	+	+	2
Oenanthe globulosa L.	.	+	1
Kickxia commutata (Bernh. ex Rchb.) Fritsch	.	1	1
Trifolium resupinatum L.	+	.	1

Table 3. *Cont.*

Other species			
Bolboschoenus maritimus (L.) Palla	+	+	2
Dipsacus fullonum L.	+	.	1
Rubus ulmifolius Schott	1	.	1
Tamarix africana Poir.	1	.	1
Schenkia spicata (L.) G. Mans.	.	2	1
Cynodon dactylon (L.) Pers.	.	2	1
Daucus carota L. subsp. *maritimus* (Lam.) Batt.	+	.	1
Xanthium italicum Moretti	.	1	1
Ranunculus trilobus Desf.	1	.	1
Mentha pulegium L.	.	1	1
Plantago media L.	.	1	1
Carex extensa Gooden.	1	.	1
Juncus acutus L.	2	.	1
Phragmites australis (Cav.) Trin. ex Steud.	.	+	1
Alisma lanceolatum With.	.	+	1
Typha domingensis (Pers.) Steud.	.	+	1

3.3.4. Dry Grasslands and Garrigues/Shrubs (Figures 6 and 7; Table S4)

The degradation of the Mediterranean maquis with *Pistacia lentiscus* allows the expansion of garrigues with *Tymbra capitata* or grassland dominated by *Hyparrhenia hirta* (habitat 6220, "Pseudo-steppe with grasses and annuals of the Thero-Brachypodietea"). From a phytosociological point of view, the grassland is very rich in thermo-xerophilous floristic elements typical of the *Hyparrhenion hirtae* alliance, such as *Hyparrenia hirta, Andropogon distachyos, Asphodelus ramosus, Ferula communis, Lathyrus clymenum, Thapsia garganica, Foeniculum vulgare* subsp. *piperitum, Carlina corymbosa, Daucus carota, Eryngium campestre, Moraea sisyrinchium, Lobularia maritima,* etc. The habitat includes arid Mediterranean grasslands, typical of shallow oligotrophic soils, characterized by a high number of hemicryptophyte species, generally mixed with annual herbaceous species. This environment is also occupied by therophytic spring vegetation, linked to small layers of soil placed on rocky ledges or in small corrosion pools, and usually forming a mosaic with the perennial herbaceous vegetation or garrigues. It is dominated by the presence of some annual Crassulaceae, such as *Sedum caeruleum, S. stellatum, Crassula tillaea,* etc. This plant community is referrable to the *Thero-Sedetum caerulei* association, a pioneer coenosis in contact with garrigue vegetation dominated by *Tymbra capitata*. On the Magnisi peninsula, the *Tymbra capitata* vegetation occupies small and limited areas, on carbonate rocky outcrops, due to the frequent summer fires which hinder the evolutionary process of these chamaephytic communities towards the evergreen Mediterranean shrub. For these reasons, there are few diagnostic species of *Cisto-Micormerietea*, such as *Micromeria graeca* subsp. *tenuifolia, Poterium spinosum,* and *Phagnalon saxatile*. The rocky slopes facing west of the Magnisi peninsula are home to communities dominated by *Artemisia arborescens* that, for their ecological, structural, and floristic features, can be referred to the *Atriplici halimi-Artemisietum arborescentis* association [48], despite the lack of the characteristic species *Atriplex halimus*. This vegetation is in contact towards the coast with the halophilous coastal vegetation of the *Crithmo-Limonieta*, while landwards, it is contact with the *Hyparrhenia* steppe grasslands. It can be considered as a permanent edaphic halo-nitrophilous community whose evolution is prevented by anthropic disturbance, combined with the action of coastal winds. Similar communities have been found on Lachea island, near Catania. Furthermore, the peninsula is affected by extensive annual meadows dominated by *Stipellula capensis*, mainly favored by summer fires and by perennial herbaceous formations dominated by *Oloptum miliaceum*, linked to grazing.

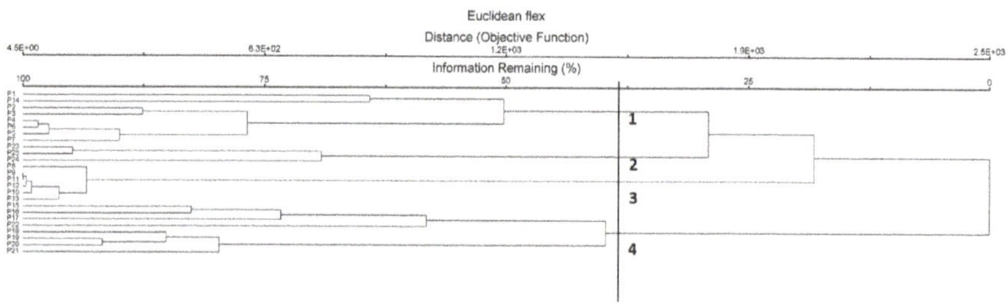

Figure 6. Cluster analysis of dry grasslands, garrigues, and shrubs. Plant communities: 1. *Artemisia arborescens* comm./*Hyparrhenietum hirto-pubescentis*/*Thymbra capitata* comm.; 2. *Thero-Sedetum caerulei*; 3. *Oloptum miliaceum* comm.; 4. *Stipellula capensis* comm.

Figure 7. Photo plate illustrating different habitat types. (**A**) Saline di Priolo during the summer; (**B**) *Juncus acutus* community (Saline di Priolo); (**C**) *Phragmites australis* community in contact with wetlands of Saline di Priolo; (**D**) *Tamarix* sp pl. vegetation; (**E**) *Bolboschoeno-Alismetum lanceolati* ass. nova (Saline di Priolo); (**F**) *Suaedo-Salicornietum patulae* (Saline di Priolo); (**G**) *Euphorbio hirsutae-Lotetum preslii* ass. nova (Saline di Priolo); (**H**) *Elymetum atherici* (Saline di Priolo).

3.4. Diacronic Analysis—Landscape Composition and Configuration

The photointerpretation of the oldest aerial photos (1955) allowed the identification of 10 main types of plant communities, which is a limited number of classes if compared to the number of types (19) derived from the photointerpretation of the recent aerial photos (2015) (Figures 8 and 9). This discrepancy is partly due to a greater spatial and thematic resolution of the aerial images than of the map in 2015, but also to the presence of artificial, synanthropic, and semi-natural classes which were not present in 1955, the landscape being completely natural at that time. In fact, the comparison of the two periods highlights that the anthropogenic transformations and the almost complete occupation of the coastal environments by industrial settlements have caused a sharp reduction in the dune and wetland systems, due to the leveling and removal of sand and to the reclamation activities (Table 4). In particular, the results show a reduction in the aquatic vegetation (*Ruppion maritimae*) of the Saline di Priolo, decreasing from 85 ha (22.6%) in 1955 to only 26 ha (7.4%) in 2015, and the disappearance of the Priolo salt-works, an ancient system for the production of sea salt from sea water, which once occupied an area of 5 hectares. As is well-known, salt-works and salt pans are anthropic modifications of coastal lagoons and, although of anthropic origin, they maintain some of the floristic and faunal peculiarities of the natural systems. The saltmarsh communities (*Juncetea maritimi, Salicornietea fruticosae*), in catenal contact with aquatic vegetation, have also undergone a strong reduction from 93 ha (37%) in 1955 to only 25 ha (8%) in 2015. The annual habitats of the coastline (*Euphorbion peplidis/Maresion nanae*) and perennial dune habitat characterized by *Elytrigia juncea* (*Elytrigienion junceae*) suffered a reduction from 40 ha (11%) to 12 ha (3.5%). The industrial transformation of the area has been a very important driver of changes affecting the coastal environments, mainly because of the massive installation, covering a surface of 96 ha (27.3%); it is the widest land-use typology of the studied area together with reforestation, at 44 ha (12.6%), and uncultivated/abandoned lands colonized by pioneer grassland vegetation (*Echio-Galactition*), with an area of 96 ha (27%). A surprising piece of data concerns the current extension of the woodland cover (about 7 ha), in the Saline di Priolo, of *Tamarix* sp. pl. vegetation, which in the past was not present, probably due to the high salt concentrations in the soil linked to the activities of salt extraction (salt-works).

Table 5 shows the results of the landscape metrics applied to the two habitat maps. Two habitat types identified in 1955 are not shown in the 2015 map, either because they disappeared (2110) or because they were so reduced that they could not be graphically represented (habitat cover less than minimum mappable area). On the other hand, habitat types 1430, 3140, 5220, and 6220 reported in the 2015 habitat map are not present in the 1955 map due to the mere fact that they often cover limited surfaces and do not have a basic cartographic product with sufficient resolution, so it was not possible to identify them. Habitat 9320, detected in 2015, was not present in 1955. For all these reasons, we decided to concentrate our considerations on the habitats strictly relevant to the coastal strip and salt-marshes (except 1310), namely 1150, 1210, 1240, 1410, and 1420.

The values for CA and NumP are shown (Figure 10A,B). It is quite evident, as already mentioned above, that there has been a strong decrease in surface area for all habitat types under consideration (particularly dramatic for habitat 1420), together with a high fragmentation (increase in the number of patches, particularly evident in 1240). This result is in accordance with the values of MPS (Figure 10C), which shows a strong decrease for all habitat types (but particularly evident in 1210, 1240, and 1420), a trend resulting from the sharp reduction and fragmentation process of coastal environments. As is well-known, the Shape index describes the ratio between the perimeter of the patch and the square root of the patch area; AWMSI equals the average shape index (SHAPE) of patches of the corresponding patch type, weighted by patch area. In general, values increase as the shapes of patches become more complex [49]. In Figure 10D,E, AWMSI values are shown; no major differences are detectable between the two years of observation, which means that, despite the ongoing dynamics, the spatial complexity in individual classes is more or less the same. MPAR is another measure of shape complexity, but because it is not standardized to a certain shape (e.g., a square), it describes the patch complexity in a straightforward way. Also, in this case, no particular changes can be evidenced between 1955 and 2015, except for habitat 1240, with a sharp increase in MPAR, probably linked to the fragmentation of the habitat into numerous small patches of irregularly linear shape. The case study of the Saline di Priolo is a paradigmatic example of how, in just a few decades, it is possible to destroy entire stretches of coastal environments with negative consequences on the structure and composition of coastal habitats and vascular flora. The comparison of aerial photos over a period of 60 years shows a dramatic change in the natural landscape, with a permanent loss of land in correspondence with the industrial settlements, the almost complete leveling of the dune system, and the extensive wetland reclamation (Figure 11). Another similar dramatic example of extensive destruction of coastal environments due to industrial installations has occurred in Macconi di Gela, along the southern coast of Sicily, with sharp changes in the dune complexes and wetlands [46].

3.5. Conservation Status

The floristic vegetation survey activities and cartographic processing have highlighted fifteen habitat types according to the Habitat Directive (Table 6), of which four are of priority conservation interest (1150, 3170, 5220, 6220). Of these fifteen, five habitat types turned out to be in the favorable conservation status, nine in the inadequate conservation status, and three in the bad conservation status, the latter one being a priority conservation habitat (5220* Zyziphus arborescent matorrals) with a decreasing trend. Furthermore, habitat 2210, "Fixed dunes of the coast (*Crucianellion maritimae*)", reported in the Natura 2000 sheet of the "Saline di Priolo" SAC, is considered to have disappeared or have been previously erroneously reported. Furthermore, the habitats 1410, "Mediterranean salt meadows (*Juncetalia maritimi*)", and 1420, "Mediterranean and thermo-Atlantic halophilous scrubs (*Sarcocornietea fruticosi*)", which once covered larger areas, are now very reduced and fragmented, resulting in an inadequate conservation status. Our results regarding the dune system, in line with Prisco et al. [50], highlighted the dramatically bad conservation status of dune habitats.

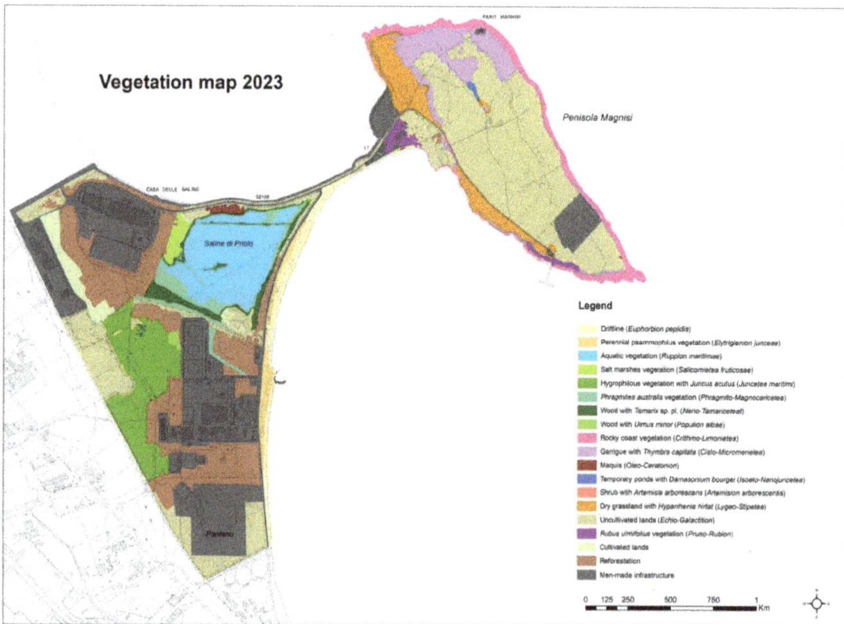

Figure 8. Current vegetation map of "Saline di Priolo and Penisola di Magnisi" (2023).

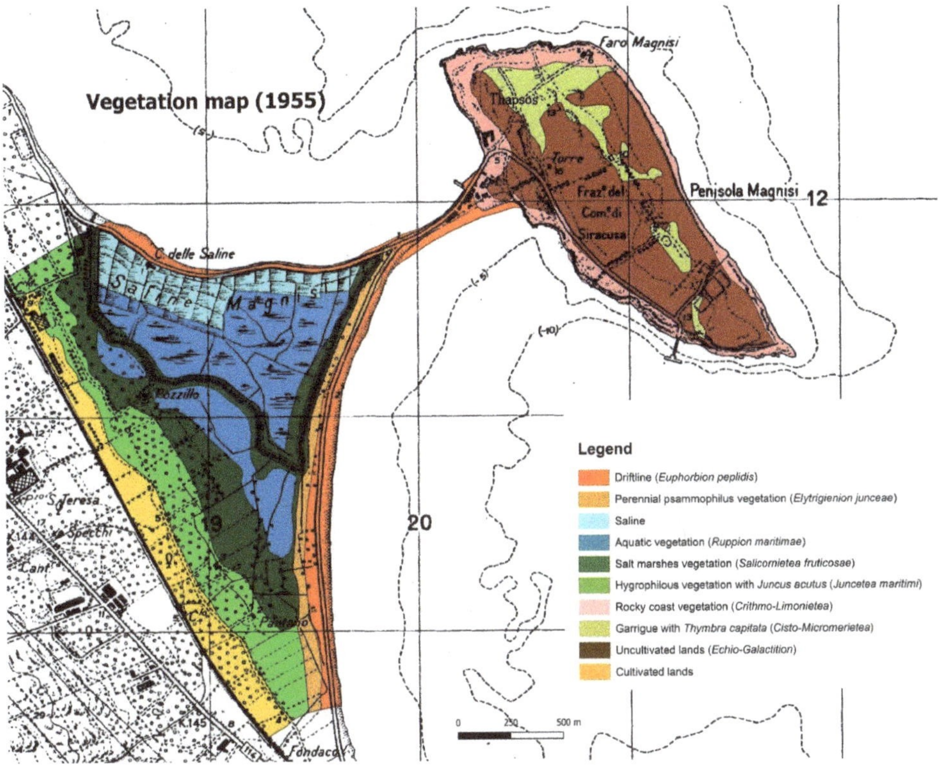

Figure 9. Historical vegetation map of "Saline di Priolo and Penisola di Magnisi" (1955).

Table 4. Surface comparison from the historical stereophoto (1955) and current aerial photos (2023).

	1955	%	2023	%
Driftline (*Euphorbion peplidis*)	29	7.8	9	2.6
Shifting dunes (*Elytrigienion junceae*)	11	2.8	4	1.2
Saline	19	5.2	0	0.0
Aquatic vegetation (*Ruppion maritimae*)	85	22.6	25.9	7.4
Salt-marsh vegetation (*Salicornietea fruticosae*)	53	14.1	2.2	0.6
Hygrophylous vegetation with *Juncus acutus* (*Juncetea maritimi*)	40	10.6	23	6.5
Phragmites australis vegetation (*Phragmito-Magnocaricetea*)	0	0.0	5.89	1.7
Woods with *Tamarix* sp. pl. (*Nerio-Tamaricetea*)	0	0.0	6.6	1.9
Woods with *Ulmus minor* (*Populion albae*)	0	0.0	0.557	0.2
Rocky coast vegetation (*Crithmo-Limonietea*)	28	7.4	11.35	3.2
Garrigue with *Thymbra capitata* (*Cisto-Micromerietea*)	16.13	4.3	14.31	4.1
Maquis with *Ziziphus lotus* (*Oleo-Ceratonion*)	0	0.0	0.87	0.2
Shrub with *Artemisia arborescens* (*Artemision arborescentis*)	0	0.0	0.117	0.0
Temporary ponds with *Damasonium bourgaei* (*Isoeto-Nanojuncetea*)	0	0.0	0.315	0.1
Hyparrhenia hirta dry grassland (*Lygeo-Stipetea*)	0	0.0	12.6	3.6
Uncultivated lands (*Echio-Galactition*)	67	17.8	96	27.3
Rubus ulmifolius vegetation (*Pruno-Rubion*)	0	0.0	3	0.8

Table 4. *Cont.*

	1955	%	2023	%
Cultivated lands	28.15	7.5	6	1.7
Reforestation	0	0.0	44.4	12.6
Man-made infrastructure	0	0.0	96	27.3
	376,274		362,261	

Table 5. Summary table of the landscape metrics applied to the different habitats in the 1955 and 2015 maps; the grey rows refer to those habitat types which it was possible to detect cartographically only on one of the two dates and which therefore were not the subject of temporal analysis.

Habitat Type	AWMSI		MSI		MPAR		MPS		NumP		CA		TLA
	1955	2015	1955	2015	1955	2015	1955	2015	1955	2015	1955	2015	2015
1150	1.74	2.23	1.62	1.88	1854.84211289.33		17.3	8.51	3	3	54.9	25.54	352.55
1210	4.74	3.68	4.74	3.92	3083.97382176.50		29.21	1.32	1	2	29.21	2.63	352.55
1240	5.31	6.23	5.31	2.82	3540.334	61161.51	27.8	1.26	1	9	27.8	11.34	352.55
1310	2.10	/	2.10	/	1676.777	/	19.45	/	1	/	19.45	/	352.55
1410	2.69	2.36	2.69	2.36	1495.051	174.30	40.16	23.04	1	1	40.16	23.04	352.55
1420	3.94	3.94	2.55	3.37	2231.83723406.60		26.06	1.09	2	2	52.12	2.19	352.55
1430	/	2.18		2.18		2264.90		0.12		1		0.12	352.55
2110	3.67	/	3.67	/	3990.6974	/	10.5	/	1	/	11.7	/	352.55
3170	/	1.50		1.50		947.90		0.31		1		0.31	352.55
5220	/	1.20		1.20		3427.80		0.02		1		0.02	352.55
5420	2.70	3.33	1.84	3.19	3998.61853336.03		5.4	1.79	3	8	16.13	14.30	352.55
6220	/	2.57		1.94		35172.79		1.62		11		17.87	352.55
9320	/	1.96		1.96		5281.10		0.02		1		0.02	352.55

Table 6. Conservation status and trend of the habitats of the "Saline di Priolo" SAC. FV—favorable; U1—inadequate; U2—bad; (=) trend stable; (-) trend decreasing; (x) trend unknown. Asterisk (*) indicates a priority habitat.

Habitat	ha	%	Conservation Status and Trend
1150 * Coastal lagoons	25.9	17.7	FV
1210 Annual vegetation of drift lines	9	6.1	FV
1240 Vegetated sea cliffs of the Mediterranean coasts with endemic Limonium spp.	11.35	7.7	U1 (-)
1310 Salicornia and other annuals colonizing mud and sand	0.1	0.1	U1 (=)
1410 Mediterranean salt meadows (Juncetalia maritimi)	23	15.7	U1 (x)
		0.0	
1420 Mediterranean and thermo-Atlantic halophilous scrubs (Sarcocornetiea fruticosi)	2.2	1.5	U1 (x)
1430 Halo-nitrophilous scrubs (Pegano-Salsoletea)	0.12	0.1	FV
2110 Embryonic shifting dunes	4	2.7	U2 (=)
2210 Crucianellion maritimae fixed beach dunes	?	0	U2 (-)
2230 Malcolmietalia dune grasslands	0.2	0.1	U1 (x)
3170 * Mediterranean temporary ponds	0.32	0.2	U1 (=)
5220 * Arborescent matorral with Zyziphus	0.1	0.1	U2 (-)
5420 Sarcopoterium spinosum phryganas	14.3	9.8	FV
6220 * Pseudo-steppe with grasses and annuals of the Thero-Brachypodietea	12.6	8.6	FV
92D0 Southern riparian galleries and thickets (Nerio-Tamaricetea and Securinegion tinctoriae)	6.6	4.5	U1 (-)
9320 Olea and Ceratonia forests	0.8	0.5	U1 (-)
	110.6		

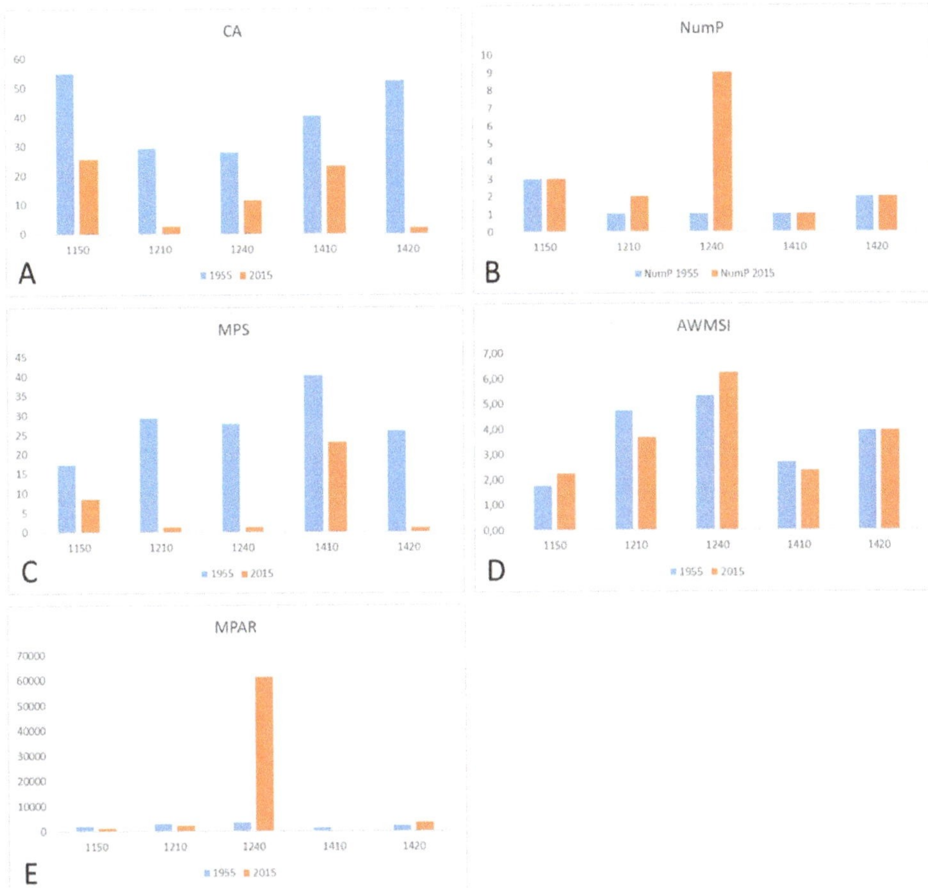

Figure 10. CA (Class Area) and NumP (Number of Patches) (**A**,**B**); MPS (Mean Patch Size) (**C**); AWMSI (Area-Weighted Mean Shape Size) and MPAR (Mean Perimeter–Area Ratio) (**D**,**E**) of habitat types 1150, 1210, 1240, 1410, and 1420 in 1955 and 2015.

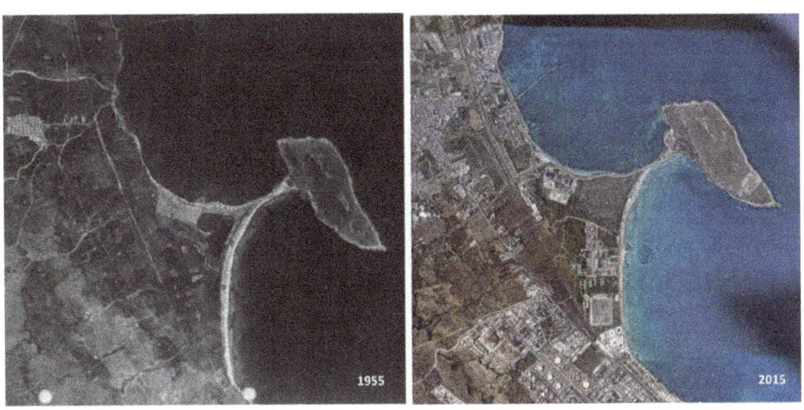

Figure 11. Aerial photos of 1955 and 2015.

4. Conclusions

Vegetation analysis with the phytosociological method is basic for detecting and assessing the terrestrial habitats in line with the objectives of the European Habitats Directive 92/43/EEC. Although the study area is strongly altered by human pressures, the study allowed us to identify notable habitat diversity and floristic richness. As demonstrated by the diachronic analysis carried out, the area has undergone notable transformations, which occurred especially in the 1950s and 1960s with the industrialization of the area. Despite that, the wetland system is still able to support specialized flora, vegetation, and fauna, and provide meaningful ecosystem services. In conclusion, the knowledge of current and past vegetation acquired in this study can represent a valid basic tool for the protected area to plan targeted conservation actions for the most threatened and endangered habitats.

Supplementary Materials: The following supporting information can be downloaded at: https://www.mdpi.com/article/10.3390/land13010062/s1. Table S1: Coastal dune vegetation; Table S2: Rocky coast vegetation; Table S3: Coastal wetland vegetation; Table S4: Dry grasslands and garrigues/shrubs.

Author Contributions: Conceptualization, S.S.; methodology, S.S. and V.T.; investigation, S.S.; data curation, S.S.; data elaboration, S.S., V.R. and V.T.; writing—original draft preparation, S.S.; writing—review and editing, S.S., V.R. and V.T. All authors have read and agreed to the published version of the manuscript.

Funding: This research was financially supported by the research program (Line 3 Starting Grant Progetto HAB-VEG cod. 22722132172) funded by the University of Catania.

Data Availability Statement: Data is contained within the article.

Acknowledgments: The authors would like to acknowledge the reviewers for their valuable suggestions and comments. The authors thank Fabio Cilea, director of the Saline di Priolo protected area, for technical support, useful advice, and for making the photographic material available.

Conflicts of Interest: The authors declare no conflict of interest.

Appendix A. Species List of the Vascular Plants Recorded from the "Saline di Priolo" SAC (SE Sicily)

Table A1. Chorology: Steno-Medit./Medit.—Mediterranean; Endem. sic.—Endemic to Sicily; Endem. Ital.—Endemic to Italy; Euro-Med.—Euro-Mediterranean; Medit.-Turan.—Mediterranean-Turanian; Medit.-Atl.—Mediterranean-Atlantic; Neotrop.—Neotropical; Paleotemp.—Paleotemperate; Cosmop.—Cosmopolitan; Eurasiat.—Eurasiantic; Circumbor.—Curcumboreal; Europ.-Caucas.—Europe-Caucasic; Paleotrop.—Paleotropical. Life form: T—Therophytes; Ch—Chamaephytes; H—Hemicryptophytes; G—Geophytes; P—Phanerophytes; NP—Nanophanerophytes; I—Hydrophytes.

N.	Family	Corology	Life form	Species	Saline Priolo	Penisola Magnisi
1	Asteraceae	Medit.-Atl.	Ch	*Achillea maritima* (L.) Ehrend. and Y.P. Guo	1	
2	Amaranthaceae	SW-Medit.	Ch	*Achyranthes sicula* (L.) All.	1	1
3	Rosaceae	Eurasiat.	H	*Agrimonia eupatoria* L.	1	
4	Lamiaceae	Euri-Medit.	T	*Ajuga chamaepitys (L.) Schreb.*	1	1
5	Alismataceae	Subcosmop.	I	*Alisma lanceolatum* With.	1	
6	Amaryllidaceae	Europ.	G	*Allium sphaerocephalon* subsp. *arvense* (Guss.) Arcang.		1
7	Amaryllidaceae	Steno-Medit.	G	*Allium roseum* L.	1	
8	Amaryllidaceae	Steno-Medit.	G	*Allium commutatum* Guss.		1
9	Amaryllidaceae	Steno-Medit.	G	*Allium subhirsutum* L.	1	
10	Asteraceae	Steno-Medit.	T	*Anacyclus clavatus* (Desf.) Pers.		1
11	Poaceae	Paleotrop.	H	*Andropogon distachyos* L.		1
12	Apiaceae	Euri-Medit.	H	*Anethum piperitum* Ucria	1	1

Table A1. *Cont.*

N.	Family	Corology	Life form	Species	Saline Priolo	Penisola Magnisi
13	Poaceae	Paleosubtrop.	T	*Anisantha rigida* (Roth) Hyl.	1	
14	Poaceae	Medit.-Turan.	T	*Anisantha rubens* (L.) Nevski	1	
15	Poaceae	Medit.-Turan.	T	*Anisantha sterilis* (L.) Nevski	1	
16	Poaceae	Paleotemp.	T	*Anisantha tectorum* (L.) Nevski	1	1
17	Asteraceae	Steno-Medit.	T	*Anthemis arvensis* L.	1	1
18	Apiaceae	Subcosmop.	T	*Torilis arvensis* (Huds.) Link	1	
19	Plantaginaceae	Endem. Ital.	Ch	*Antirrhinum siculum* Mill.		1
20	Asteraceae	S-Medit.	NP	*Artemisia arborescens* (Vaill.) L.		1
21	Asteraceae	Steno-Medit.	H	*Carlina corymbosa* L.		
22	Lamiaceae	Steno-Medit.	T	*Stachys major* (L.) Bartolucci and Peru		
23	Amaranthaceae	Medit.	Ch	*Arthrocaulon meridionale* Es.Ramírez, Rufo, Sánchez Mata, V.Fuente	1	1
24	Araceae	Steno-Medit.	G	*Arum italicum* Mill.	1	
25	Poaceae	Subcosmop.	G	*Arundo donax* L.	1	
26	Asparagaceae	Steno-Medit.	G	*Asparagus acutifolius* L.	1	1
27	Asphodelaceae	Subtrop.	H	*Asphodelus fistulosus* L.	1	1
28	Asphodelaceae	Steno-Medit.	G	*Asphodelus ramosus* L.	1	1
29	Asteraceae	Steno-Medit.	T	*Asteriscus aquaticus* (L.) Less.		1
30	Fabaceae	S-Medit.	T	*Astragalus epiglottis* L.		1
31	Fabaceae	S-Medit.	T	*Astragalus boeticus* L.	1	1
32	Fabaceae	Medit.-Turan.	T	*Astragalus hamosus* L.		1
33	Fabaceae	Steno-Medit.	T	*Astragalus pelecinus* (L.) Barneby		1
34	Amaranthaceae	Circumbor.	T	*Atriplex prostrata* Boucher ex DC.	1	
35	Poaceae	Medit.-Turan.	T	*Avena barbata* Pott ex Link	1	1
36	Poaceae	Medit.-Turan.	T	*Avena sterilis* L.		1
37	Orobanchaceae	Euri-Medit.	T	*Bellardia trixago* (L.) All.	1	
38	Asparagaceae	Endem. Ital.	G	*Bellevalia dubia* (Guss.) Rchb.		1
39	Asteraceae	Steno-Medit.	T	*Bellis annua* L.		1
40	Amaranthaceae	Euri-Medit.	H	*Beta vulgaris* subsp. *maritima* (L.) Arcang.	1	1
41	Brassicaceae	Medit.-Turan.	T	*Biscutella didyma* L.	1	
42	Cyperaceae	Cosmop.	G	*Bolboschoenus maritimus* (L.) Palla	1	
43	Poaceae	Medit.-Turan.	T	*Brachypodium distachyon* (L.) P. Beauv.	1	1
44	Poaceae	Europ.-Caucas.	T	*Bromus racemosus* L.		1
45	Crassulaceae	Subatl.	T	*Bulliarda vaillantii* (Willd.) DC.		1
46	Apiaceae	Euri-Medit.	T	*Bupleurum tenuissimum* L.	1	
47	Brassicaceae	Medit.-Atl.	T	*Cakile maritima* Scop.	1	
48	Capparaceae	Steno-Medit.	NP	*Capparis orientalis* Veill.	1	1
49	Brassicaceae	Cosmop.	T	*Cardamine hirsuta* L.	1	
50	Asteraceae	Steno-Medit.	T	*Carduus argyroa* Biv.		1
51	Asteraceae	Medit.-Turan.	H	*Carduus pycnocephalus* L.	1	
52	Cyperaceae	Steno-Medit.	G	*Carex hispida* Willd. ex Schkuhr	1	
53	Cyperaceae	Medit.-Atl.	H	*Carex extensa* Gooden.	1	
54	Cyperaceae	Atl.	H	*Carex otrubae* Podp.	1	
55	Cyperaceae	Eurosiber.	H	*Carex vulpina* L.		1
56	Cyperaceae	Euri-Medit.	H	Carex distans L.	1	
57	Asteraceae	Endem. Ital.	H	*Carlina hispanica* Lam.	1	1
58	Asteraceae	Euri-Medit.	T	*Carthamus lanatus* L.	1	1
59	Poaceae	Medit.-Atl.	T	*Catapodium balearicum* (Willk.) H. Scholz		1
60	Asteraceae	SW-Medit.	H	*Centaurea sicula* L.		1
61	Asteraceae	Steno-Medit.	H	*Centaurea sphaerocephala* L.	1	
62	Gentianaceae	Paleotemp.	T	*Centaurium tenuiflorum* (Hoffmanns. and Link) Fritsch	1	
63	Characeae		I	Chara sp.	1	
64	Asparagaceae	Steno-Medit.	G	*Squilla maritima* (L.) Steinh.	1	1

Table A1. *Cont.*

N.	Family	Corology	Life form	Species	Saline Priolo	Penisola Magnisi
65	Asteraceae	Cosmop.	H	*Cichorium intybus* L.		1
66	Asteraceae	Orof. NE-Medit.	H	*Cirsium creticum* subsp. *triumfettii* (Lacaita) K. Werner	1	
67	Boraginaceae	Euri-Medit.	H	*Cynoglossum creticum* Mill.	1	1
68	Lamiaceae	Medit.-Mont.	Ch	*Clinopodium nepeta* (L.) Kuntze		1
69	Convolvulaceae	Steno-Medit.	H	*Convolvulus althaeoides* L.	1	1
70	Convolvulaceae	Cosmop.	G	*Convolvulus soldanella* L.	1	
71	Asteraceae	Medit.	H	*Crepis bursifolia* L.	1	
72	Convolvulaceae	Cosmop.	Ch	*Cressa cretica* L.	1	
73	Apiaceae	Euri-Medit.	Ch	*Crithmum maritimum* L.	1	1
74	Convolvulaceae	Eurasiat.	T	*Cuscuta epithymum* (L.) L.		1
75	Poaceae	Steno-Medit.	T	*Cutandia maritima* (L.) Benth. ex Barbey	1	
76	Poaceae	Steno-Medit.	T	*Cutandia divaricata* (Desf.) Barbey	1	
77	Poaceae	Cosmop.	G	*Cynodon dactylon* (L.) Pers.	1	1
78	Cyperaceae	Steno-Medit.	G	*Cyperus capitatus* Vand.	1	
79	Poaceae	Steno-Medit.	H	*Dactylis glomerata* subsp. *hispanica* (Roth) Nyman		1
80	Poaceae	Paleotrop.	T	*Dactyloctenium aegyptium* (L.) Willd.		1
81	Alismataceae	Atl.	I	*Damasonium bourgaei* Coss.		1
82	Apiaceae	Paleotemp.	H	*Daucus carota* L.	1	1
83	Apiaceae	Cosmop.	H	*Daucus carota* L. subsp. *carota*	1	
84	Apiaceae	W-Medit.	H	*Daucus carota* L. subsp. *maritimus* (Lam.) Batt.	1	
85	Brassicaceae	W-Medit.	T	*Diplotaxis erucoides* (L.) DC.		1
86	Dipsacaceae	Euri-Medit.	H	*Dipsacus fullonum* L.	1	
87	Asteraceae	Euri-Medit.	H	*Dittrichia viscosa* (L.) Greuter	1	1
88	Apiaceae	Euri-Medit.	H	*Echinophora spinosa* L.	1	
89	Boraginaceae	Steno-Medit.	H	*Echium arenarium* Guss.	1	
90	Boraginaceae	Endem. Sic.	H	*Echium italicum* subsp. *siculum* (Lacaita) Greuter and Burdet	1	
91	Boraginaceae	Steno-Medit.	H	*Echium parviflorum* Moench	1	
92	Cyperaceae	Subcosmop.	G	*Eleocharis palustris* (L.) Roem. and Schult.	1	
93	Geraniaceae	Subcosmop.	T	*Erodium cicutarium* (L.) L'Hér.		1
94	Geraniaceae	Steno-Medit.	T	*Erodium laciniatum* (Cav.) Willd.	1	
95	Apiaceae	Euri-Medit.	H	*Eryngium campestre* L.		1
96	Apiaceae	SW-Medit.	H	*Eryngium dicothomum* Desf.		1
97	Apiaceae	Medit.-Atl.	G	*Eryngium maritimum* L.	1	
98	Euphorbiaceae	Cosmop.	T	*Euphorbia helioscopia* L.		1
99	Euphorbiaceae	Medit.	G	*Euphorbia hirsuta* L.	1	
100	Euphorbiaceae	Cosmop.	T	*Euphorbia peplus* L.		1
101	Euphorbiaceae	W-Medit.	H	*Euphorbia segetalis* L.		1
102	Euphorbiaceae	Steno-Medit.	H	*Euphorbia terracina* L.	1	
103	Apiaceae	Euri-Medit.	H	*Ferula communis* L.		1
104	Poaceae	Subcosmop.	T	*Festuca danthonii* Asch. and Graebn.	1	1
105	Poaceae	Subcosmop.	T	*Festuca myuros L.*		1
106	Poaceae	Medit.-Atl.	T	*Festuca fasciculata* Forssk.	1	
107	Moraceae	Medit.-Turan.	P	*Ficus carica* L.	1	1
108	Asteraceae	Steno-Medit.	T	*Filago pygmaea* L.		1
109	Frankeniaceae	Steno-Medit.	Ch	*Frankenia hirsuta* L.		1
110	Frankeniaceae	Steno-Medit.	T	*Frankenia pulverulenta* L.		1
111	Asteraceae	Steno-Medit.	H	*Galactites tomentosus* Moench	1	1
112	Rubiaceae	Eurasiat.	T	*Galium aparine* L.	1	
113	Geraniaceae	Cosmop.	T	*Geranium dissectum* L.		1
114	Geraniaceae	Paleotemp.	T	*Geranium rotundifolium* L.	1	1
115	Iridaceae	Euri-Medit.	G	*Gladiolus italicus* Mill.	1	1

Table A1. *Cont.*

N.	Family	Corology	Life form	Species	Saline Priolo	Penisola Magnisi
116	Papaveraceae	Euri-Medit.	H	*Glaucium flavum* Crantz		1
117	Asteraceae	Steno-Medit.	T	*Glebionis coronaria* (L.) Spach		
118	Amaranthaceae	Circumbor.	Ch	*Halimione portulacoides* (L.) Aellen	1	
119	Asteraceae	Steno-Medit.	T	*Hedypnois rhagadioloides* (L.) F.W. Schmidt	1	1
120	Asteraceae	Euri-Medit.	T	*Helminthotheca echioides* (L.) Holub	1	
121	Brassicaceae	Subatl.	H	*Hirschfeldia incana* (L.) Lagr.-Foss.	1	1
122	Poaceae	Circumbor.	H	*Holcus lanatus* L.	1	
123	Poaceae	Euri-Medit.	T	*Hordeum murinum* subsp. *leporinum* (Link) Arcang.	1	1
124	Asteraceae	Steno-Medit.	H	*Hyoseris radiata* L.		1
125	Poaceae	Paleotrop.	H	*Hyparrhenia hirta* (L.) Stapf		1
126	Hypericaceae	Steno-Medit.	H	*Hypericum triquetrifolium* Turra		1
127	Asteraceae	Steno-Medit.	T	*Hypochaeris achyrophorus* L.	1	1
128	Cyperaceae	Subcosmop.	H	*Isolepis cernua* (Vahl) Roem. and Schult.	1	
129	Juncaceae	Euri-Medit.	H	*Juncus acutus* L.	1	1
130	Juncaceae	Circumbor.	G	*Juncus articulatus* L.	1	
131	Juncaceae	Cosmop.	T	*Juncus bufonius* L.		1
132	Juncaceae	Paleosubtrop.	G	*Juncus fontanesii* J.Gay	1	
133	Juncaceae	Euri-Medit.	T	*Juncus hybridus* Brot.	1	1
134	Juncaceae	Subcosmop.	G	*Juncus maritimus* Lam.	1	
135	Juncaceae	S-Medit.	G	*Juncus subulatus* Forssk.	1	
136	Plantaginaceae	Steno-Medit.	H	*Kickxia commutata* (Bernh. ex Rchb.) Fritsch	1	
137	Poaceae	Euri-Medit.	T	*Lagurus ovatus* L.	1	
138	Fabaceae	Euri-Medit.	T	*Lathyrus annuus* L.	1	
139	Fabaceae	Euri-Medit.	T	*Lathyrus aphaca* L.	1	1
140	Fabaceae	Euri-Medit.	T	*Lathyrus cicera* L.		1
141	Fabaceae	Steno-Medit.	T	*Lathyrus clymenum* L.		1
142	Fabaceae	Steno-Medit.	T	*Lathyrus ochrus* (L.) DC.	1	1
143	Malvaceae	Steno-Medit.	H	*Malva arborea (L.) Webb and Berthel.*		
144	Brassicaceae	Euri-Medit.	T	*Lepidium coronopus* (L.) Al-Shehbaz		1
145	Asteraceae	Medit.Atl.	Ch	*Limbarda crithmoides* (L.) Dumort.	1	1
146	Plumbacinaceae	Euri-Medit.	H	*Limonium narbonense* Mill.	1	
147	Plumbaginaceae	Endem. Sic.	Ch	*Limonium syracusanum* Brullo	1	1
148	Plantaginaceae	W-Medit.	T	*Linaria triphylla* (L.) Mill.		1
149	Linaceae	Euri-Medit.	H	*Linum usitatissimum* subsp. *angustifolium* (Huds.) Thell.		1
150	Brassicaceae	Steno-Medit.	Ch	*Lobularia maritima* (L.) Desv.	1	1
151	Poaceae	Paleotemp.	H	*Lolium arundinaceum* (Schreb.) Darbysh.	1	
152	Fabaceae	Euri-Medit.	H	*Lotus corniculatus* L. subsp. *preslii* (Ten.) P.Fourn.	1	
153	Fabaceae	Steno-Medit.	Ch	*Lotus cytisoides* L.	1	1
154	Fabaceae	Steno-Medit.	T	*Lotus edulis* L.		1
155	Fabaceae	Steno-Medit.	T	*Lotus ornithopodioides* L.		1
156	Primulaceae	Cosmop.	T	*Lysimachia arvensis* (L.) U. Manns and Anderb.	1	1
157	Lythraceae	Subcosmop.	T	*Lythrum hyssopifolia* L.		1
158	Lythraceae	Steno-Medit.	H	*Lythrum junceum* Banks and Sol.	1	
159	Malvaceae	Subcosmop.	H	*Malva sylvestris* L.	1	
160	Solanaceae	Steno-Medit.	H	*Mandragora autumnalis* Bertol.		1
161	Lamiaceae	Euri-Medit.	H	*Marrubium vulgare* L.	1	1
162	Brassicaceae	Steno-Medit.	T	*Matthiola tricuspidata* (L.) W.T. Aiton	1	1
163	Fabaceae	Euri-Medit.	T	*Medicago littoralis* Rohde ex Loisel.	1	1
164	Fabaceae	Euri-Medit.	Ch	*Medicago marina* L.	1	

Table A1. *Cont.*

N.	Family	Corology	Life form	Species	Saline Priolo	Penisola Magnisi
165	Fabaceae	Euri-Medit.	T	*Medicago minima* (L.) L.	1	1
166	Fabaceae	Euri-Medit.	T	*Medicago polymorpha* L.		1
167	Fabaceae	Steno-Medit.	T	*Medicago truncatula* Gaertn.		1
168	Lamiaceae	Euri-Medit.	H	*Mentha pulegium* L.	1	1
169	Euphorbiaceae	Paleotemp.	T	*Mercurialis annua* L.	1	1
170	Aizoaceae	S-Medit.	T	*Mesembryanthemum nodiflorum* L.		1
171	Lamiaceae	Endem. Ital.	Ch	*Micromeria graeca subsp. tenuifolia* (Ten.) Nyman	1	1
172	Lamiaceae	S-Medit.	Ch	*Micromeria nervosa* (Desf.) Benth.		1
173	Plantaginaceae	Euri-Medit.	T	*Misopates orontium* (L.) Raf.	1	
174	Iridaceae	Steno-Medit.	G	*Moraea sisyrinchium* (L.) Ker Gawl.		1
175	Asparagaceae	Euri-Medit.	G	*Muscari comosum* (L.) Mill.		1
176	Asparagaceae	Steno-Medit.	G	*Muscari parviflorum* Desf.		1
177	Myrtaceae	Steno-Medit.	P	*Myrtus communis* L.	1	
178	Ranunculaceae	Euri-Medit.	T	*Nigella damascena* L.		1
179	Asteraceae	Steno-Medit.	T	*Notobasis syriaca* (L.) Cass.		1
180	Apiaceae	Steno-Medit.	H	*Oenanthe globulosa* L.	1	
181	Oleaceae	Steno-Medit.	P	*Olea europaea* L. var. *sylvestris* (Mill.) Lehr	1	1
182	Poaceae	Medit.-Turan.	H	*Oloptum miliaceum* (L.) Röser and H.R. Hamasha	1	
183	Fabaceae	Steno-Medit.	T	*Onobrychis caput-galli* (L.) Lam.	1	1
184	Fabaceae	Euri-Medit.	H	*Ononis natrix* subsp. *ramosissima* (Desf.) Batt.	1	1
185	Fabaceae	Medit.-Turan.	T	*Ononis reclinata* L.	1	1
186	Fabaceae	Steno-Medit.	T	*Ononis variegata* L.	1	
187	Asteraceae	Steno-Medit.	H	*Onopordum illyricum* L.		1
188	Orchidaceae	Steno-Medit	G	*Ophrys bertolonii* Moretti	1	
189	Cactaceae	Americ.	P	*Opuntia dillenii* (Ker Gawl.) Haw.	1	
190	Cactaceae	Neotrop.	P	*Opuntia ficus-indica* (L.) Mill.		1
191	Asparagaceae	Steno-Medit.	G	*Ornithogalum gussonei* Ten.		1
192	Oxalidaceae	Africana	G	*Oxalis pes-caprae* L.	1	1
193	Asteraceae	Euri-Medit.	H	*Pallenis spinosa* (L.) Cass.		1
194	Amaryllidaceae	Steno-Medit.	G	*Pancratium maritimum* L.	1	
195	Poaceae	Euri-Medit.	T	*Parapholis cylindrica* (Willd.) Romero Zarco		1
196	Poaceae	Medit.-Atl.	T	*Parapholis filiformis* (Roth) C.E. Hubb.	1	
197	Poaceae	Medit.-Atl.	T	*Parapholis incurva* (L.) C.E. Hubb.	1	1
198	Urticaceae	Euri-Medit.	H	*Parietaria judaica* L.	1	
199	Asteraceae	Steno-Medit.	Ch	*Phagnalon saxatile* (L.) Cass.		1
200	Poaceae	Steno-Medit.	H	*Phalaris coerulescens* Desf.	1	
201	Poaceae	Subcosmop.	G	*Phragmites australis* (Cav.) Trin. ex Steud.	1	
202	Verbenaceae	Pantrop.	H	*Phyla nodiflora* (L.) Greene	1	
203	Anacardiaceae	S-Medit.	P	*Pistacia lentiscus* L.		1
204	Plantaginaceae	Euri-Medit.	H	*Plantago coronopus* L.		1
205	Plantaginaceae	Steno-Medit.	T	*Plantago lagopus* L.	1	1
206	Plantaginaceae	Steno-Medit.	H	*Plantago macrorhiza* Poir.		1
207	Plantaginaceae	Eurasiat.	H	*Plantago media* L.	1	
208	Plantaginaceae	Steno-Medit.	H	*Plantago serraria* L.		1
209	Poaceae	Euri-Medit.	T	*Poa infirma* Kunth	1	1
210	Poaceae	Paleosubtrop.	T	*Polypogon monspeliensis* (L.) Desf.	1	
211	Rosaceae	Paleotemp.	H	*Potentilla reptans* L.	1	
212	Rosaceae	Steno-Medit.	NP	*Poterium spinosum* L.		1
213	Asparagaceae	Steno-Medit.	G	*Prospero autumnale* (L.) Speta		1
214	Asteraceae	Steno-Medit.	T	*Pulicaria sicula* (L.) Moris	1	
215	Asteraceae	Steno-Medit.	H	*Pulicaria dysenterica* (L.) Bernh.	1	

Table A1. *Cont.*

N.	Family	Corology	Life form	Species	Saline Priolo	Penisola Magnisi
216	Rosaceae	Eurasiat.	P	*Pyrus spinosa* Forssk.		1
217	Ranunculaceae	W-Medit.	T	*Ranunculus trilobus* Desf.	1	
218	Asteraceae	Steno-Medit.	H	*Reichardia picroides* (L.) Roth		1
219	Resedaceae	Steno-Medit.	H	*Reseda alba* L.	1	
220	Rhamnaceae	Steno-Medit.	P	*Rhamnus alaternus* L.	1	1
221	Iridaceae	Steno-Medit.	G	*Romulea ramiflora* Ten.		1
222	Rosaceae	Steno-Medit.	NP	*Rosa sempervirens* L.	1	
223	Poaceae	Paleotemp.	T	*Rostraria cristata (L.) Tzvelev*		1
224	Rubiaceae	Steno-Medit.	P	*Rubia peregrina* L.	1	
225	Rosaceae	Euri-Medit.	NP	*Rubus ulmifolius* Schott	1	
226	Polygonaceae	Euri-Medit.	H	*Rumex pulcher* L.	1	1
227	Polygonaceae	Eurasiat.	H	*Rumex conglomeratus* Murray		
228	Polygonaceae	Subcosmop.	H	*Rumex crispus L.*	1	
229	Polygonaceae	W-Medit.	H	*Rumex thyrsoides* Desf.		1
230	Ruppiaceae	Cosmop.	I	*Ruppia maritima* L.	1	
231	Amaranthaceae	Euri-Medit.	Ch	*Salicornia fruticosa* (L.) L.		
232	Amaranthaceae	W-Europ.	T	*Salicornia perennans* Willd.	1	
233	Amaranthaceae	Steno-Medit.	Ch	*Salicornia perennis* subsp. *alpini* (Lag.) Castrov.	1	
234	Salicaceae	Steno-Medit.	P	*Salix pedicellata* Desf.	1	
235	Amaranthaceae	Paleotemp.	T	*Salsola tragus* L.	1	
236	Lamiaceae	Euri-Medit.	H	*Salvia verbenaca* L.		1
237	Gentianaceae	Euri-Medit.	T	*Schenkia spicata* (L.) G. Mans.	1	
238	Cyperaceae	Euri-Medit.	G	*Scirpoides holoschoenus* (L.) Soják	1	
239	Asteraceae	SW-Medit.	H	*Scolymus grandiflorus* Desf.		1
240	Asteraceae	Euri-Medit.	H	*Scolymus hispanicus* L.	1	
241	Fabaceae	S-Medit.	T	*Scorpiurus vermiculatus L.*		1
242	Crassulaceae	SW-Medit.	T	*Sedum caeruleum* L.	1	1
243	Asteraceae	Cosmop.	T	*Senecio vulgaris* L.	1	
244	Caryophyllaceae	S-Medit.	T	*Silene bellidifolia* Jacq.		1
245	Caryophyllaceae	Steno-Medit.	T	*Silene colorata* Poir.	1	1
246	Caryophyllaceae	Steno-Medit.	T	*Silene niceensis* All.	1	
247	Caryophyllaceae	Steno-Medit.	T	*Silene sedoides* Poir.		1
248	Caryophyllaceae	Paleotemp.	H	*Silene vulgaris* (Moench) Garcke		1
249	Asteraceae	Medit.-Turan.	H	*Silybum marianum* (L.) Gaertn.	1	1
250	Brassicaceae	E-Medit.	T	*Sinapis alba L.*	1	
251	Dipsacaceae	Steno-Medit.	H	*Sixalix atropurpurea* (L.) Greuter and Burdet	1	1
252	Apiaceae	Medit.-Atl.	H	*Smyrnium olusatrum* L.	1	
253	Asteraceae	Eurasiat.	H	*Sonchus asper* (L.) Hill		1
254	Asteraceae	Steno-Medit.	G	*Sonchus bulbosus* (L.) Kilian and Greuter	1	
255	Asteraceae	Cosmop.	H	*Sonchus oleraceus* L.	1	1
256	Asteraceae	Steno-Medit.	H	*Sonchus tenerrimus* L.		1
257	Caryophyllaceae	Subcosmop.	T	*Spergularia marina* (L.) Besser	1	1
258	Poaceae	Subtrop.	G	*Sporobolus virginicus* (L.) Kunth	1	
259	Lamiaceae	Steno-Medit.	Ch	*Stachys major* (L.) Bartolucci and Peruzzi		1
260	Lamiaceae	Steno-Medit.	T	*Stachys romana* (L.) E.H.L. Krause	1	1
261	Poaceae	Steno-Medit.	T	*Stipellula capensis* (Thunb.) Röser and H.R. Hamasha	1	1
262	Amaranthaceae	Cosmop.	T	*Suaeda maritima* (L.) Dumort.	1	
263	Amaranthaceae	Cosmop.	NP	*Suaeda vera* J. F. Gmelin	1	1
264	Asteraceae	Neotrop.	H	*Symphyotrichum squamatum* (Spreng.) G. L. Nesom	1	
265	Tamaricaceae	W-Medit.	P	*Tamarix africana* Poir.	1	1
266	Tamaricaceae	S-Medit.	P	*Tamarix arborea* (Ehrenb.) Bunge	1	

Table A1. *Cont.*

N.	Family	Corology	Life form	Species	Saline Priolo	Penisola Magnisi
267	Tamaricaceae	W-Medit.	P	*Tamarix gallica* L.	1	
268	Lamiaceae	Europ.-Caucas.	H	*Teucrium scordium* L.	1	
269	Apiaceae	S-Medit.	H	*Thapsia garganica* L.		1
270	Poaceae	Submedit.	H	*Thinopyrum flaccidifolium* (Boiss. and Heldr.) Moustakas	1	
271	Poaceae	Euri-Medit.	G	*Thinopyrum junceum* (L.) Á. Löve	1	
272	Lamiaceae	Steno-Medit.	Ch	*Thymbra capitata* (L.) Cav.		1
273	Thymelaeaceae	S-Medit.	Ch	*Thymelaea hirsuta* (L.) Endl.	1	
274	Apiaceae	Steno-Medit.	T	*Tordylium apulum* L.		1
275	Apiaceae	Subcosmop.	T	*Torilis arvensis* (Huds.) Link	1	
276	Asteraceae	Euri-Medit.	H	*Tragopogon porrifolius* L.		1
277	Fabaceae	Paleotemp.	T	*Trifolium campestre* Schreb.	1	1
278	Fabaceae	Euri-Medit.	T	*Trifolium cherleri* L.	1	1
279	Fabaceae	Paleotemp.	H	*Trifolium fragiferum* L.	1	
280	Fabaceae	Euri-Medit.	T	*Trifolium nigrescens* Viv.	1	1
281	Fabaceae	Paleotemp.	H	*Trifolium resupinatum* L.		1
282	Fabaceae	Euri-Medit.	T	*Trifolium scabrum* L.	1	1
283	Fabaceae	Euri-Medit.	T	*Trifolium stellatum* L.		1
284	Fabaceae	Paleotemp.	T	*Trifolium tomentosum* L.		1
285	Fabaceae	S-Medit.	T	*Trigonella sulcata* (Desf.) Coulot and Rabaute		1
286	Poaceae	Steno-Medit.	T	*Trisetaria aurea* (Ten.) Banfi and Galasso		1
287	Poaceae	Medit.-Turan.	T	*Triticum vagans* (Jord. and Fourr.) Greuter		1
288	Typhaceae	Pantrop.	G	*Typha domingensis* (Pers.) Steud.	1	
289	Typhaceae	Circumbor.	G	*Typha angustifolia* L.	1	
290	Ulmaceae	Europ.-Caucas.	P	*Ulmus minor* Mill.	1	
291	Asteraceae	Euri-Medit.	T	*Urospermum picroides* (L.) Scop. ex F.W. Schmidt	1	1
292	Urticaceae	S-Medit.	T	*Urtica membranacea* Poir.	1	
293	Asteraceae	Neotrop.	T	*Symphyotrichum squamatum* (Spreng.) G.L. Nesom		
294	Rubiaceae	Steno-Medit.	T	*Valantia muralis* L.		1
295	Scrophulariaceae	Euri-Medit.	H	*Verbascum sinuatum* L.	1	1
296	Plantaginaceae	Cosmop.	H	*Veronica anagallis-aquatica* L.	1	
297	Fabaceae	Euri-Medit.	T	*Vicia hybrida* L.		1
298	Fabaceae	S-Europ.	T	*Vicia melanops* Sm.	1	1
299	Fabaceae	Steno-Medit.	T	*Vicia sativa* L.		1
300	Fabaceae	Steno-Medit.	H	*Vicia villosa* Roth	1	1
301	Fabaceae	Euri-Medit.	T	*Ervum gracile* DC. (=Vicia tenuissima (Bieb.) Sch. and Th.)	1	1
302	Lamiaceae	Medit.-Turan.	P	*Vitex agnus-castus* L.	1	
303	Asteraceae	S-Europ.	T	*Xanthium italicum* Moretti	1	
304	Rhamnaceae	S-Medit.	P	*Ziziphus lotus* (L.) Lam.		1

Appendix B. Syntaxonomical Scheme of the Vegetation Units Recorded from the "Saline di Priolo" SAC (SE Sicily)

CAKILETEA MARITIMAE Tüxen and Preising ex Br.-Bl. and Tüxen 1952
EUPHORBIETALIA PEPLIS Tüxen 1950
EUPHORBION PEPLIS Tüxen 1950
1. *Salsolo-Cakiletum maritimae* Costa and Mansanet 1981 corr. Rivas-Martínez et al., 1992
EUPHORBIO PARALIAE-AMMOPHILETEA AUSTRALIS Géhu and Rivas-Martínez in Rivas Martínez, Asensi, Díez-Garretas, Molero, Valle, Cano, Costa and Díaz 2011

AMMOPHILETALIA AUSTRALIS Br.-Bl. 1933
AMMOPHILION AUSTRALIS Br.-Bl. 1921 corr. Rivas-Martìnez, Costa Izco in Rivas-Martìnez, Lousa, T.
E.Diaz, Fernandez-Gonzalez and J.C.Costa 1990
2. *Cypero capitati-Agropyretum juncei* Kühnholtz-Lordat (1923) Br.-Bl. 1933
ONONIDION RAMOSISSIMAE Pignatti 1952
3. *Centaureo sphaerocephalae-Ononidetum ramosissimae* Br.-Bl. e Frei in Frei, 193
HELIANTHEMETEA GUTTATI (Br.-Bl. in Br.-Bl., Roussine and Nègre 1952) Rivas Goday and Rivas-Martìnez 1963 em. Rivas-Martìnez 1978
CUTANDIETALIA MARITIMAE Rivas- Martìnez, Díez-Garretas, and Asensi 2002
ALKANNO-MARESION NANAE Rivas Goday ex Rivas Goday and Rivas-Martínez 1963 corr. Diaz-Garretas et al., 2001
4. *Sileno coloratae-Ononidetum variegatae* Gèhu and Gèhu-Franck 1986
RUPPIETEA MARITIMAE Tuxen ex Den Hartog and Segal 1964
RUPPIETALIA MARITIMAE Tuxen ex Den Hartog and Segal 1964
RUPPION MARITIMAE Br.-Bl. ex Br.-Bl., Roussine and Nègre 1952
5. *Enteromorpho intestinalidis-Ruppietum maritimae* Westhoff ex R.Tx. and Böckelmann 1957
PHRAGMITO-MAGNOCARICETEA Klika in Klika and Novák 1941
PHRAGMITETALIA Koch 1926
PHRAGMITION Koch 1926
6. *Phragmitetum communis* (Koch 1926) Schmale 1939
7. *Typhetum domingensis* Brullo, Minissale and Spamp. 1994
NASTURTIO-GLYCERIETALIA Pign. 1954
GLYCERIO-SPARGANION Br.-Bl. and Sissing in Boer 1942
8. *Eleocharido-Alismetum lanceolati* Minissale and Spampinato 1987
9. *Bolboschoeno maritimi-Alismetum lanceolati* ass. nov. hoc loco
MOLINIO-ARRHENATHERETEA R.Tx.1937
PASPALO-AGROSTION SEMIVERTICILLATI Br.-Bl. in Br.-Bl. Roussine and Negre 1952
PASPALO-HELEOCHLOETALIA Br.-Bl. ex Rivas Goday 1956
10. *Euphorbio hirsutae-Lotetum preslii* ass. nov. hoc loco
SALICORNIETEA FRUTICOSAE Br.-Bl. et Tx. ex A. Bolòs y Vayreda et. O. de Bolòs in A. Bolòs et Vayreda 1950
SALICORNIETALIA FRUTICOSAE Br.-Bl. 1933
SALICORNION FRUTICOSAE Br.-Bl. 1933
11. *Junco subulati-Sarcocornietum alpini* Brullo et Sciandrello in Giusso et al., 2008
ARTHROCNEMION GLAUCI Rivas-Mart. et Costa M. 1984
12. *Arthrocaulo meridionalis-Juncetum subulati* Brullo et Furnari 1976 *nom. corr.* Sciandrello et al., 2019
13. *Limonio virgati-Arthrocnemetum macrostachyi* Biondi, Casavecchia and Guerra 2006
SUAEDION BREVIFOLIAE Br.-Bl. et O. de Bolòs 1958 (= *Suaedion verae* Brullo et Furnari 1988)
14. *Halimiono-Suaedetum verae* Molinier et Tallon 1970 corr. Géhu 1984
INULION CRITHMOIDIS Brullo et Furnari 1988
15. *Agropyro scirpei-Inuletum crithmoidis* Brullo in Brullo et al., 1988
THERO-SUAEDETEA SPLENDENTIS Rivas-Martínez 1972
THERO-SALICORNIETALIA Tüxen in Tüxen et Oberdorfer ex Géhu et Géhu-Franck 1984
Salicornion patulae Géhu et Géhu-Franck ex Rivas-Martínez 1990
16. *Suaedo-Salicornietum patulae* Brullo et Furnari ex Géhu et Géhu-Franck 1984
JUNCETEA MARITIMI Br.Bl. in Br.-Bl., Roussine and Nègre 1952
JUNCETALIA MARITIMI Br.Bl. ex Horvatic 1934
JUNCION MARITIMI Br.Bl. ex Horvatic 1934
17. *Juncetum maritimo-acuti* Horvatic 1934 (*Juncus acutus* comm.)
HALO-ARTEMISION COERULESCENTIS Pignatti 1953
18. *Elymetum atherici* Pellizzari, Merloni et Piccoli 1998 (=*Thinopyrum acutum* (DC.) Banfi)
SAGINETEA MARITIMAE Westhoff, Van Leeuwen et Adriani 1962
SAGINETALIA MARITIMAE Westhoff, Van Leeuwen et Adriani 1962
SILENO SEDOIDIS-CATAPODION BALEARICI de Foucault and Bioret 2010 corr. Tomaselli et al., 2020
19. *Parapholido incurvae-Spergularietum marinae* ass. nov. hoc loco
ISOËTO-NANOJUNCETEA Br.-Bl. and R. Tx. ex Westhoff, Dijk and Passchier 1946
Isoëtetalia Br.-Bl. 1936

Isoëtion Br.-Bl. 1936
20. Lythro hyssopifoliae-Crassuletum vaillantii Bagella et al., 2009
NERIO-TAMARICETEA Br.-Bl. et O.Bolòs 1958
TAMARICETALIA AFRICANAE Br.-Bl. et O. Bolòs 1958
TAMARICION AFRICANAE Br.-Bl. et O.Bolòs 1958
21. Inulo crithmoidis-Tamaricetum africanae Gamisans 1992 (=*Limbardo crithmoidis-Tamaricetum africanae* Sciandrello et al., 2019)
CRITHMO-LIMONIETEA Br.-Bl. in Br- Bl., Roussine and Nègre1952
CRITHMO-LIMONIETALIA Molinier 1934
CRITHMO-LIMONION MOLINIER 1934
22. Limonietum syracusani Bartolo, Brullo and Marcenò 1982
PEGANO HARMALAE-SALSOLETEA VERMICULATAE Br-Bl and O.Bolòs 1958
SALSOLO VERMICULATAE-PEGANETALIA HARMALAE Br.-Bl. and O. Bolòs 1954
Artemision arborescentis Géhu et al., 1986
23. Atriplici halimi-Artemisietum arborescentis Biondi 1988 (*Artemisia arborescens* comm.)
CISTO-MICROMERIETEA Oberd. 1954
CISTO-ERICETALIA Horvatic 1958
CISTO-ERICION Horvatic 1958
24. Thymbra capitata comm.
LYGEO SPARTI-STIPETEA TENACISSIMAE Rivas-Martinez 1978
CYMBOPOGONO-BRACHYPODIETALIA RAMOSI Horvatić 1963
HYPARRHENENION HIRTAE Brullo, Minissale and Spamp. in C. Brullo et al., 2010
25. Hyparrhenietum hirto-pubescentis A.and O. Bolòs and Br.-Bl. in A.and O. Bolòs 1950
BROMO-ORYZOPSION MILIACEAE O.Bolòs 1970
26. *Oloptum miliaceum* comm.
STIPO-TRACHYNIETEA DISTACHYAE Brullo in Brullo, Scelsi and Spampinato 2001
TRACHYNIETALIA DISTACHYAE Rivas-Martínez 1978
TRACHYNION DISTACHYAE Rivas-Martínez 1978
27. *Thero-Sedetum caerulei* Brullo 1975
CHENOPODIETEA Br.-Bl. in Br.-Bl. et al., 1952
BROMETALIA RUBENTI-TECTORUM (Rivas Goday et Rivas-Martinez 1973) Rivas-Martinez and Izco 1977
ECHIO PLANTAGINEI-GALACTITION TOMENTOSAE O. Bolòs and Molinier 1969
28. *Stipellula capensis* comm.

Appendix C. Localities and Dates of Phytosociological Relevés

Table 1: *Parapholido incurvae-Spergularietum marinae* ass. nova (Rel. 17–22, Scogliera Penisola Magnisi, 30.03.2021, Cambria, Minissale, Sciandrello, Rel. 23–27 Scogliera Penisola Magnisi, 09.04.2021 (Cambria, Minissale, Sciandrello, Tavilla).

Table 2: *Bolboschoeno-Alismetum lanceolati* ass. nova (Rel. 22–24, 26–27, 30–32, pozze Saline di Priolo, 21.04.2021, Cambria, Minissale, Ranno, Sciandrello, Tavilla)

Table 3: *Euphorbio hirsutae-Lotetum preslii* ass. nova (Rel. 40–41, Saline Priolo, pozza lunga 21.09.2021, Sciandrello).

References

1. Leberger, R.; Geijzendorffer, I.R.; Gaget, E.; Gwelmami, A.; Galewski, T.; Pereira, H.M.; Guerra, C.A. Mediterranean wetland conservation in the context of climate and land cover change. *Reg. Environ. Chang.* **2020**, *20*, 67. [CrossRef]
2. Taylor, N.G.; Grillas, P.; Al Hreisha, H.; Balkız, Ö.; Borie, M.; Boutron, O.; Catita, A.; Champagnon, J.; Cherif, S.; Çiçek, K.; et al. The future for Mediterranean wetlands: 50 key issues and 50 important conservation research questions. *Reg. Environ. Chang.* **2021**, *21*, 33. [CrossRef] [PubMed]
3. Scott, D.B.; Frail-Gauthier, J.; Mudie, P.J. *Coastal Wetlands of the World: Geology, Ecology, Distribution and Applications*; Cambridge University Press: New York, NY, USA, 2014.
4. Garbutt, A.; De Groot, A.; Smit, C.; Pétillon, J. European salt marshes: Ecology and conservation in a changing world. *J. Coast. Conserv.* **2017**, *21*, 405–408. [CrossRef]
5. Sciandrello, S.; Musarella, C.M.; Puglisi, M.; Spampinato, G.; Tomaselli, V.; Minissale, P. Updated and new insights on the coastal halophilous vegetation of southeastern Sicily (Italy). *Plant Sociol.* **2019**, *56*, 81–98.
6. De Pietro, R.; Ientile, R.; Puccia, S.; Sabella, G. Birds of Gelsari and Lentini marshes, special protection area for the protection and maintenance of aquatic avifauna in central Mediterranean. *Ocean Coast. Manag.* **2019**, *169*, 96–103. [CrossRef]

7. Frei, M. Studi fitosociologici su alcune associazioni litorali in Sicilia (Ammophiletalia e Salicornietalia). *Nuovo G. Bot. Ital.* **1937**, *44*, 273–294. [CrossRef]
8. Brullo, S.; Di Martino, A. Vegetazione dell'Isola Grande dello Stagnone (Marsala). *Boll. Studi Inf. Giard. Colon. Palermo* **1974**, *26*, 15–62.
9. Brullo, S.; Furnari, F. Le associazioni vegetali degli ambienti palustri costieri della Sicilia. *Not. Soc. Ital. Fitosociol.* **1976**, *11*, 1–43.
10. Brullo, S.; Furnari, F. La vegetazione costiera della Cirenaica. *Boll. Dell'accademia Gioenia Sci. Nat.* **1988**, *21*, 31–117.
11. Minissale, P.; Sciandrello, S. Flora e vegetazione terrestre della Riserva Naturale di Vendicari (Sicilia sud-orientale). *Ente Fauna Sicil.* **2010**, *12*, 145–208.
12. Sciandrello, S. Coastal saltmarsh vegetation in Sicily (Italy): Phytosociological insights and plant diversity. *Plant Biosyst.* **2020**, *154*, 860–876. [CrossRef]
13. Benadusi, M. Oil in Sicily: Petrocapitalist imaginaries in the shadow of old smokestacks. *Econ. Anthropol.* **2018**, *5*, 45–58. [CrossRef]
14. Brullo, S.; Furnari, F. *Vegetazione dei Pantani Litoranei della Sicilia Sud-Orientale e Problema della Conservazione Dell'ambiente*; Istituto di Botanica dell'Università di Catania: Catania, Italy, 1971.
15. Fagotto, F.; Baglieri, S. Ornitofauna e vegetazione delle saline di Siracusa (un luogo umido costiero della Sicilia orientale). *Animalia* **1976**, *3*, 81–103.
16. La Mantia, A.; Scuderi, L. Distribuzione presente e passata di *Ziziphus lotus* (L.) Lam. subsp. lotus (Rhamnaceae) in Sicilia. *Nat. Sicil.* **2021**, *45*, 7–28.
17. Turner, M.G.; Gardner, R.H.; O'Neill, R.V. *Landscape Ecology in Theory and Practice: Pattern and Process*; Springer: New York, NY, USA, 2001.
18. Uuemaa, E.; Mander, Ü.; Marja, R. Trends in the use of landscape spatial metrics as landscape indicators: A review. *Ecol. Indic.* **2013**, *28*, 100–106. [CrossRef]
19. Tomaselli, V.; Mantino, F.; Tarantino, C.; Albanese, G.; Adamo, P. Changing landscapes: Habitat monitoring and land transformation in a long-time used Mediterranean coastal wetland. *Wetl. Ecol. Manag.* **2022**, *31*, 1–28. [CrossRef]
20. Ben-Avraham, Z.; Grasso, M. Collisional zone segmentation in Sicily and surrounding areas in the Central Mediterranean. *Ann. Tectonicae* **1990**, *4*, 131–139.
21. Garfì, G.; Carimi, F.; Pasta, S.; Rühl, J.; Trigila, S. Additional insights on the ecology of the relic tree Zelkova sicula Di Pasquale, Garfì & Quézel (Ulmaceae) after the finding of a new population. *Flora* **2011**, *206*, 407–417.
22. Guarino, R.; Raimondo, F.M.; Domina, G. A new species of Anthemis sect. Hiorthia (Asteraceae) SE Sicily. *Plant Biosyst.* **2013**, *147*, 821–825. [CrossRef]
23. Zampino, D.; Duro, A.; Piccione, V.; Scalia, C. Termoudogrammi secondo Walter & Lieth [Thermoudogram according to Walter & Lieth]. In *Atti 5° Workshop Progetto Strategico Clima Ambiente e Territorio nel Mezzogiorno*; CNR: Rome, Italy, 1997; Volume 2, pp. 7–54.
24. Pesaresi, S.; Biondi, E.; Casavecchia, S. Bioclimate of Italy. *J. Maps* **2017**, *13*, 955–960. [CrossRef]
25. Braun-Blanquet, J. *Pflanzensoziologie. Grundzüge der Vegetationskunde [Plant Sociology. Basic Course of Vegetation Science]*; Springer: New York, NY, USA, 1964.
26. Pignatti, S. Volume 1: Flora d'Italia & Flora Digitale. In *Flora d'Italia: In 4 Volumi*, 2nd ed.; Edagricole-Edizioni Agricole di New Business Media: Milano, Italy, 2017.
27. Pignatti, S. Volume 2: Flora d'Italia & Flora Digitale. In *Flora d'Italia: In 4 Volumi*, 2nd ed.; Edagricole-Edizioni Agricole di New Business Media: Milano, Italy, 2017.
28. Pignatti, S. Volume 3: Flora d'Italia & Flora Digitale. In *Flora d'Italia: In 4 Volumi*, 2nd ed.; Edagricole-Edizioni Agricole di New Business Media: Milano, Italy, 2018.
29. Pignatti, S.; Guarino, R.; La Rosa, M. Volume 4: Flora d'Italia & Flora Digitale. In *Flora d'Italia: In 4 Volumi*, 2nd ed.; Edagricole-Edizioni Agricole di New Business Media: Milano, Italy, 2019.
30. Portal to the Flora of Italy. Available online: http:/dryades.units.it/floritaly (accessed on 5 April 2023).
31. Martellos, S.; Bartolucci, F.; Conti, F.; Galasso, G.; Moro, A.; Pennesi, R.; Peruzzi, L.; Pittao, E.; Nimis, P.L. FlorItaly—The portal to the Flora of Italy. *PhytoKeys* **2020**, *156*, 5571. [CrossRef] [PubMed]
32. Giardina, G.; Raimondo, F.M.; Spadaro, V. Checklist of the vascular plants growing in Sicily. *Bocconea* **2007**, *20*, 5–582.
33. McCune, B.; Mefford, M.J. *PC-ORD. Multivariate Analysis of Ecological Data*, Version 4.0; MjM Software Design: Gleneden Beach, OR, USA, 1999.
34. Lengyel, A.; Podani, J. 12 Assessing the relative importance of methodological decisions in classifications of 13 vegetation data. *J. Veg. Sci.* **2015**, *26*, 804–815. [CrossRef]
35. Biondi, E.; Blasi, C.; Allegrezza, M.; Anzellotti, I.; Azzella, M.M.; Carli, E.; Casavecchia, S.; Copiz, R.; Del Vico, E.; Facioni, L.; et al. Plant communities of Italy: The Vegetation Prodrome. *Plant Biosyst.* **2014**, *148*, 728–814. [CrossRef]
36. Mucina, L.; Bültmann, H.; Dierßen, K.; Theurillat, J.P.; Raus, T.; Čarni, A.; Šumberová, K.; Willner, W.; Dengler, J.; García, R.G.; et al. Vegetation of Europe: Hierarchical floristic classification system of vascular plant, bryophyte, lichen, and algal communities. *Appl. Veg. Sci.* **2016**, *19*, 3–264. [CrossRef]
37. Biondi, E.; Blasi, C. *Italian Interpretation Manual of the Habitats (92/43/EEC Directive)*; Ministero dell'Ambiente e della Tutela del Territorio e del Mare: Roma, Italy, 2009.

38. Cushman, A.; McGarigal, K.; Neel, M.C. Parsimony in landscape metrics: Strength, universality and consistency. *Ecol. Indic.* **2008**, *8*, 691–703. [CrossRef]
39. Brullo, S. Taxonomic and nomenclatural notes on the genus Limonum in Sicily. *Bot. Not.* **1980**, *133*, 281–293.
40. Rossi, G.; Orsenigo, S.; Gargano, D.; Montagnani, C.; Peruzzi, L.; Fenu, G.; Abel, T.; Alessandrini, A.; Astuti, G.; Bacchetta, G.; et al. *Lista Rossa della Flora Italiana. 2 Endemiti e Altre Specie Minacciate;* Ministero dell'Ambiente e della Tutela del Territorio e del Mare: Roma, Italy, 2020.
41. Conti, F.; Manzi, A.; Pedrotti, F. *Liste Rosse Regionali delle Piante D'italia;* WWF–SBI: Camerino, Italy, 1997.
42. Orsenigo, S.; Fenu, G.; Gargano, D.; Montagnani, C.; Abeli, T.; Alessandrini, A.; Bacchetta, G.; Bartolucci, F.; Carta, A.; Castello, M.; et al. Red list of threatened vascular plants in Italy. *Plant Biosyst.* **2021**, *155*, 310–335. [CrossRef]
43. Zodda, G. Nuove aggiunte alla flora di Siracusa. *Ann. R. Liceo Sc. "O. M. Corbino" Siracusa* **1930**, *4*, 69–74.
44. Martinoli, G. *Poterium spinosum* L. elemento del Mediterraneo Orientale a Calamosca (Capo S. Elia, Sardegna). *Plant Biosyst.* **1969**, *103*, 325–340.
45. Gargano, D.; Fenu, G.; Medagli, P.; Sciandrello, S.; Bernardo, L. The status of *Sarcopoterium spinosum* (Rosaceae) at the western periphery of its range: Ecological constraints lead to conservation concerns. *Isr. J. Plant Sci.* **2007**, *55*, 1–13. [CrossRef]
46. Sciandrello, S.; Tomaselli, G.; Minissale, P. The role of natural vegetation in the analysis of the spatio-temporal changes of coastal dune system: A case study in Sicily. *J. Coast. Conserv.* **2015**, *19*, 199–212. [CrossRef]
47. Biondi, E.; Casavecchia, S.; Guerra, V. Analysis of vegetation diversity in relation to the geomorphological characteristics in the Salento coasts (Apulia-Italy). *Fitosociologia* **2006**, *43*, 25–38.
48. Biondi, E.; Allegrezza, M.; Filigheddu, R. Su alcune formazioni ad *Artemisia arborescens* L. della Sard. Settentrionale. *Boll. Soc. Sarda Sci. Nat.* **1988**, *26*, 177–185.
49. McGarigal, K.; Marks, B.J. *Fragstats: Spatial Pattern Analysis Program for Quantifying Landscape Structure;* Reference Manual; Forest Science Department Oregon State University: Corvallis, OR, USA, 1994; p. 60.
50. Prisco, I.; Angiolini, C.; Assini, S.; Buffa, G.; Gigante, D.; Marcenò, C.; Sciandrello, S.; Villani, M.C.; Acosta, A.T.R. Conservation status of Italian coastal dune habitats in the light of the 4th Monitoring Report (92/43/EEC Habitats Directive). *Plant Sociol.* **2020**, *57*, 55–64. [CrossRef]

land

MDPI

Article

Assessment of Coastal Cultural Ecosystem Services and Well-Being for Integrating Stakeholder Values into Coastal Planning

Kristina Veidemane [1,*], Agnese Reke [1], Anda Ruskule [1] and Ivo Vinogradovs [2]

[1] Baltic Environmental Forum—Latvia, LV-1050 Riga, Latvia; agnese.reke@bef.lv (A.R.); anda.ruskule@bef.lv (A.R.)
[2] Faculty of Geography and Earth Sciences, University of Latvia, LV-1589 Riga, Latvia; ivo.vinogradovs@lu.lv
* Correspondence: kristina.veidemane@bef.lv

Abstract: Coastal areas provide ecosystem services (ES), including a wide range of cultural ecosystem services (CES). This study aims to operationalize the ES approach for integrated assessment and mapping of coastal CES through the case of the eastern Baltic Sea coast in Latvia. It explores an interdisciplinary approach to enhance coastal planning, leveraging the strengths of plural disciplines to ensure a more holistic representation of coastal CES. A set of methods and techniques from landscape ecology (e.g., landscape characterization, quality assessment, biophysical mapping) and social sciences (participatory GIS, stakeholder engagement events, nationwide survey) are developed and tested, particularly demonstrating links and correlations between landscape character and CES values and well-being dimensions. The results illuminate the main perceived well-being benefits that people gain from the coastal areas, highlighting the different perspectives of stakeholders. Finally, the integrated assessment results helped to construct proposals for sustainable tourism development in the area. The outcomes of the study are intended to assist planners and decision-makers in evaluating the potential for development and trade-offs in coastal regions. This research contributes to the advancement of coastal spatial planning methodologies, emphasizing the importance of stakeholder engagement and ES assessment for informed decision-making.

Keywords: ecosystem services; stakeholders; social values; perceived well-being benefits; coastal spatial planning; integrated assessment; Baltic Sea; Latvia

1. Introduction

Coastal regions represent a spatial realm where terrestrial and marine environments interact, giving rise to distinctive ecosystems on land and at sea. Within the European context, coastal areas have historically and contemporarily played a pivotal role in the socio-economic and cultural development of communities [1]. Due to the multidimensional nature of coastal areas, the ecosystems therein exhibit diverse and specific features and qualities, contributing to the provision of various ecosystem services (ES). However, the coast has experienced escalating pressures from human activities over the last decades, accompanied by an expanding demand for space [2,3]. The growing human influence poses a significant threat to the uninterrupted supply of ES unless a more sustainable approach to the management of coastal natural capital is promptly implemented [4–6]. This is particularly evident in the context of coastal recreational and tourism development. European coastal tourism grew until the onset of COVID-19, then subsequently experienced a dramatic decline, and is now gradually recovering, with an increasing number of people, both within the European Union and internationally, allocating larger portions of their income to travel and leisure [7].

The concept of ES has gained recognition for its application in addressing challenges and controversial issues during the planning of nature conservation and the sustainable use of ecosystems [5,8]. Maritime spatial planning, integrated coastal zone planning, and marine strategy represent key governance tools employed to progressively incorporate the outcomes of ES assessments [9–11]. The ES concept encompasses ecological, economic, and social systems, thereby establishing a nexus between the environmental dimension and political decision-making processes [12,13]. The advancement of indicators, encompassing both biophysical and socio-cultural dimensions, is encouraged to elucidate the intricate relationships between humans and nature in a quantified format [14,15].

Cultural ecosystem services (CES) provide opportunities for social interactions and intangible benefits through active or passive engagement with nature. They involve the reception of artistic and scenic appreciation, contribute to educational development, and foster spiritual fulfillment. They are based on cultural diversity and identity, thus linking cultural landscapes and heritage values [16]. CES can be combined with cultural landscape research [17–19]; in particular, studies related to aesthetics or cultural heritage could provide valuable input for a comprehensive assessment of CES [20] and implications for management [21]. Choosing a landscape perspective could serve as a practical approach, as numerous researchers either have not tried to establish connections between CES and ecosystems or have encountered challenges in doing so [22].

When assessing CES, assigning intangible values to marine and coastal cultural ecosystem services has been challenging [23,24]. It has become essential to establish an appropriate assessment unit by considering ecosystem types (e.g., beach or forest), as well as habitats (e.g., coastal lagoons or wooded dunes), and other distinctive features of each individual coastal and marine region. Another challenge is the interface between land and sea, the characteristics of direct shoreline and adjacent land, and the identification of the proper biophysical and socio-cultural assessment criteria and indicators [25–27].

Coastal spatial planning processes necessitate the incorporation of robust participatory procedures to ensure the recognition of the multiple interests of stakeholders and plural values, as well as support to mitigate conflicts and manage trade-offs [28]. Different societal groups perceive ecosystems, including marine and coastal systems, through the prism of their unique economic, cultural, and societal perspectives. The review conducted by Kosanic and Petzold [29] emphasizes the presence of knowledge gaps in how different categories of CES affect various societal groups and different types of well-being. A pertinent strategy for assessing CES involves engaging stakeholders and other beneficiaries, especially in recognizing and using values in a spatially explicit manner [30,31].

Participatory GIS (PGIS) has emerged as a valuable tool for mapping CES by pinpointing specific locations or delineating areas of significance to local stakeholders [32]. This process involves identifying preferred areas for deriving benefits from ecosystems and landscapes, as well as assessing attitudes towards land and space usage [33]. While conventional methods such as paper maps were once prevalent, modern digital or online GIS tools facilitate surveys conducted during events or remotely [34,35]. However, the mere incorporation of spatial technology into a singular participatory exercise does not guarantee influence on decisions regarding land and marine use [33]. To achieve a substantial impact, multiple interactions with stakeholders are indispensable to ensure that social values are integrated throughout the planning process [33]. This inclusive approach helps to overcome barriers to meaningful public engagement in sustainable planning.

Integration of biophysical expert-based CES mapping with participatory approaches is increasingly considered best practice for eliciting meaningful values for planning and decision making [36]. Narrowing the assessment scope to exclusively encompass natural science or biophysical measurements will significantly diminish the efficacy of ES indicators in providing valuable and pertinent information for decision-makers regarding a specific policy objective [15]. Integration of multiple sources of data and assessment results can be achieved by using established decision-support frameworks such as multi-criteria analysis [37]. This approach has been tested in the CES assessment of the Latvian coastline as an input to maritime spatial planning by prioritizing the suitability of the coastal areas for recreation [38].

The integrated assessment approach represents different combinations of methods and underlines the necessity for collaboration among scientists from different disciplines [24]. Joint efforts between economists and ecologists have been firmly established in the development of valuation studies of CES, while the contributions of social sciences and humanities have been relatively neglected [24]. Understanding the human dimensions of the sea, including its impact on health, inspiration, childhood experiences, cultural connection, and place identity, is crucial to ensuring a sustainable future for the coastal areas and the well-being of coastal communities. Addressing this complexity could be facilitated by integrating social science methods into assessments and valuations of marine cultural ecosystem services [39]. A lack of understanding of interlinks between coastal ecosystem structures and CES has been identified by recent marine and coastal studies [40].

The aim of the study is to enhance the integrated assessment of CES to support sustainable coastal development planning, particularly in terms of incorporating expert-based and stakeholder engagement approaches. By operationalizing biophysical mapping and stakeholder engagement, the study addresses key questions regarding coastal CES assessment, stakeholder involvement, and sustainable tourism development. Key research questions are as follows:

- How does the landscape characterization approach, and landscape quality assessment, contribute to biophysical mapping and assessment of coastal CES?
- How can social methods such as participatory GIS contribute to interpretation of expert based biophysical mapping and assessment of CES?
- What are the public's perceived benefits and links to CES? What values are assigned to CES by different stakeholders?
- In which way can CES assessment be taken up to plan level sustainably?

The application of this approach is demonstrated through a study conducted in a coastal area within Europe—the southwestern coast of Latvia, the eastern part of the Baltic Sea.

2. Materials and Methods

2.1. Study Area

The study area—the Southwestern Kurzeme coastal area—is located in the Kurzeme Region in the southwestern part of Latvia (Figure 1). It stretches for about 120 km along the coast from the Latvian-Lithuanian border until the border of Ventspils municipality. The total size of the study area is around 1200 km^2. The study area covers the terrestrial part up to 10 km inland from the shoreline.

Administratively, it includes Liepaja city and the coastal part of Dienvidkurzeme municipality. According to the National Statistical System of Latvia [41], the largest settlements in the area are Liepaja City (around 67,000 inhabitants) and Pavilosta town (around 820 registered inhabitants; however, the population of Pavilosta increases dramatically during the summer months). The habitation comprises also 17 coastal and inland villages, as well as smaller settlements. The rural part of the study area is characterized by a low population density—9.18 individuals/km^2 [41].

The study area consists of a highly diverse land-use mosaic [42]. It encompasses coastal ecosystems (sandy beaches, dune habitats), vast coniferous forest areas, wetlands, lakes and rivers, grasslands, and agricultural lands, including polders. The study area also plays an important role in nature conservation—some parts of it are included in nature protection areas (e.g., Pape Nature Park and Bernati Nature Park) (MoEPRD, 2016). The adjacent coastal waters are important for fish spawning and nursery, as well as migrating birds and wintering waterfowl [43].

Key maritime economic activities in the study area include coastal tourism, fishing, and shipping [41]. A multimodal harbor operates in Liepaja, while Pavilosta serves for fisheries and expanding maritime tourism, particularly in the realm of yachting. The inland part of the study area is used for agriculture, forestry, and, more recently, wind energy production. Currently, wind turbines are operational in the onshore part, but there is an emerging interest among developers in building offshore wind farms in the adjacent territorial waters and Exclusive Economic Zone [26].

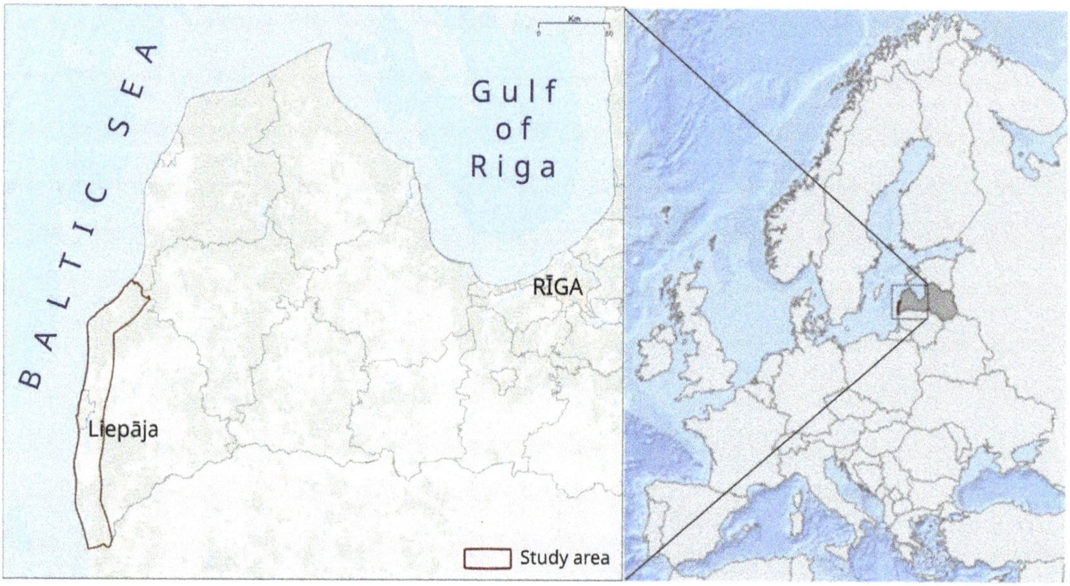

Figure 1. Location of the study area (ESRI base map).

The main stakeholder groups directly engaged in the study are representatives of local authorities, national and regional environmental and nature conservation authorities, and other governmental institutions, as well as representatives of niche tourism and renewable energy production sectors. The anticipated offshore wind energy development is raising concerns among local stakeholders regarding the potential impact on landscape qualities and coastal tourism. At the same time, stakeholders are worried about expansive, uncontrolled tourism development and insufficient tourism infrastructure, resulting in damage to the fragile coastal ecosystems and landscape.

2.2. Methods

The study's framework (Figure 2) aligns with the recommended key planning steps for coastal zone and ecosystem-based maritime spatial planning. The process commences with framing the policy theme, followed by stocktaking, assessment, analysis, and the development of proposals for decision-making [41,42]. The innovative methodological aspect of this approach lies within the robust integration of mapping and assessment of ecosystem services, coupled with the evaluation of perceived well-being benefits, providing additional perspectives for this type of planning. Furthermore, the study incorporates both expert-based (biophysical methods) and stakeholder engagement (social methods) approaches. This comprehensive and transdisciplinary approach offers additional perspectives for this type of planning.

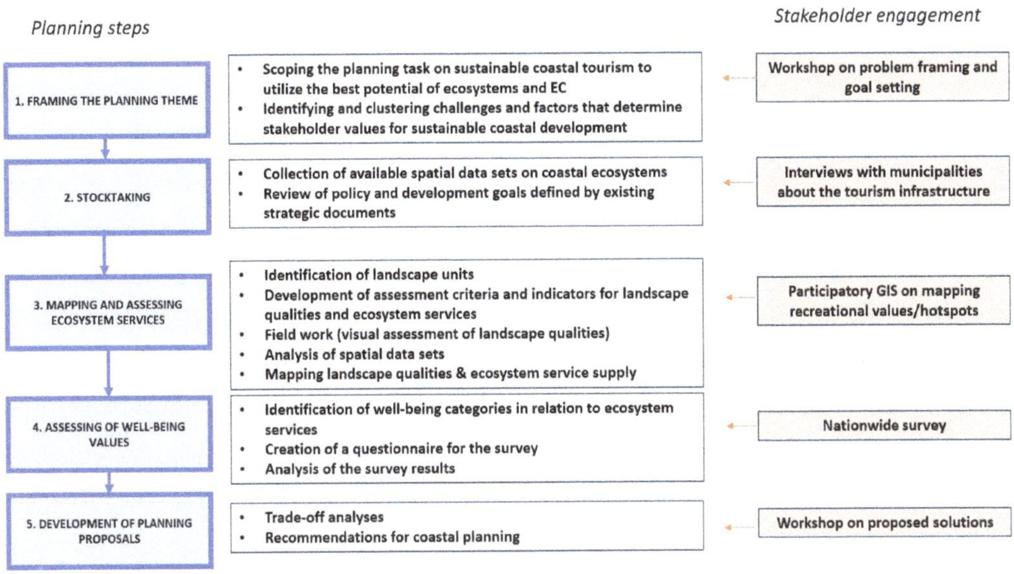

Figure 2. Framework and workflow of the study.

The subsequent thematic task bundles are designed to develop and showcase a methodology aimed at ensuring sustainable coastal development planning:

- **Biophysical mapping and assessing** coastal **CESs** involve applying a landscape characterization approach to identify and delineate landscape areas as fundamental units for CES. This process also includes the development of criteria and indicators for both quantitative and qualitative assessment, along with the comprehensive mapping and assessment of CES.
- **Stakeholder engagement** in assessing coastal development potential and priorities is facilitated through workshops, operationalizing the PGIS method for mapping recreational hotspots and incorporating local perspectives, as well as implementing a nationwide survey to evaluate the contribution of coastal ecosystems to human well-being. This combination of methods aims to contribute to the analysis of trade-offs by gathering insights from a diverse range of participants.
- **Development of planning proposals** for sustainable tourism development grounded in landscape qualities and CES, with a specific focus on addressing the trade-off between "naturalness vs. amenities".

2.2.1. Biophysical Mapping and Assessment of CES

The study involves biophysical mapping [37] of CES, adapting a landscape characterization approach developed in England and Scotland [44,45]. This approach includes two main tasks: (i) landscape characterization—conducting a desk study of available spatial data to identify key biophysical and socio-cultural features that determine landscape character, represented types, and landscape areas as an assessment unit; followed by performing a field survey to assess aesthetic and perceptual aspects of the landscape; and classifying, mapping and delineating the landscape character types and areas; (ii) landscape assessment—selecting assessment criteria and indicators; quantifying the assessment scale for defined indicators; scoring the landscape quality value for each indicator in each landscape area by analysis of the spatial data sets or expert judgment.

We used the landscape area as the spatial unit for assessing CES (i.e., service-providing area). The landscape area, defined as a discrete geographical area that is structurally and functionally relatively homogeneous, exhibiting similar land-use patterns and development history [44,46], is recognized as a suitable unit for defining landscape quality objectives [47], communicating planning solutions [48], and mapping ES [46]. We developed two typologies for delineating coastal landscape areas—one for inland and another for shoreline—due to the distinctive ecosystem structures present in these two landscape domains. The boundaries of landscape areas were delineated using spatial data on dominant land cover/land use and relief (topography maps in scale 1:50,000 and 1: 10,000, orthophoto images, Corine Land Cover maps, a digital relief model and maps of shoreline character and sediments). Additional boundaries were verified during a field survey in the summer of 2019, when each landscape area was visited to get an in-situ impression of its characteristics through photo documentation, filling a fact sheet, and expert-based assessment (with a pre-defined scoring scale) for features that cannot be calculated from spatial datasets.

Each landscape area was assessed in relation to six landscape quality criteria: aesthetic quality, uniqueness, landscape diversity, naturalness, cultural heritage, and recreational potential (Table 1). For the assessment of landscape qualities, we adapted a set of indicators developed by Zariņa et al. [49], adjusting the indicators and their quantification scale (1–5) to suit the context of the coastal landscape. The set of indicators allows the combination of the assessment of the biophysical and socio-cultural characteristics of landscape as well as objectively measurable and perceptual features. The indicators are related to the relevant CES categories according to the CICES v.5.1: "Characteristics of living systems that enable activities promoting health, recuperation or enjoyment through active or immersive interactions" (3.1.1.1); "Characteristics of living systems that enable activities promoting health, recuperation or enjoyment through passive or observational interactions" (3.1.1.2); "Characteristics of living systems that are resonant in terms of culture or heritage" (3.1.2.3); and "Characteristics of living systems that enable aesthetic experiences" (3.1.2.4).

In this assessment scheme, a score of 1 refers to the lowest value, while a score of 5 represents the highest value of the landscape characteristics along the entire Latvian coastline. The score was calculated either based on landscape metrics (e.g., the share of land cover types), number of specific features (e.g., cultural history monuments, accommodation possibilities, tourism attractions, visitors), or using expert judgment derived from field surveys' observations. Finally, we aggregated the indicator scores of each landscape area at the level of landscape quality criteria and normalized them on a scale of 0–1 (with 1 representing the highest rating). The indicators, assessment scale, and data sources used for each indicator are available in Appendix A (Table A1—for inland landscape assessment indicators and Table A2—for shoreline landscape assessment indicators).

The maps of landscape quality value at the criteria level were created using ArcGIS Arcgis Desktop 10.8 software. These maps were then published in an online platform Land-Sea-Act Map Explorer[1] to communicate the CES assessment results with stakeholders and facilitate their involvement in the development of planning solutions. The online platform was developed using ArGIS Online tools which involved combining the feature layers into "Web maps" and creating the platform with "Experience Builder".

Statistical analyses of correlations between landscape quality indicators were performed using the R programming language (version 4.3.2). The 'stats' and 'corrr' packages were employed to compute correlation coefficients and construct the correlation matrix. The Pearson correlation coefficient, along with a significance level, was computed to assess linear relationships between continuous variables (landscape quality indicators) both for shoreline and inland landscapes. Correlation coefficients were calculated using metric values for the 17 shoreline landscape area units and 55 inland landscape area units. The correlation matrices were then visually inspected to identify any strong positive or negative correlations, which were color-marked in the final table.

Table 1. Landscape quality assessment criteria and indicators and related CES (according to CICES v.5.1 [50]).

CICES Code	Landscape Quality Criteria	Domain	Indicators	Abbreviations of Indicator
3.1.2.4	Landscape aesthetic quality (LAQ)	Land	Openness of landscape	LAQ_open
		Land/Shore	Scenic views	LAQ_views
		Land/Shore	Attractive landscape elements (inland: gardens, alleys, farmstead clusters, fishermen villages, small forest patches; shoreline: fishermen boats, ships, yacht, big stones/rocks)	LAQ_attr.el.
		Land	Share of scenically attractive roads	LAQ_roads
		Land/Shore	Visual pollution and degrading elements (abandoned sites, demolished buildings)	LAQ_vis.pol.
3.1.2.4	Uniqueness of the landscape (UN)	Land/Shore	Unique landscapes on a national/regional scale	UN_unique
3.1.2.4	Landscape diversity (LD)	Land	Diversity of land use types	LD_land.use
		Shore	Shoreline diversity depending on its character (bays, capes etc.) and exposition	LD_shoreline
		Land/Shore	Small-scale landscape elements (inland: rows of trees/alleys, single large trees, piles of stones; shoreline: lighthouses, piers, ships, boats, rocks, fallen/washed out trees)	LD_small.el.
		Land	Settlement structure	LD_settl.str.
3.1.1.2.	Landscape naturalness (LN)	Land Land	Presence of natural microrelief Naturalness of water streams and rivers	LN_micro.rel. LN_water
		Land	Share of semi-natural habitats	LN_habitat
		Shore	Share of natural shoreline stretches	LN_shoreline
		Land/Shore	Share of protected areas	LN_prot.area
		Land	Presence of industrial objects (big agroindustry farms, wind-parks etc.)	LN_industr.
3.1.2.3	Cultural heritage (CH)	Land/Shore	Number of cultural history monuments	CH_No.mon.
		Land	Diversity of cultural history monuments	CH_div.mon.
		Shore	Presence of traditional land/sea use practices (e.g., fishing) in the landscape	CH_trad.use
		Land	Presence of historical spatial structures (traditional landscape, historical roads or their sections, old villages, historic distribution pattern of farmsteads)	CH_sp.str.
		Land/Shore	Visually outstanding cultural history objects (manor houses, churches, mills, bridges, mounds, fishing villages, lighthouses etc.)	CH_outst.ob.
3.1.1.1	Recreation potential (RP)	Land/Shore	Density of tourism attractions/objects	RP_den.tour.ob
		Land	Accommodation possibilities	RP_accom.
		Land	Presence of nature parks, national parks and nature monuments	RP_nat.park
		Land/Shore	Diversity of tourism infrastructure	RP_div.tour.inf.
		Land	Cycling Routes	RP_cycling
		Land/Shore	Accessibility	RP_access
		Shore	Number of visitors	RP_visitors

2.2.2. Stakeholder Engagement Methods

We employed a variety of social methods to gather insights into stakeholder concerns, values, and perceptions while implementing the planning steps presented in the framework (Figure 2). Face-to-face interaction remains crucial, especially in the initial stages of coastal or marine spatial planning [10,51]. Therefore, we organized a facilitated stakeholder workshop on problem framing and goal setting. Thematic clusters focusing on the challenges, benefits, and losses associated with different uses of coastal space and the development of coastal tourism emerged from the workshop, providing valuable input for further stakeholder engagement activities. Additionally, interviews were conducted with representatives from local municipalities responsible for spatial and development planning. Another facilitated workshop was organized in the final step of the study to discuss the development proposals for coastal tourism. The interactions between the research team and stakeholders served as a valuable forum for sharing experiences and debating the relevance of the coastal ecosystems and landscapes to their lives.

A methodological advancement was dedicated to the following two tasks: (i) operationalizing Participatory GIS to collect local knowledge on the spatial distribution and significance of CES and (ii) implementing a nationwide survey to evaluate the contribution of coastal ecosystems to human well-being.

Participatory GIS (PGIS)

The PGIS method supports planners and researchers in engaging with people in the generation of spatially explicit information for a variety of planning and decision-making purposes [33]. In this study, it was applied to (i) collect local knowledge to articulate the CES values, (ii) to verify the biophysical mapping results on recreational potential, and (iii) to support the development of proposals for coastal tourism development.

The online PGIS survey was created using the ArcGIS Online Tool "Survey123" and launched in the autumn of 2020. Respondents of the survey were asked to pinpoint locations they deemed critical for recreation and tourism and then to complete a brief questionnaire detailing the functionality or suitability for concrete activities of the identified site. The respondents were allowed to allocate several functions to one site. The PGIS survey utilized in the study is available in Appendix C.

The primary target audience for the survey was stakeholders of the local communities in the study area. The survey was disseminated through workshops, local municipality websites, and social media channels. During the surveying process, data on 82 sites were collected. The data on site functionality was used to associate the sites with CES during the formulation of tourism and recreation proposals linked to the method described in Section 2.2.3.

National scale social survey

A nationwide scale Computer-Assisted Web Interviewing (CAWI) social survey (n = 1000) was conducted in May 2021. The main aim of the survey was to obtain data on the perceived well-being benefits provided by the ecosystems of Southwestern Kurzeme coastal area both to local residents and visitors.

The survey content was developed by the authors of the study. The survey was distributed by an outsourced sociology company, which ensured a representative sample of respondents from the general national population (aged 18–74) by gender, age, region, type of settlement, and nationality. Survey respondents were selected using random stratified sampling.

Out of the total 1000 survey respondents, 90 were local inhabitants of the study area. 224 respondents stated that they have never visited the Southwestern Kurzeme coastal area and therefore were excluded from the analysis of perceived well-being benefits provided by the area. Thus, the responses represent the views of 686 visitors and 90 local inhabitants.

The survey consisted of four distinct sets of questions:

- Respondent's connection to the Southwestern Kurzeme coastal area: in this section, respondents indicated their connection to the area (local resident/visitor), primary objectives of their visits, and other relevant visitation details.
- Perceived well-being benefits associated with the Southwestern Kurzeme coastal area: this section was exclusively presented to respondents who previously indicated that they had visited the area. The study encompassed seven well-being dimensions derived from a literature analysis [14,49,50] and then tailored to the study context. The seven well-being dimensions included in this study were: (i) safety and sense of security, (ii) financial/material security, (iii) leisure, (iv) health and life satisfaction, (v) social relations, (vi) education, and (vii) ecological contentment. Respondents were asked to rate statements (Appendix B, Q5) related to the various well-being dimensions on a scale from 1 to 4 where 1—completely disagree, 2—partially disagree, 3—partially agree, and 4—completely agree. To accommodate situations where participants may not have a well-formed opinion on the statement, an additional response option, "I don't know" was included in the potential response list. These responses were not included in the final analysis of perceived well-being benefits. As a result, some of the statements have varying numbers of responses.
- Preferences for coastal recreation: This section focused on identifying factors that influence the selection of a location for coastal recreation, both ecosystem and amenity-related. Respondents were asked to rate the importance of various features (Appendix B, Q10) on a scale from 1 to 4 where 1—completely unimportant, 2—minor importance, 3—important but not decisive, and 4—very important. To accommodate situations where participants may not have a well-formed opinion on a feature, an additional response option, "I don't know" was included in the potential response list. These responses were not included in the final analysis of the preferences.
- Demographics: The final section was dedicated to the collection of demographic data from the respondents.

The questionnaire is available in Appendix B.

The statistical analysis of the survey results was performed in JASP 0.18.3 software [52]. The Mann-Whitney U-test—a non-parametric statistical test that is often utilized to evaluate if statistically significant differences between two independent groups exist—was employed to test for potential differences in perceived well-being benefits between local inhabitants of the Southwestern Kurzeme coastal area and visitors to this area.

2.2.3. Development of Planning Proposals

The development of planning proposals for coastal tourism development served as the final planning step within the framework and workflow outlined in the study. Based on landscape character assessment, landscape areas were segmented into clusters based on high values in certain landscape quality criteria. The aim of the clustering is to highlight the existing high values (above average in the given landscape criteria group) of the landscape areas and to use them as the main driver for tourism development. A single landscape area may correspond to several clusters in terms of its characteristics, e.g., it may have both high natural and aesthetic values. In such case, tourism development should consider all values present in the landscape area, ranking them in order of importance (from highest priority): (1) naturalness values, (2) cultural and historical values, and (3) aesthetic values. The gradation of importance is based on the normative weight of the given values. The protection of nature values is linked to EU-level legislation (e.g., Habitats Directive). Cultural heritage is protected under national or local regulations (e.g., National Law on Protection of Cultural Monuments). In addition, landscape areas were assessed in relation to the use of their recreational potential–an area with high scores for indicators representing tourism interest ("Density of tourism attractions/objects" and "Presence of nature parks, national parks, and nature monuments"), but low scores for indicators of tourism infrastructure and accessibility were indicated as "areas with underused recreational potential". The results of the PGIS complemented the clustering outcomes during the development of solutions.

For each cluster, recommendations for tourism development were proposed. General recommendations relevant to planning tourism development in all landscape areas were summarised separately. The protection regime of aesthetic values could potentially be included in municipal spatial or development planning documents. Identified landscape area clusters and recommendations for tailoring tourism development to site-specific values/landscape qualities were elaborated and discussed with stakeholders during a dedicated workshop.

3. Results

3.1. Biophysical Mapping and Assessment of Cultural Ecosystem Service

3.1.1. Identification of Landscape Areas

For landscape characterization, we have delineated 17 relatively homogeneous seashore landscape areas and 55 inland landscape areas in the study area. The landscape areas were classified into seven shoreline landscape types (based on seashore width; material texture—sand, gravel, pebbles; and morphology) and 11 inland landscape types (based on land-cover type, land-use and function patterns, historical development, and visual features) (Figure 3). The dominant landscape type on shore is sandy beach (covering 62% of the shoreline in the study area), followed by gravel-sand-pebble beach (with boulders) with cliff (10%), sand-gravel-pebble beach (7%) and sandy beach with cliff (6%). Other types—sandy beaches with variable pebble content with (and without) cliff as well as anthropogenic type—ports covered 5% of the shoreline each. The inland area is dominated by forest landscapes in a flat terrain (covering 44%), followed by agricultural landscapes in a flat terrain (12%), polder landscape (10%), wetland and lake landscapes (9%), and agricultural landscape in an undulated terrain (6%). Other inland landscape types are formed by wooded dunes, mires, rivers, villages and urban landscapes, as well as mixed forest-rural landscapes in flat terrain, each covering 2–5%.

3.1.2. Assessment of Landscape Qualities

Inland landscape qualities were assessed by aggregating and normalizing the indicator value of each landscape area on the level six quality criteria (Figure 4). Landscape aesthetic quality indicators include "Openness of landscape", "Scenic views", "Attractive landscape elements", "Share of scenically attractive roads" as well as "Visual pollution and degrading elements" (the last one was assessed using the reverse scale).

Thus, the highest scores mostly received open landscape types dominated by agricultural land and coastal settlements, wooded dunes, wetlands, and lakes. Uniqueness was represented only by one indicator—"Unique landscapes on a national/regional scale", derived from the national scale survey of "Landscape treasures"[2], highlighting an urban landscape—Liepaja city, one village landscape, as well as a few areas representing outstanding agricultural landscape features. Landscape diversity scores were calculated based on the diversity of land use types, small-scale landscape elements, and settlement structure. Accordingly, the highest value again was received by agricultural lands and coastal settlements, mixed forest-rural landscapes, and river valleys. The landscape naturalness was assessed by the "Presence of natural microrelief", "Naturalness of water streams and rivers", "Share of semi-natural habitats" and "Protected areas", as well as the "Presence of industrial objects" (the last one was assessed using the reverse scale). Accordingly, the highest values were received by wooded dunes, wetlands and lakes, and few forest landscape areas. The cultural heritage value was assessed by the number and diversity of "cultural history monuments", "Presence of historical spatial structures" (like traditional landscapes and historic roads) and "Visually outstanding cultural history objects". The highest values were received by landscape areas with coastal villages and Liepaja City, as well as a few agricultural landscape areas. The recreational potential was assessed by "Density of tourism attractions", "Accommodation possibilities", "Presence of nature parks, national parks and nature monuments", "Diversity of tourism infrastructure", "Cycling Routes" as well as "Accessibility". The highest scores were received by the majority of the landscape areas next to the shoreline, as well as the ones including water bodies with recreational facilities.

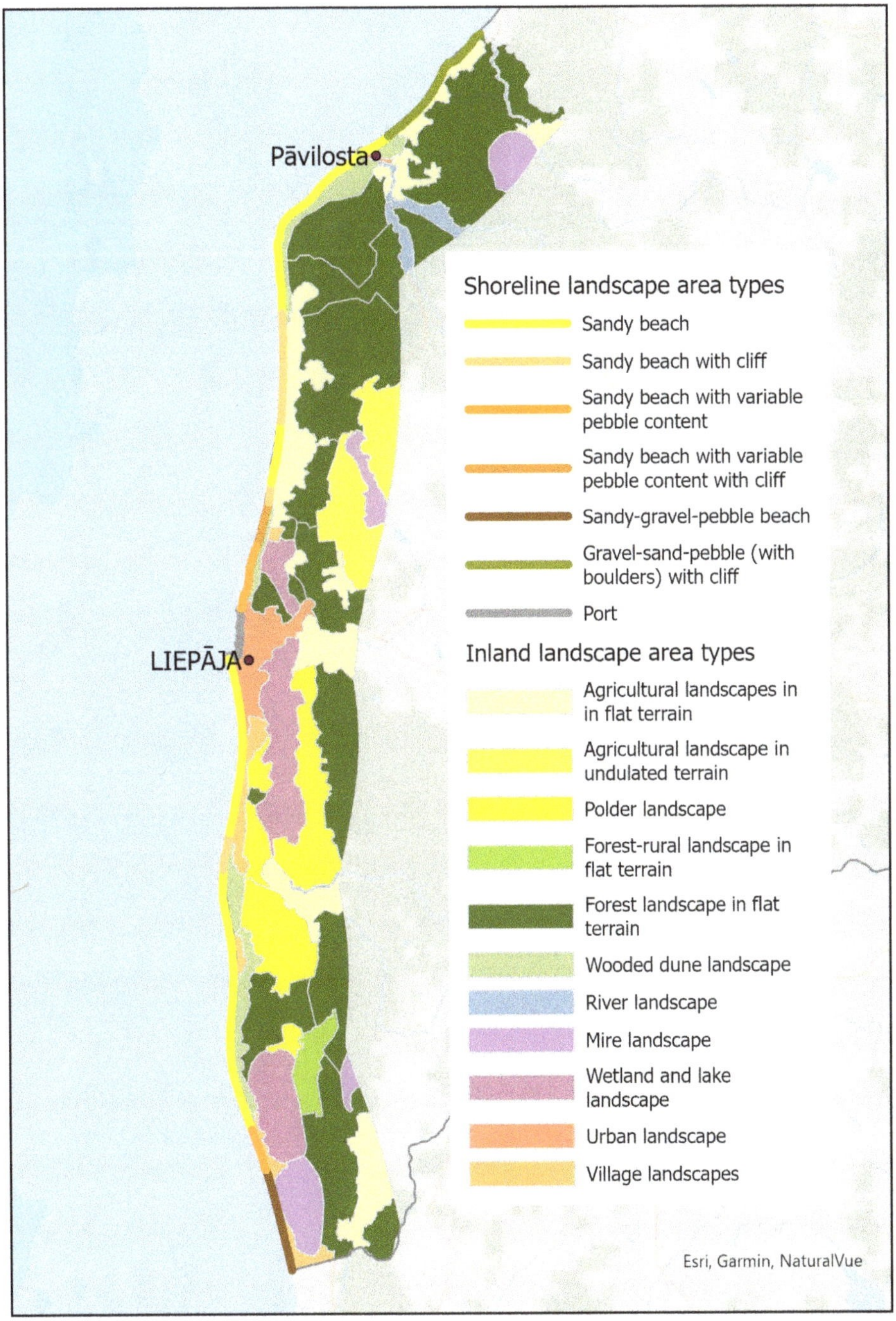

Figure 3. Identified inland and shoreline landscape areas and their classification in landscape types.

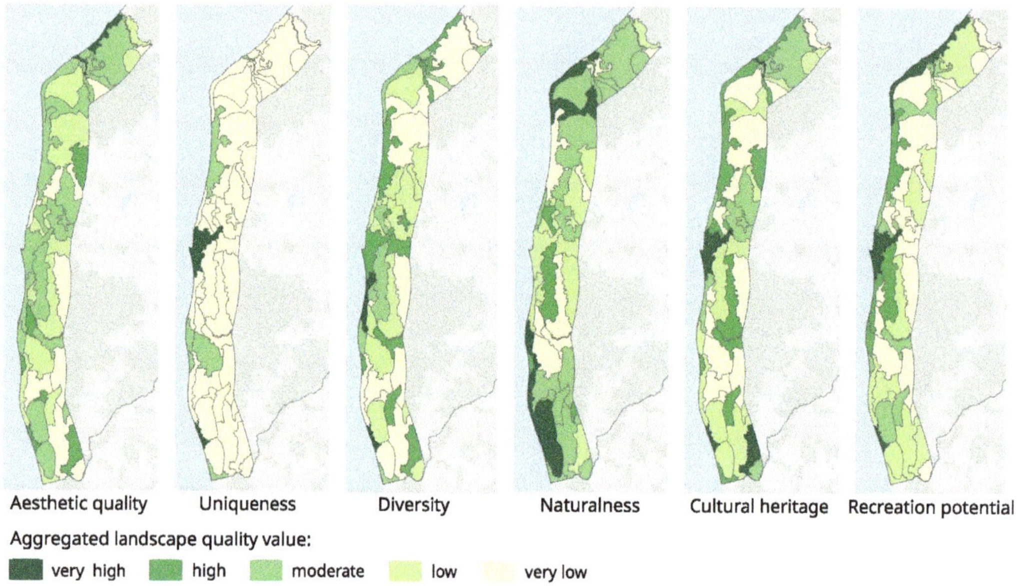

Figure 4. Aggregated value of landscape qualities in inland landscape areas (darker color—higher value).

Shoreline landscape qualities (Figure 5) were assessed using the same criteria, while the indicators were partly adjusted to shoreline specifics. For aesthetic quality assessment on shore, we used only "Scenic views", "Attractive landscape elements" and "Visual pollution and degrading elements".

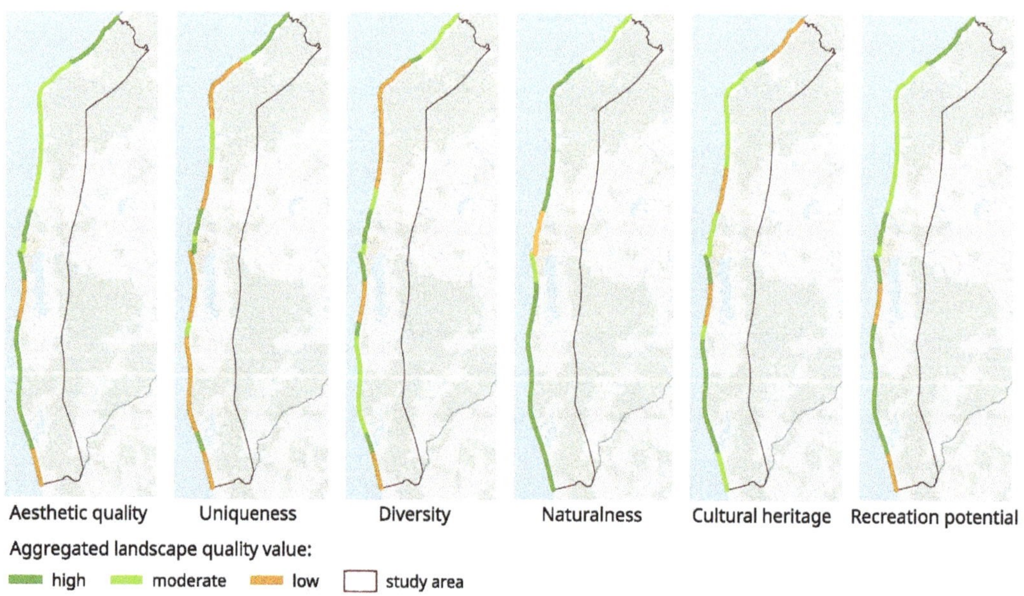

Figure 5. Aggregated value of landscape qualities in shoreline landscape areas (dark green color—very high value; light green—high value; orange—medium value).

Over half of the shoreline was assessed to have high aesthetic value, capturing coastal stretches with cliffs and the stretches next to populated areas. The uniqueness value was highest in the shoreline stretches with cliffs near Liepaja city and a traditional fishermen village near Pape. The shoreline diversity was assessed using its character (bays, capes, etc.) and exposure, as well as small-scale landscape elements (e.g., lighthouses, piers, ships, boats, rocks, fallen/washed-out trees, etc.), also resulting in the highest scores in stretches next to populated areas. Naturalness included two indicators—"Share of natural shoreline stretches" and "Share of protected areas", revealing that most of the shoreline can be considered natural. Cultural heritage was assessed by indicators "Number of cultural history monuments", "Presence of traditional land/sea use practices" (e.g., fishing) and "Visually outstanding cultural history objects", again revealing higher values next to populated areas. Recreation potential (assessed by "Density of tourism attractions/objects", "Diversity of tourism infrastructure", "Accessibility" and "Number of visitors") was also higher in populated areas but more related to the distribution of aesthetic qualities.

3.1.3. Statistical Analysis of Landscape Quality Trade-Offs

The relationships between shoreline landscape quality indicators were explored through the construction of a correlation matrix (Table 2), providing insights into the strength and direction of the associations among certain landscape quality indicators. Strong positive correlations (r > 0.5) were observed between variables "Uniqueness of landscape" (UN_unique) and "Attractive landscape elements" (LAQ_Attr.el) (r = 0.63, $p < 0.01$), between "Attractive landscape elements" and "Shoreline diversity" (LD_shorline) (r = 0.76, $p < 0.001$), presence of "Small landscape elements" (LD_small.el) (r = 0.89, $p < 0.001$), and "Visually outstanding cultural history objects" (CH_outst.ob.) (r = 0.58, $p < 0.05$), between "Small landscape elements" and "Visually outstanding cultural history objects" (r = 0.62, $p < 0.01$), between "Shoreline naturalness" (LN_shoreline) and "Share of protected areas" (LN_prot.area) (r = 0.57, $p < 0.05$), between presence of "Number of cultural history monuments" (CH_No.mon.) and "Presence of traditional land/sea use practices" (CH_trad.use) (r = 0.58, $p < 0.05$). Also, positive correlations were observed among all tourism-related landscape quality indicators. Strong negative correlations (r < −0.5) were observed among "Shoreline naturalness" and the presence of "Attractive landscape elements" (r = −0.70, $p < 0.005$), "Shoreline diversity" (r = −0.57, $p < 0.05$) and "Visually outstanding cultural history objects" (r = 0.75, $p < 0.001$). Possible multicollinearity (*r* value as close to 1) could be observed among such variables as "Diversity of tourism infrastructure" and "Density of tourism attractions/objects" and between "Attractive landscape elements" and "Small landscape elements".

Table 2. Correlation matrix of landscape quality indicators in shoreline landscape, with cells highlighted in green indicating strong positive correlations (r > 0.5) and cells highlighted in red indicating strong negative correlations (r < −0.5).

	LAQ views	LAQ Attr.el.	LAQ vis.pol.	UN unique	LD shoreline	LD small.el.	LN shoreline	LN prot.area	CH No.mon.	CH. trad.use	CH outst.ob.	RP den.tour.ob	RP div.tour.inf.	RP access.	RP visitors
LAQ_views	1														
LAQ_attr.el.	0.33	1													
LAQ_vis.pol.	0.41	−0.31	1												
UN_unique	0.63	0.37	0.22	1											
LD_shoreline	0.44	0.76	−0.01	0.57	1										
LD_small.el.	0.30	0.89	−0.30	0.49	0.78	1									
LN_shoreline	−0.25	−0.70	0.33	−0.51	−0.57	−0.79	1								
LN_prot.area	−0.22	−0.29	0.34	−0.44	−0.20	−0.41	0.57	1							
CH_No.mon.	0.15	0.25	0.02	0.20	0.21	0.47	−0.39	0.11	1						
CH_trad.use	0.36	0.32	0.31	0.20	0.38	0.45	−0.13	0.26	0.58	1					
CH_outst.ob.	0.22	0.58	−0.37	0.11	0.45	0.62	−0.75	−0.43	0.30	−0.05	1				
RP_den.tour.ob	0.44	0.20	0.25	0.20	0.49	0.39	−0.17	−0.01	0.45	0.39	0.41	1			
RP_div. tour.inf.	0.44	0.27	0.24	0.18	0.47	0.40	−0.28	−0.11	0.42	0.29	0.50	0.92	1		
RP_access	0.23	0.23	0.04	0.11	0.32	0.37	−0.04	−0.38	−0.03	0.03	0.29	0.64	0.58	1	
RP_visitors	0.028	0.21	0.05	−0.07	0.46	0.26	−0.06	0.02	−0.04	−0.03	0.38	0.68	0.65	0.75	1

Through the creation of a correlation matrix, the relationships between inland landscape quality indicators were investigated, offering insights into the direction and intensity of linkages among certain landscape quality indices. Strong positive correlations (r > 0.5) were observed between more than 40 pairs of variables (Table 3). Variable "Small landscape elements" (LD_small.el) composed seven pairs of strong correlation with variables mostly from recreational landscape qualities (i.e., "RP_accom" (r = 0.79, $p < 0.001$) and "Accessibility" (RP_access) (r = 0.65, $p < 0.001$), as well as landscape qualities associated with cultural heritage (i.e., "Presence of historical spatial structures" (CH_sp.str) (r = 0.71, $p < 0.001$)). Variables associated with recreational landscape qualities displayed a strong correlation among themselves (i.e., "Accessibility" (RP_access) and "Diversity of tourism objects" (RP_div.tour.ob) (r = 0.86, $p < 0.001$)) and among landscape qualities associated with cultural heritage, for instance variable "Visually outstanding cultural history objects" (CH_outst.ob.) showed strong correlation with variables "Accommodation" (r = 0.58, $p < 0.001$), "Diversity of tourism object" (r = 0.64, $p < 0.001$), "Cycling routes" (RP_cycling) (r = 63, $p < 0.001$) and "Accessibility" (r = 0.62, $p < 0.001$). Variable "Scenic views" had strong positive relations with indicators from all landscape quality groups—"Attractive landscape elements" (r = 0.75, $p < 0.001$), "Small landscape elements" (r = 0.54, $p < 0.001$), "Scenic views" and "Presence of natural microrelief" (LN_mikro.rel) (r = 0.58, $p < 0.001$), "Presence of historical spatial structures" (r = 0.65, $p < 0.05$) and "Diversity of tourism objects" (r = 0.57, $p < 0.001$). There was revealed only one strong negative correlation between "Visual pollution and degrading elements" (LAQ_vis.pol) and "Presence of historical spatial structures" (r = −0.63, $p < 0.001$). Possible multicollinearity was observed between "Diversity of tourism objects" and "Accessibility".

Table 3. Correlation matrix of landscape quality indicators in inland landscape areas, with cells highlighted in green indicating strong positive correlations (r > 0.5) and cells highlighted in red indicating strong negative correlations (r < −0.5).

	LAQ_open	LAQ_views	LAQ_attr.el.	LAQ_roads	LAQ_vis.pol.	UN_unique	LD_land.use	LD_small.el	LD_settl.str.	LN_micro.rel.	LN_water	LN_habitat	LN_prot.area	LN_industr.	CH_no.mon.	CH_div.mon.	CH_sp.str.	CH_outst.ob.	RP_den.tour.ob	RP_accom.	RP_nat.park	RP_div.tour.ob	RP_cycling	RP_access
LAQ_open	1.00																							
LAQ_views	0.06	1.00																						
LAQ_attr.el.	0.13	0.75	1.00																					
LAQ_roads	0.35	0.42	0.33	1.00																				
LAQ_vis.pol.	−0.44	−0.10	−0.10	−0.41	1.00																			
UN_unique	0.09	0.20	0.44	0.17	−0.29	1.00																		
LD_land.use	0.36	0.16	0.10	0.13	−0.43	−0.13	1.00																	
LD_small.el	0.16	0.54	0.69	0.39	−0.41	0.65	0.08	1.00																
LD_settl.str.	0.33	0.24	0.34	0.60	−0.63	0.44	0.27	0.55	1.00															
LN_micro.rel.	0.25	0.58	0.37	0.15	0.00	0.13	0.03	0.18	−0.05	1.00														
LN_water	−0.39	0.12	0.20	−0.22	0.38	−0.05	−0.34	−0.08	−0.30	0.31	1.00													
LN_habitat	−0.17	0.33	0.34	−0.20	0.11	0.10	−0.03	0.19	−0.17	0.42	0.41	1.00												
LN_prot.area	0.12	0.04	0.23	−0.23	0.17	0.14	−0.01	0.05	−0.24	0.19	0.21	0.64	1.00											
LN_industr.	−0.44	0.00	−0.07	−0.21	0.65	−0.26	−0.27	−0.32	−0.47	0.31	0.44	0.14	0.25	1.00										
CH_no.mon.	0.22	0.21	0.14	0.47	−0.40	0.31	0.18	0.30	0.40	0.21	−0.23	−0.10	−0.05	−0.19	1.00									
CH_div.mon.	0.25	0.38	0.34	0.45	−0.35	0.36	−0.01	0.40	0.44	0.21	−0.10	−0.01	−0.08	−0.32	0.69	1.00								
CH_sp.str.	−0.01	0.65	0.74	0.21	−0.22	0.42	0.13	0.71	0.36	0.39	0.21	0.37	0.16	−0.10	0.13	0.20	1.00							
CH_outst.ob.	0.02	0.46	0.49	0.36	−0.32	0.40	0.13	0.55	0.55	0.12	0.01	0.07	−0.26	−0.31	0.33	0.51	0.44	1.00						
RP_den.tour.ob	0.03	0.50	0.60	0.24	−0.13	0.51	−0.02	0.65	0.33	0.29	0.02	0.48	0.24	−0.17	0.28	0.29	0.55	0.48	1.00					
RP_accom.	0.16	0.41	0.46	0.35	−0.33	0.63	0.12	0.79	0.57	0.07	−0.13	0.13	−0.04	−0.30	0.37	0.40	0.52	0.58	0.62	1.00				
RP_nat.park	0.02	0.25	0.30	0.16	0.12	0.17	−0.14	0.24	−0.09	0.26	0.26	0.36	0.51	0.17	0.35	0.24	0.18	0.18	0.40	0.21	1.00			
RP_div.tour.ob	0.08	0.57	0.65	0.21	−0.10	0.26	−0.01	0.64	0.20	0.19	0.16	0.33	0.14	−0.10	0.12	0.27	0.57	0.64	0.51	0.45	0.44	1.00		
RP_cycling	−0.04	0.40	0.43	0.31	−0.11	0.47	0.00	0.55	0.39	0.28	0.00	0.20	−0.03	−0.09	0.21	0.33	0.24	0.63	0.60	0.57	0.26	0.42	1.00	
RP_access	0.08	0.49	0.61	0.15	−0.25	0.36	0.03	0.65	0.28	0.15	0.00	0.29	0.12	−0.22	0.15	0.30	0.57	0.62	0.52	0.42	0.22	0.86	0.42	1.00

3.2. *Assessment of the Perceived Well-Being Benefits Provided by Southwestern Kurzeme COASTAL Area*

We focused our analysis of the nationwide survey responses on two key aspects: (i) the main perceived well-being benefits provided by the Southwestern Kurzeme coastal area in general and (ii) the distinction between regional and national perspectives (locals vs. visitors) on the values of well-being benefits.

Well-being benefits were divided into seven broad dimensions: safety and sense of security; financial/material security; leisure; health and life satisfaction; social relations; education; and ecological contentment.

3.2.1. Perceived Well-Being Benefits from Coastal Ecosystems

In the survey, respondents were asked to rate several statements (Table 4, column "Statement included in the survey") related to perceived well-being benefits provided by the Southwestern Kurzeme coastal area on a scale from 1 to 4 where: 1—completely disagree, 2—partially disagree, 3—partially agree, 4—completely agree. To accommodate situations where participants may not have a well-formed opinion on the statement, an additional response option, "I don't know" was included in the potential response list. However, these responses were not included in the analysis of perceived well-being benefits. As a result, some of the statements have varying numbers of responses.

Table 4. Results of perceived well-being benefit assessment.

Well-Being Dimension	State-Ment No.	Statement Included in the Survey	No of Responses (Excl. "I Don't Know")		Mean	*p*-Value
Health and life satisfaction	1	Visiting the Southwestern Kurzeme coastal area provides me with an opportunity to relax and clarify my thoughts	Locals Visitors	87 564	3.506 3.041	<0.001
Health and life satisfaction	2	I feel healthier, more energetic after visiting the Southwestern Kurzeme coastal area	Locals Visitors	86 546	3.512 3.046	<0.001
Financ./ material security	3	The natural resources provided by the Southwestern Kurzeme coastal area significantly contribute to my financial well-being (my work is related to the coast—hospitality, fishing, tourism, etc.)	Locals Visitors	79 564	2.051 1.711	0.002
Financ./ material security	4	The opportunity to fish, pick berries, mushrooms and other goods for self-consumption on the Southwestern Kurzeme coastal area is important for me	Locals Visitors	87 546	3.540 2.489	<0.001
Social relations	5	The Southwestern Kurzeme coastal area plays a crucial role in my social life: I spend time with my family and/or friends here, engage in social activities	Locals Visitors	87 543	3.264 2.223	<0.001
Education	6	Exploring the Southwestern Kurzeme coastal area has enabled me to acquire new knowledge (e.g., about nature, the sea)	Locals Visitors	81 537	3.136 2.784	<0.001
Leisure	7	The Southwestern Kurzeme coastal area holds a significance for me in terms of recreation opportunities	Locals Visitors	86 579	3.651 2.926	<0.001
Safety and sense of security	8	The role of the Southwestern Kurzeme coastal area in strengthening the national military security holds importance for me	Locals Visitors	70 485	2.457 2.233	0.095
Safety and sense of security	9	The potential energy independence that can be achieved through the establishment of wind farms along the Southwestern Kurzeme coastal area is vital for me	Locals Visitors	79 516	2.734 2.659	0.376
Safety and sense of security	10	The Southwestern Kurzeme coastal area provides me with a sense of security—I feel protected from the effects of climate change, natural disasters (storms, floods etc.)	Locals Visitors	78 470	2.808 2.398	<0.001
Ecological content-ment	11	I hold the nature diversity of the Southwestern Kurzeme coastal area in high regard, and its preservation brings me a sense of satisfaction	Locals Visitors	86 590	3.547 3.197	<0.001

According to the survey results, in general people in Latvia value ecological contentment the highest, which refers to the joy derived from the awareness of the diverse ecosystems on the Southwestern Kurzeme coastal area and their protection, followed by perceived leisure and health benefits (Figure 6). The lowest rates were allocated to perceived financial benefits and safety and security.

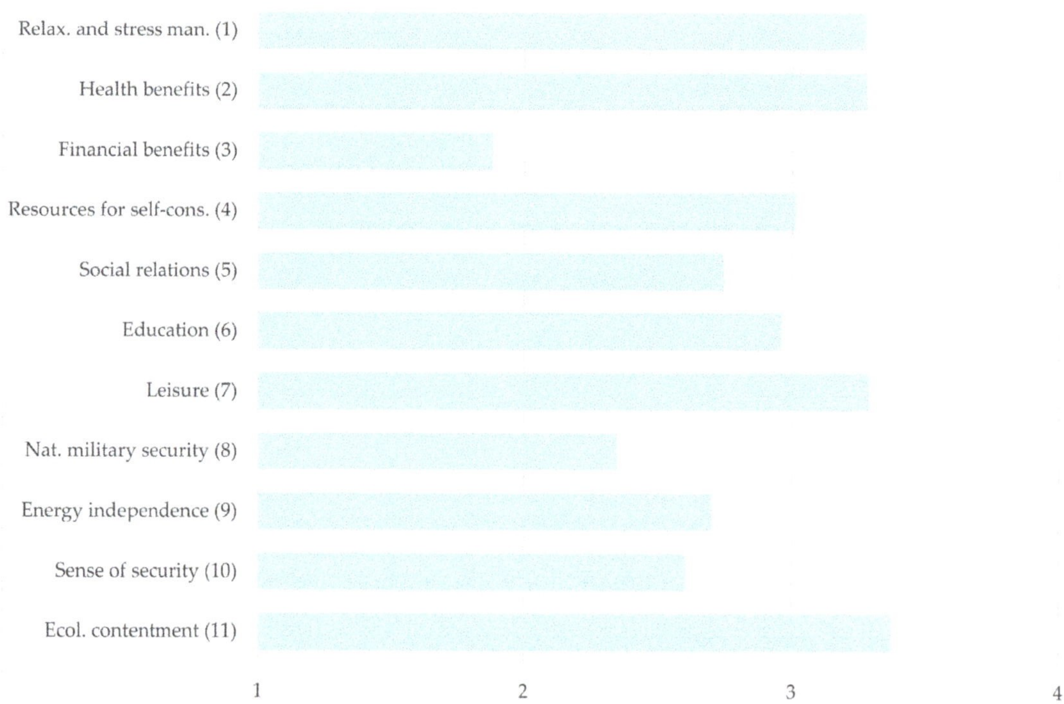

Figure 6. Mean ratings of the perceived well-being benefits based on the survey results. Values of the rating scale: 1—completely disagree, 2—partially disagree, 3—partially agree, 4—completely agree. For full statements included in the survey see Table 4—numbers in brackets correspond to table column "Statement No".

3.2.2. Perceptions of Inhabitants versus Visitors (Regional versus National Benefits)

The survey results elucidated significant differences between the valuation of perceived well-being benefits between respondents who indicated they live in the study area (locals) and the respondents who only visit the area. The locals value all perceived well-being benefits provided by the Southwestern Kurzeme coastal area more than visitors. The statistical analysis (Table 4) indicates that the differences are statistically significant ($p < 0.005$) for all well-being dimensions included in the survey, except energy independence and national military security (however, both are still rated slightly higher by the locals).

Leading preferences in both groups were similar. Both allocated the highest rates to nature diversity and protection, which is linked to the well-being dimension of ecological contentment. It was closely followed by perceived health and recreational benefits. The highest differences between groups were linked to resources for self-consumption and social relations that were valued significantly higher by local inhabitants.

3.2.3. Necessity for Amenities for Coastal Recreation

Another objective of the survey was to illuminate the preferences of Latvian residents concerning coastal recreation sites, with a particular focus on pinpointing the factors that influence the choice of coastal recreation sites and determining preferences for beach types, distinguishing between natural beaches and those equipped with amenities. In the context of this study, the term "natural beach" refers to shoreline stretches that are largely unaltered by human interventions and are characterized by more natural settings—minimal to no human-built structures, natural landscapes, relatively natural and undisturbed ecosystems. The term "beaches with amenities" is used for such stretches where natural beaches have been significantly modified by human-made elements to cater to the needs and comfort of visitors—e.g., cafes, changing rooms, children's playgrounds, benches, recreational equipment rentals.

Based on the results of the survey, approximately 64% express a preference for beaches characterized by more natural features, while 36% favor beaches equipped with amenities for recreational purposes. The assumption that Latvian residents prefer the natural coast is further substantiated by the importance ratings assigned to factors influencing the selection of coastal recreation sites in the survey results (Figure 7). The highest importance was associated with the cleanliness of the site, followed by the attractiveness of the landscapes and natural diversity. The availability of amenities and services and public transport received the lowest rates.

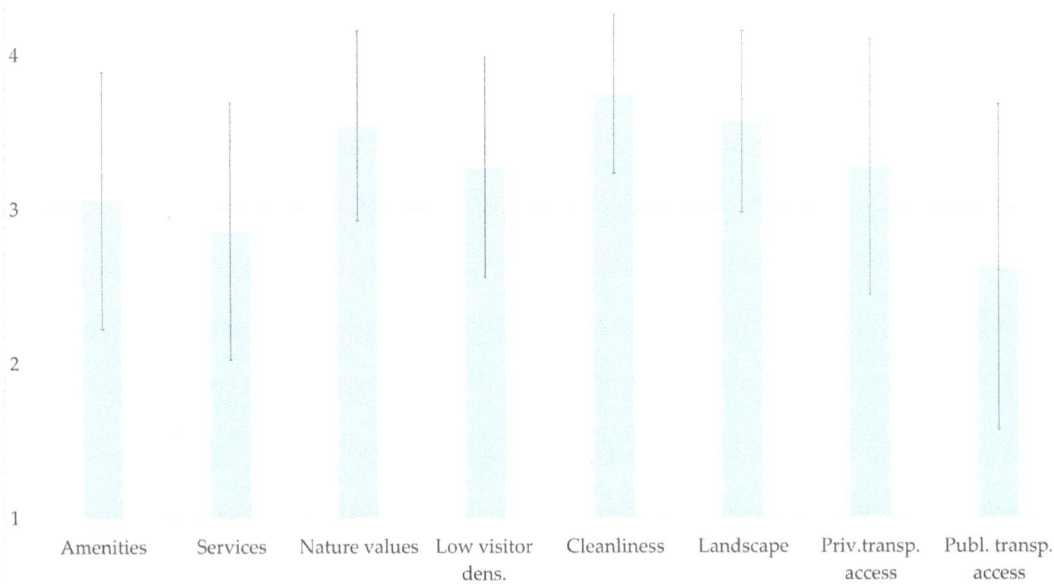

Figure 7. Mean ratings including Standard deviation (SD) of the factors that influence the choice of coastal recreation site. Values of the rating scale: 1—completely unimportant, 2—minor importance, 3—important but not decisive, 4—very important.

3.3. *Proposals for Coastal Tourism Development*

The solutions for sustainable tourism development in the Southwestern Kurzeme coastal area were constructed based on the results of landscape quality assessment and the evaluation of current tourism and recreational potential in each landscape area. We identified 31 landscape areas with high naturalness value, 24 landscape areas with high cultural heritage value, and 30 landscape areas with high aesthetic value (Figure 8).

For the creation of the solutions, data from the PGIS exercise were analyzed to draw attention to favored places and their features. In total, the PGIS survey identified 82 points of tourism and recreational interest. The results highlight the multifunctionality of the coast, with 80 of the surveyed points being associated with the potential to perform several socio-cultural activities. The predominant functions were landscape appreciation, followed by hiking and walking and relaxation by the sea (Table 5).

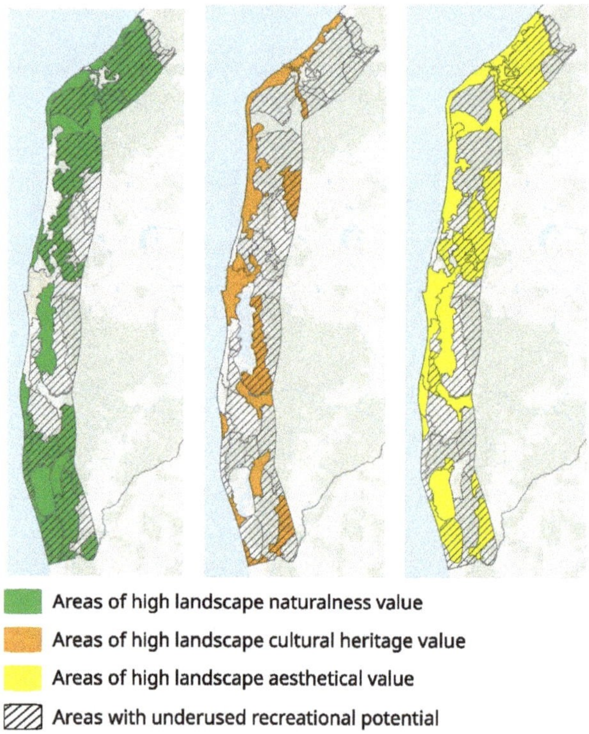

Figure 8. Landscape based solutions for sustainable tourism development: colors indicate areas of specific value and the striped pattern—areas with underused recreational potential.

Table 5. The number of mapped and described locations suitable for conducting identified socio-cultural activities in the study areas.

Socio-Cultural Activities	No of PGIS Points in the Study Area
Enjoying landscape/seascape	73
Hiking and walking	70
Relaxation by the sea	60
Overnighting	31
Bird watching	28
Exploring cultural history	27
Fishing	26
Public entertainment events	25
Sports	19
Water sports	18
Yacht and boat docks	14
Jet skis	12
Hunting	3

The proposed solutions are based on the premise that tourism development should not only focus on the economic interests of the tourism industry but also respect nature values, landscapes, place identity, and cultural heritage of the area—factors that have provided the resources for tourism development potential. In addition, the proposed tourism development solutions for the Southwestern Kurzeme coastal area would facilitate and ensure that the increased tourism intensity does not harm (and even improve) the well-being of the local communities, as it aims at highlighting and preserving the values that are important to them.

The proposition for the direction of tourism development was constructed for each identified cluster of landscape values (Table 6). In the cluster of landscape areas with high naturalness values, the direction of tourism development should be linked to highlighting the naturalness of the landscape while considering that tourism development activities that conflict with the preservation of naturalness or nature conservation objectives should be avoided. In these areas, particular attention should be paid to monitoring and prevention of anthropogenic pressures. In the cluster of landscape areas with high cultural heritage values, the development of tourism should be linked to the sustainable integration of cultural heritage into the tourism offer. Appropriate management of the heritage is essential to ensure that the area's tourism development potential is not undermined. In the cluster of landscape areas with high aesthetic values, the direction of tourism development in this cluster is related to the integration of the aesthetic value of landscapes into the tourism offer. In this cluster, the development of public infrastructure should be carried out in a way that does not conflict with the aesthetic value. Information on landscape areas with underused recreational potential serves as additional guidance for planners, identifying priority areas for infrastructure improvements, but infrastructure solutions should be designed according to the recommendations for each cluster.

Table 6. Recommendations to planners for sustainable coastal tourism development.

Cluster	Characteristics of the Cluster	Recommendations
Areas of high aesthetic value	scenic views, open landscape with attractive landscape elements and roads, no visual pollution	Landscape experts should be involved in identification of valuable landscapes and integration of the landscape qualities in spatial planning/development documents. Access to the attractive landscape sites/scenic views should be improved, including suitable public infrastructure and information for tourism marketing.
Areas of high naturalness value	high share of protected areas, natural or semi-natural land cover and natural (not modified) streams	Nature experts should be involved in assessment of anthropogenic pressure and developing measures for risk prevention as well as ensure appropriate integration of the nature values in the spatial planning/development documents. Suitable public infrastructure should be developed or improved for reducing anthropogenic pressure to vulnerable habitats and species, not contrasting with landscape character, and including nature education elements
Areas of high cultural heritage value	high number and diversity of cultural heritage monuments, historic land use features and visually impressive historic objects	Cultural heritage experts should be involved in identification of cultural heritage values and ensuring their appropriate integration in the spatial planning/development documents. Action plans should be developed for better integration of cultural heritage in tourism offer. Connection between nature and cultural heritage should be highlighted (e.g., coastal meadows, grey dunes etc.). Public infrastructure should be harmonic with the cultural historic identity of the site.

4. Discussion

4.1. Landscape Character and Qualities for Enhanced Coastal CES Assessment

In our study, we use a landscape characterization approach in mapping and assessment of CES. The landscape, characterized by its fusion of biophysical and socio-cultural dimensions—provides a more comprehensive and robust understanding of CES [53]. Medeiros et al. [54] have highlighted the need for this comprehensive approach, particularly in the integration of socioeconomic, perceptual, and historical-cultural indicators. Landscape quality indicators play a crucial role in sustainable planning, as they help identify resources, qualities, and criticalities of a particular area [55]. These indicators should be flexible and universal, encompassing environmental, socio-economic, and aesthetic criteria [56].

The construction of correlation matrices of landscape quality indicators provided certain insights and valuable information for spatial planning. Strong positive correlations between certain indicators signify potential synergies that can be leveraged to design integrated spatial and development plans that address multiple indicators simultaneously, promoting holistic and efficient landscape management [57]. Strong negative correlations suggest potential trade-offs between certain indicators [58] and could guide planners when navigating the trade-offs by carefully balancing conflicting aspects and ensuring that improvements in one dimension do not come at the expense of another [59]. In our study, strong negative correlations were found among naturalness indicators and indicators associated with human use. To address this issue, we prioritized the values associated with naturalness in the planning proposals. High correlations between landscape quality indicators may suggest redundancy [60], allowing planners to allocate resources more effectively by focusing on unique aspects of landscape quality. Even though we argue that possible multicollinearity revealed in our analysis is not a sign of redundancy of certain indicators, it is rather spatial contingency. For instance, in the case of shoreline landscape quality assessment, the diversity of tourism objects and the presence of tourism infrastructure exhibited possible multicollinearity (as well as the diversity of tourism objects and accessibility in inland landscape quality, implying that both indicators capture the same aspect of landscape quality). However, it is conceivable that a situation might arise where the presence of tourism attractions coincides with low values of infrastructure quality or accessibility, thereby indicating the necessity for its development. As in the case of possible multicollinearity between attractive landscape elements and small landscape elements, these two indicators are not redundant as they represent different landscape quality criteria (landscape aesthetic quality and landscape diversity), but their definitions should be more specific to avoid possible double-counting.

4.2. Role of Public Preferences and Well-Being Aspects in Planning of Coastal Development

Coastal and marine ecosystems are increasingly recognized for their multifaceted contributions to human well-being. For, recent research by Severin et al. [61] underscores the role of coastal settings in stress reduction and energy restoration, while Ashbullby et al. [62] associate coastal visits with diverse health and social benefits. Despite these findings, the incorporation of non-material values, such as perceived well-being benefits and intangible social values, remains limited in spatial or development planning processes, as highlighted by various scholars investigating the perceived benefits of coastal environments [63].

The nationwide survey in Latvia delivers the priorities of the well-being benefits by residents, revealing a distinct emphasis on the value of diverse ecosystems, followed closely by perceived leisure and health benefits associated with coastal areas. Similar priorities were also discovered in a study conducted to explore the relationship between CES, socio-cultural values, and well-being in Ireland [64].

Employing a range of social research methods, including interactive workshops, interviews, and surveys, their findings underscored the high esteem in which individuals hold intangible coastal values, particularly wilderness experiences, aesthetic appreciation, and health benefits. Distinctly, leisure activities such as coastal walks emerged as significant contributors to the perceived well-being, with participants consistently expressing the profound impact of coastal environments on their overall quality of life.

Complementing these insights, a study by Ahtiainen et al. [65] on CES provided by the Baltic Sea further elucidates the specific coastal ecosystem services deemed crucial in the Latvian context. Recreation, habitats, and landscapes emerge as focal points, underscoring the nuanced interplay between ecosystem functionality and human well-being within coastal regions.

Inhabitants versus visitors (regional versus national benefits)

The study results indicate that residents value all perceived well-being benefits offered by the Southwestern Kurzeme coastal area more than visitors. This observation is consistent with findings from research examining the perspectives of various stakeholder groups. Residents tend to attach greater value to local assets and demonstrate a stronger commitment to protecting resources, often due to the absence of readily available alternative sources of benefits. When studying natural resource-dependent communities, those whose material benefits depend on provisioning ES, the findings revealed that they simultaneously perceived the high importance of CES, such as scenic beauty, which can serve as a strong incentive for people to safeguard ecosystems [21]. Conversely, visitors may exhibit an increased awareness and appreciation of these distinctive environments, but they may also intentionally seek out alternative ways of enjoying coastal areas [66].

Naturalness versus amenities in coastal recreation planning

Coastal regions offer a diverse array of recreational opportunities, ranging from water-based activities to nature exploration and cultural experiences. Understanding the preferences of visitors regarding different types of coastal recreation is vital for municipal planners, developers, and managers to tailor offerings and enhance satisfaction among both locals and visitors. One notable finding from our study suggests that planners should not necessarily aim to equip all beaches and coastal areas with extensive amenities and services, as a considerable proportion of respondents in Latvia express a preference for untouched beaches with no human-made elements.

This result resonates with previous research conducted in Latvia [67] and the broader Baltic Sea region [65,68], where respondents prioritize natural habitats and their elements, such as sandy beaches, clean water, dunes, pine forest landscapes over built amenities and entertainment options. Similar trends have been observed in studies conducted in various other regions as well. For instance, Dimmock's study [66] uncovered a strong preference among residents and visitors for the natural aspects of marine landscapes along the coast of NSW, Australia. Furthermore, Bell et al. [69] stressed the significance of preserving coastline diversity to accommodate the diverse and individual ways in which people engage with the coast.

The segmentation of beach stretches and the management of visitor flows prove advantageous not only for stakeholders but also for the coastal ecosystem, which is particularly susceptible to the impacts of various natural and anthropogenic factors. A mutual connection between undertaking an activity, environmental awareness and appreciation has been articulated in previous studies [63], a connection that resonates with our study's findings, including the outcomes of PGIS.

While people highly value the naturalness of coastal landscapes, clean water emerges as a crucial prerequisite for a significant portion of respondents in Latvia, as well as in other coastal countries in Europe and globally [64,70,71]. Therefore, establishing certain minimum facilities, such as toilets, becomes necessary to uphold water quality and prevent degradation.

4.3. *Integrated Approach in CES Assessment to Support Coastal Planning*

The necessity of integrating biophysical and social methods of ES assessment to support development planning and decision-making has been recognized by several authors [36,37,39]. Our study presents an integrated framework for coastal CES assessment, including biophysical mapping of service supply, as well as social methods for revealing public preferences in the use of CES and their contribution to perceived well-being, which allowed us to develop evidence-based spatial solutions and recommendations for coastal tourism development. A similar approach was applied by Inácio et al. [72] in mapping and assessing coastal CES in a neighboring country—Lithuania, where various spatial data sets were used to model the natural components (e.g., naturalness, relief heterogeneity, tree density, water bodies, and other landcover types, silent areas, protected areas) and cultural components (e.g., religious sites, points of interest, sports facilities, accommodation, gastronomy, tourism infrastructure) contributing to CES supply; whereas flow and demand of the services were assessed using the online PGIS survey, identify the sites visited for recreation and points of origin of respondents as well as using this information also for verification of the supply assessment. Our approach differs in terms of the assessment unit—landscape area, which encompasses relatively homogeneous biophysical and socio-cultural characteristics as well as place identity [26], therefore being more recognizable/perceivable by local stakeholders and robust for decision-making. The applicability of landscape areas in decision-making and conveying planning solutions was highlighted by Hazeau et al. [48] and Fish et al. [73] and demonstrated in our study by using the CES assessment at the scale of landscape areas to guide the costal tourism development recommendations and their communication to stakeholders.

In this study, the data collected with participatory methods had a complementary character in CES mapping and assessment (e.g., PIGIS data were used to verify the biophysical mapping results on recreational potential as well as provide insights into the use of the areas for different recreational activities). The contribution of social methods was crucial when designing the solutions. Public demand for the preservation of natural landscape qualities as a resource for well-being was taken as the basis for targeting coastal tourism development by our method of clustering of landscape areas (with priority for maintaining natural, cultural heritage or aesthetic values) and thus providing a foundation for sustainable development planning in coastal regions.

The active involvement of stakeholders throughout the planning process is fundamental to our framework. Understanding their needs and perceptions is crucial for informed decision-making and the successful implementation of planning initiatives [74,75]. Incorporating stakeholder input enhances the relevance and acceptance of planning outcomes, contributing to more sustainable and inclusive development practices [51].

The framework and methods demonstrated in this study offer a practical approach to addressing the complexities of planning processes associated with manifold ecosystem values. Our findings illustrate the applicability of this approach across various planning contexts, including thematic planning of coastal tourism development, maritime spatial planning, and the promotion of sustainable blue economy initiatives. By considering land-sea interactions, our framework enriches the planning process, ensuring a holistic approach that accounts for the interconnectedness of coastal ecosystems (shoreline, terrestrial) and human activities.

Practically, our approach serves as valuable input for the development of guidance tailored to respond to the needs of coastal planning practitioners [76,77]. By operationalizing the concept of CES through our framework, coastal communities can align their development goals with national governance ambitions, while addressing the priorities and concerns of local stakeholders.

5. Conclusions

In conclusion, our study presents a significant contribution to the field of coastal planning and management by offering a comprehensive framework for evaluating development potentials and associated trade-offs in coastal regions. By integrating coastal landscape ecology and social science to operationalize the concept of ecosystem services, our research provides valuable insights that can inform decision-making processes for marine and coastal planners.

Our findings emphasize the importance of adopting an interdisciplinary approach to coastal planning, which leverages the strengths of plural disciplines and assessments to ensure a more holistic representation of coastal CES. Through this approach, we can better recognize and value the well-being benefits provided by coastal ecosystems and use this information in development of planning solutions.

Moreover, our study highlights the necessity of a transdisciplinary approach that actively engages stakeholders and acknowledges their diverse perspectives and valuations of well-being. By systematically involving stakeholders throughout the planning process, we can foster a deeper understanding of local contexts, preferences, and benefits, ultimately leading to more inclusive and effective coastal management strategies.

While our proposed coastal planning framework has been tested at a regional level, its scalability to national and cross-border contexts holds high potential for addressing the complex challenges facing coastal areas globally. By enhancing the management of trade-offs through smart spatial segmentation, informed by interdisciplinary insights and stakeholder engagement, we envision a future where coastal regions and communities can achieve a balanced and sustainable trajectory of development for equitable well-being. Moving forward, it is essential to continue building upon this research by addressing multilevel and multi-scalar perspectives in the assessment of CES and well-being.

Author Contributions: Conceptualization, all; methodology, all; validation, all; formal analysis, I.V. and A.R. (Agnese Reke); investigation, A.R. (Anda Ruskule) and K.V.; data curation, I.V. and A.R. (Agnese Reke); writing—original draft preparation, all; writing—review and editing, all; visualization, I.V., A.R. (Anda Ruskule) and A.R. (Agnese Reke); supervision, K.V.; project administration, K.V.; funding acquisition, K.V. All authors have read and agreed to the published version of the manuscript.

Funding: This research was financially supported by a project of the Baltic Sea Region Interreg program, "Land-sea interactions advancing Blue Growth in Baltic Sea coastal areas" (Land-Sea-Act, #R098).

Institutional Review Board Statement: Not applicable.

Data Availability Statement: Data is contained within the article.

Acknowledgments: The authors would like to thank Marta Stube for her contributions to the study by supporting the team with her GIS knowledge and technical skills.

Conflicts of Interest: The authors declare no conflicts of interest. The funders had no role in the design of the study; in the collection, analyses, or interpretation of data; in the writing of the manuscript; or in the decision to publish the results.

Appendix A

Table A1. Landscape quality assessment criteria, indicators, and data sources, applied to inland part of the case study area.

Landscape Quality	Indicators	Indicator Assessment Scale	Data Sources
Landscape aesthetic quality	Openness of landscape	1—closed landscape (forest cover > 90%) 3—partly open landscape (mosaic type) 5—open landscape (forest < 20%)	Land cover maps Topography maps Field survey
	Scenic views	1—no scenic views (plain terrain) 2—very few scenic views (almost plain terrain) 3—several scenic views but mostly plain terrain 4—several scenic views, undulating terrain 5—many outstanding scenic views (e.g., sea view)	Field survey
	Attractive landscape elements (gardens, alleys, farmstead clusters, fishermen villages, small forest patches)	1—no attractive landscape elements/landscape is visually homogeneous 3—few attractive landscape elements 5—many attractive landscape elements/landscape is visually diverse	Field survey
	Share of scenically attractive roads	1—no scenically attractive roads or only short sections 3—few scenically attractive roads 5—many and diverse scenically attractive roads	Field survey
	Visual pollution and degrading elements (abandoned sites or agricultural land, demolished buildings)	1—visual pollution dominates the landscape 3—few visual degrading elements 5—no visual degrading elements	Field survey
Uniqueness of the landscape	Unique landscapes on a national/regional scale	1—no unique landscapes of national or regional scale 3—at least one unique landscape 5—several unique landscapes of national or regional scale	Map of national/regional "landscape treasures"

Table A1. *Cont.*

Landscape Quality	Indicators	Indicator Assessment Scale	Data Sources
Landscape diversity	Diversity of land use (LU) types	1—low (dominated by one LU type) 2—medium low (in some parts of the area many LU types can be found) 3—medium (dominated by at least two LU types) 4—medium high (several LU types distributed all over the area) 5—high (several LU types well represented; many different LU type concentration areas)	Corine Land Cover Topography map
	Small-scale landscape elements (rows of trees/alleys, single large trees, piles of stones, etc.)	1—very few small-scale landscape elements 2—few significant small-scale landscape elements 3—several significant small-scale landscape elements, but their diversity is low 4—many different small-scale landscape elements, but not evenly distributed 5—many different small-scale landscape elements, distributed evenly all over the area	Topography map Field survey
	Settlement structure	1—area not populated 2—very few settlements (farmsteads) 3—settlements distributed unevenly (e.g., along the roads) 4—several concentration areas of the settlements 5—settlement distributed all over the area	Topography map
Landscape naturalness	Presence of natural microrelief	1—no microrelief (levelled area) 3—microrelief in some parts maintained 5—microrelief maintained	Topography map Field survey
	Naturalness of water streams and rivers	1—almost completely regulated water streams (>90%) 3—water streams partly regulated 5—natural water streams	Topography map Map of drainage systems
	Share of semi-natural habitats	1—no semi-natural habitats (<1%) 2—only few areas with semi-natural habitats (<20%) 3—medium distribution of semi-natural habitats (20–50%) 4—area dominated by semi-natural habitats (>50%) 5—almost only semi-natural habitats (>90%)	Habitat mapping data
	Share of protected areas	1—no or negligible share of protected areas 3—relatively high share of protected areas (>20%) 5—protected areas are dominating (>50%)	Map of protected areas
	Presence of industrial objects (big agroindustry farms, wind-parks etc.)	1—lot of different industrial objects are present throughout the landscape 3—few industrial objects are present at some locations 5—no industrial objects are visible in landscape	Field survey
Cultural heritage	Number of cultural history monuments	1—only few cultural heritage objects at some locations 3—relatively high number of cultural heritage objects, but distributed unevenly 5—high density of cultural heritage objects distributed evenly throughout the area	Data base of cultural history monuments Cultural heritage maps of the municipality spatial planning documents
	Diversity of cultural history monuments	1—no diversity of cultural heritage objects 3—medium diversity of cultural heritage objects (at least two types of objects), 5—high diversity—many different types of cultural heritage objects (at least three)	Data base of cultural history monuments Cultural heritage maps of the municipality spatial planning documents
	Presence of historical spatial structures (traditional landscape, historical roads or their sections, old villages, historic distribution pattern of farmsteads)	1—very little evidence of historic spatial structures at some locations (landscape structure is significantly transformed) 3—medium diversity of historic spatial structures and uneven distribution 5—high diversity of historic spatial structures evenly distributed throughout the area	Field survey and other data (e.g., municipality spatial planning documents)
	Visually outstanding cultural history objects (manor houses, churches, mills, bridges, mounds, fishing villages, etc.)	1—only few outstanding cultural heritage objects at some locations 3—relatively high number of outstanding cultural heritage objects, but distributed unevenly 5—high density of outstanding cultural heritage objects distributed evenly throughout the area	Field survey and other data (e.g., municipality spatial planning documents)

Table A1. *Cont.*

Landscape Quality	Indicators	Indicator Assessment Scale	Data Sources
Recreation potential	Density of tourism attractions/objects	1—only few tourism objects at some locations 3—medium density of tourism objects, concentrated only in few locations 5—high density of tourism objects concentrated at several locations	Municipality spatial planning documents, thematic plans etc.
	Accommodation possibilities	1—very few (1–2 sites) or no accommodation possibilities 3—relatively good accommodation possibilities but only in a few locations 5—accommodation possibilities are more or less evenly distributed all over the area and concentrated near significant tourism objects	Data collected by tourism experts (A. Klepers), Accommodation websites, e.g., booking.com, airbnb.com (Accessed on 31 August 2020)
	Presence of nature parks, national parks and nature monuments	1—no nature parks, national parks or nature monuments 3—one nature park, national park or nature monument 5—two or more nature parks, national parks or nature monument	Map of protected areas
	Diversity of tourism infrastructure	1—no diversity of tourism infrastructure, since only one or few tourism objects are present in the area 3—medium diversity of tourism infrastructure 5—high diversity of tourism infrastructure	Municipality spatial planning documents, thematic plans etc. Field survey
	Cycling Routes	1—no cycling routes 3—there are 1–2 cycling routes crossing the area 5—many cycling rotes, well distributed in the area	Municipality spatial planning documents, thematic plans etc.
	Accessibility	1—access to tourism attractions is not provided or very limited/difficult 3—tourist attractions are relatively well reachable, but the infrastructure or quality of access roads is poor 5—free and convenient access to tourist attractions with appropriate infrastructure	Field survey Municipality spatial planning documents, thematic plans etc

Table A2. Landscape quality assessment criteria, indicators and data sources, applied to shoreline part of the study area.

Landscape Quality	Indicators	Expert Assessment Scale	Data Sources
Landscape aesthetic quality	Scenic views (cliffs, bays, capes etc.)	1—Monotonous shoreline (or landscape when viewed from shore), different shore types cannot be distinguished 3—moderately monotonous shoreline/landscape 5—spectacular/attractive shoreline, diverse landscape/shore types	Field survey
	Attractive landscape elements (fishermen boats, ships, yacht, big stones/rocks)	1—no attractive landscape elements/landscape is visually homogeneous 3—few attractive landscape elements 5—many attractive landscape elements/landscape is visually diverse	Field survey
	Visual pollution and degrading elements (abandoned sites, demolished buildings)	1—visual pollution dominates the landscape 3—few visual degrading elements 5—no visual degrading elements	Field survey
Uniqueness of the landscape	Unique landscapes on a national/regional scale	1—no unique landscapes of national or regional scale 3—at least one unique landscape 5—several unique landscapes of national or regional scale	Map of national/regional "Landscape treasures"
Landscape diversity	Shoreline diversity depending on its character (bays, capes etc) and exposition	1—low (smooth shoreline, dominated by one shore type) 2—medium low (some parts of the shoreline stretch have a different character/shore type) 3—medium (two shore types dominate equally) 4—medium high (several shore types are present) 5—high (great variety of shore types, many bays, estuaries)	Corine Land Cover Topography map Map of shore types Field survey

Table A2. *Cont.*

Landscape Quality	Indicators	Expert Assessment Scale	Data Sources
	Small-scale landscape elements—presence of natural and human made elements in landscape (e.g., lighthouses, piers, ships, boats, rocks, fallen/washed out trees)	1—very few small-scale landscape elements 2—few significant small-scale landscape elements 3—several significant small-scale landscape elements, but their diversity is low 4—many different small-scale landscape elements, but not evenly distributed 5—many different small-scale landscape elements, distributed evenly all over the area	Topography map Field survey
Landscape naturalness	Share of natural shoreline stretches	1—artificial/transformed shore—port 3—partly natural shore 5—distinctly natural shore	Field survey
	Share of protected areas	1—no or negligible share of protected areas 3—relatively high share of protected areas (>20%) 5—protected areas are dominating (>50%)	Map of protected areas
Cultural heritage	Number of cultural history monuments (e.g., wrecks, lighthouses etc.)	1—no cultural history monuments 3—few (1–2) cultural history monuments 5—many cultural history monuments	Data base of cultural history monuments Cultural heritage maps of the municipality, spatial planning documents and MSP
	Presence of traditional land/sea use practices (e.g., fishing) in the landscape	1—No signs of traditional practices 2—Some elements (e.g., few fisherman boats) 3—Many fisherman boats, fishing nets or other features	Field survey
	Visually outstanding cultural history objects (manor houses, churches, mills, bridges, mounds, fishing villages, lighthouses etc.)	1—only few outstanding cultural heritage objects at some locations 3—relatively high number of outstanding cultural heritage objects, but distributed unevenly 5—high density of outstanding cultural heritage objects distributed evenly throughout the area	Field survey and other data (municipality spatial plans, etc.)
Recreation potential	Density of tourism attractions/objects	1—only few tourism objects at some locations 3—medium density of tourism objects, concentrated only in few locations 5—high density of tourism objects concentrated at several locations	Density of tourism attractions/objects
	Diversity of tourism infrastructure	1—no diversity of tourism infrastructure, since only one or few tourism objects are present in the area 3—medium diversity of tourism infrastructure 5—high diversity of tourism infrastructure	Municipality spatial planning documents, thematic plans etc. Field survey
	Accessibility	1—access to tourism attractions is not provided or very limited/difficult 3—tourist attractions are relatively well reachable, but the infrastructure or quality of the access roads is poor 5—free and convenient access to tourist attractions with appropriate infrastructure	Field survey Assessment of coastal infrastructure
	Number of visitors	1—negligible number of visitors (<100) 2—low number of visitors (100–300) 3—average number of visitors (300–1000) 4—high number of visitors (1000–5000) 5—intensively visited beach (>5000)	Visitor counts

Appendix B

Questionnaire for assessing the perceived well-being benefits provided by the South-western Kurzeme coastal area

Please note—*originally, the questionnaire was developed and distributed in Latvian and Russian languages. The English translation is generated for the purpose of this publication only. Questions that were not utilized in this study are excluded from the translated version.*

INTRODUCTION

Hello!

We are currently conducting a social survey about the importance of Southwestern Kurzeme coastal area for residents of Latvia. The survey is anonymous, and all your answers will be used only in an aggregated form. The survey will take ~10 min.

QUESTIONNAIRE

D1 Gender

D2 Age

D3 Nationality

D4 Place of residence

In case the place of the residence indicated by the respondent is geographically part of Southwestern Kurzeme coasal area, the respondent is automatically classified as local inhabitant of Southwestern Kurzeme coastal area and therefore excluded from several questions directed to visitors (see "FILTER").

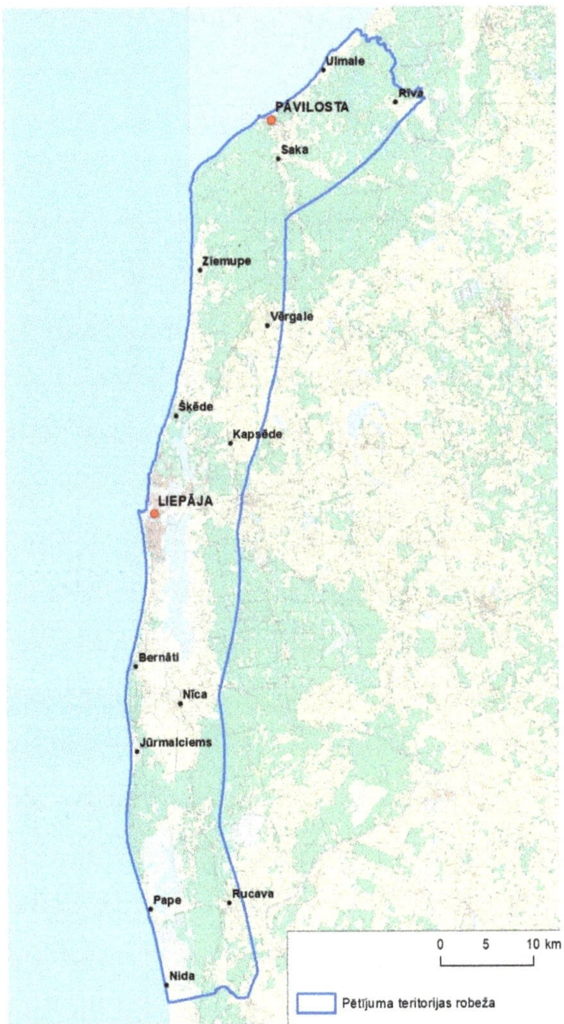

INTRO1: Please take a look at this map. This is the area we identify as the "Southwestern Kurzeme coastal area" in this study. The marked territory includes Liepāja,

Pāvilosta, Grobiņa, Grobiņa Parish, Medzes Parish, Nīcas Parish, Otaņki Parish, Sakas Parish, Vērgales Parish, Rucavas Paris.

FILTER: *Display only to "Not a permanent resident of the Southwestern Kurzeme coastal area"*

Q1 When was the last time you visited the Southwestern Kurzeme coastal area?

1. In the last three years (2018–2021)
2. 4–10 years ago (2011–2017)
3. Earlier than 10 years ago (before 2011)
4. I have never been to the Southwestern Kurzeme coastal area -> *excluded from Q2–Q5*

FILTER: *Display only to "Not a permanent resident of the Southwestern Kurzeme coastal area"*

Q2 Please select which of the following statements best describes your connection to the Southwestern Kurzeme coastal area? You can select multiple answers.

1. I live on the Southwestern coast occasionally/seasonally (e.g., I own or rent a summer house/apartment/cottage there)
2. I visit the Southwestern Kurzeme coastal area for private business (e.g., recreation, visiting relatives/friends, cultural and sport events)
3. I visit the Southwestern Kurzeme coastal area for work
4. I visit the Southwestern Kurzeme coastal area when passing through (i.e., it is not my main destination)

FILTER: *Display to "Not a permanent resident of the Southwestern Kurzeme coastal area" and has visited the Southwestern Kurzeme in the last three years (Q1 = 1)*

Q3 How many times have you visited the Southwestern Kurzeme coastal area in the last 3 years?

1. 1–2 times
2. 3–5 times
3. 6–10 times
4. 11–30 times
5. 31 or more times
6. *I don't know*

FILTER: *Display only to "Not a permanent resident of the Southwestern Kurzeme coastal area"*
Q4 Which seasons of the year do you usually visit the Southwestern Kurzeme coastal area? You can select multiple answers.

1. Spring
2. Summer
3. Autumn
4. Winter

FILTER: *Display to all respondents, except the ones who have never been in Soutwestern Kurzeme coastal area (Q1 = 4)*

Q5 Please rate the following statements!

		Compl. Dis-Agree	Part. Dis-Agree	Part. Agree	Compl. Agree	I Don't Know
1.	Visiting the Southwestern Kurzeme coastal area provides me with an opportunity to relax and clarify my thoughts	1	2	3	4	n/a
2.	I feel healthier, more energetic after visiting the Southwestern Kurzeme coastal area	1	2	3	4	n/a
3.	The natural resources provided by the Southwestern Kurzeme coastal area significantly contribute to my financial well-being (my work is related to the coast—hospitality, fishing, tourism, etc.)	1	2	3	4	n/a

		Compl. Dis-Agree	Part. Dis-Agree	Part. Agree	Compl. Agree	I Don't Know
4.	The opportunity to fish, pick berries, mushrooms and other goods for self-consumption on the Southwestern Kurzeme coastal area is important for me	1	2	3	4	n/a
5.	The Southwestern Kurzeme coastal area plays a crucial role in my social life: I spend time with my family and/or friends here, engage in social activities	1	2	3	4	n/a
6.	Exploring the Southwestern Kurzeme coastal areas has enabled me to acquire new knowledge (e.g., about nature, the sea)	1	2	3	4	n/a
7.	The Southwestern Kurzeme coastal area holds a significance for me in terms of recreation opportunities	1	2	3	4	n/a
8.	The role of the Southwestern Kurzeme coastal area in strenghtening the national military security holds importance for me	1	2	3	4	n/a
9.	The potential energy independence that can be achieved through the establishment of wind farms along the Southwestern Kurzeme coastal area is vital for me	1	2	3	4	n/a
10.	The Southwestern Kurzeme coastal area provides me with a sense of security—I feel protected from the effects of climate change, natural disasters (storms, floods etc.)	1	2	3	4	n/a
11.	I hold the nature diversity of the Southwestern Kurzeme coastal area in high regard, and its preservation brings me a sense of satisfaction	1	2	3	4	n/a

FILTER: *All respondents*

Q9 Which of these picture collages best fits your idea of a pleasant environment for recreation?

1. A
2. B
3. *Difficult to say*

FILTER: *All respondents*

Q10 Please rate how important are these factors are to you when selecting a recreation site by the sea!

		Compl. Un-Import.	Minor Import.	Import. but Not Decisive	Very Import.	I Don't Know
1.	Amenities are available (e.g., benches, changing rooms, access paths, children's playgrounds)	1	2	3	4	n/a
2.	Services are available nearby (e.g., shops, restaurants, cafes)	1	2	3	4	n/a
3.	The site has special nature values (e.g., beautiful beaches, dunes, many bird species)	1	2	3	4	n/a
4.	There are not many other visitors on the site	1	2	3	4	n/a
5.	The site is clean	1	2	3	4	n/a
6.	The site has beautiful landscapes/seascapes	1	2	3	4	n/a
7.	Easy access by private transport	1	2	3	4	n/a
8.	Easy access by public transport	1	2	3	4	n/a

Appendix C

PGIS survey: Recreation on the Southwestern Kurzeme coast

Please note—*originally the PGIS survey was developed and distributed in Latvian language. The English translation is generated for the purpose of this publication only.*

Introduction

We would like to find out what you think are the most important recreation/tourism sites on the Southwestern Kurzeme coastal area. The information is being gathered within the framework of the Interreg project Land-Sea-Act to find solutions to balance the development of offshore wind farms with the interests of the local community and tourism industry.

FILTER: *All respondents*

Q1 Please mark on the interactive map a place of interest for tourism/recreation! Please select a place within the marked project area.

FILTER: *All respondents*

Q2 Please type the name of the place: ____________
FILTER: *All respondents*

Q3 The marked place is suitable for:
You can select multiple answers.

1. Relaxation by the sea
2. Water sports
3. Jet skis
4. Yacht and boat decks
5. Fishing
6. Hunting
7. Bird watching
8. Hiking and walking
9. Exploring cultural history
10. Public entertainment events (festivals, performances etc.)
11. Sports
12. Overnighting
13. Enjoying landscape/seascape
14. Other ____ (please specify)

FILTER: *Optional question*

Q4 Additional comments about the place: __________
Demographic questions
D1 Place of residence (city, parish or municipality)
D2 Occupation
D3 Age

Appendix D

Table A3. Descriptive statistics for the rating of the factors influencing the choice of coastal recreation sites. Note that responses of "I don't know" were excluded from the analysis, resulting in varying number of responses (N) across factors.

Factor *	N	Mean	SD	Coefficient of Variation
Availability of amenities	981	3.059	0.833	0.272
Availability of services	976	2.858	0.835	0.292
Presence of nature values	977	3.540	0.616	0.174
Low visitor density	973	3.270	0.714	0.218
Cleanliness of the site	981	3.744	0.514	0.137
Beautiful landscapes	978	3.566	0.587	0.164
Private transport access	974	3.270	0.830	0.254
Public transport access	967	2.620	1.056	0.403

* For full statements used in the survey see Appendix B, Q10.

Notes

1 **Land-Sea-Act Map Explorer:** https://experience.arcgis.com/experience/2447e76e306a4e68bf82323e33b72b26/ (accessed on 31 January 2024).

2 https://ainavudargumi.lv/en/ (accessed on 31 January 2024).

References

1. European Environment Agency. *Marine Messages II: Navigating the Course towards Clean, Healthy and Productive Seas through Implementation of an Ecosystem-Based Approach*; Publications Office: Luxembourg, 2019.
2. Petrişor, A.-I.; Hamma, W.; Nguyen, H.D.; Randazzo, G.; Muzirafuti, A.; Stan, M.-I.; Tran, V.T.; Aştefănoaiei, R.; Bui, Q.-T.; Vintilă, D.-F.; et al. Degradation of Coastlines under the Pressure of Urbanization and Tourism: Evidence on the Change of Land Systems from Europe, Asia and Africa. *Land* **2020**, *9*, 275. [CrossRef]
3. Dahlet, L.I.; Selim, S.A.; Van Putten, I. A Review of How We Study Coastal and Marine Conflicts: Is Social Science Taking a Broad Enough View? *Marit. Stud.* **2023**, *22*, 29. [CrossRef]

4. Turner, R.K. Integrating Natural and Socio-Economic Science in Coastal Management. *J. Mar. Syst.* **2000**, *25*, 447–460. [CrossRef]
5. Bennett, E.M.; Chaplin-Kramer, R. Science for the Sustainable Use of Ecosystem Services. *F1000Research* **2016**, *5*, 2622. [CrossRef]
6. Smart, L.S.; Vukomanovic, J.; Sills, E.O.; Sanchez, G. Cultural Ecosystem Services Caught in a 'Coastal Squeeze' between Sea Level Rise and Urban Expansion. *Glob. Environ. Chang.* **2021**, *66*, 102209. [CrossRef]
7. Eurostat Data Browser. Available online: https://ec.europa.eu/eurostat/databrowser/product/page/TOUR_OCC_NINATDC (accessed on 31 January 2024).
8. Custodio, M.; Moulaert, I.; Asselman, J.; Van Der Biest, K.; Van De Pol, L.; Drouillon, M.; Hernandez Lucas, S.; Taelman, S.E.; Everaert, G. Prioritizing Ecosystem Services for Marine Management through Stakeholder Engagement. *Ocean Coast. Manag.* **2022**, *225*, 106228. [CrossRef]
9. Frederiksen, P.; Morf, A.; Von Thenen, M.; Armoskaite, A.; Luhtala, H.; Schiele, K.S.; Strake, S.; Hansen, H.S. Proposing an Ecosystem Services-Based Framework to Assess Sustainability Impacts of Maritime Spatial Plans (MSP-SA). *Ocean Coast. Manag.* **2021**, *208*, 105577. [CrossRef]
10. Galparsoro, I.; Pınarbaşı, K.; Gissi, E.; Culhane, F.; Gacutan, J.; Kotta, J.; Cabana, D.; Wanke, S.; Aps, R.; Bazzucchi, D.; et al. Operationalisation of Ecosystem Services in Support of Ecosystem-Based Marine Spatial Planning: Insights into Needs and Recommendations. *Mar. Policy* **2021**, *131*, 104609. [CrossRef]
11. Veidemane, K.; Ruskule, A.; Strake, S.; Purina, I.; Aigars, J.; Sprukta, S.; Ustups, D.; Putnis, I.; Klepers, A. Application of the Marine Ecosystem Services Approach in the Development of the Maritime Spatial Plan of Latvia. *Int. J. Biodivers. Sci. Ecosyst. Serv. Manag.* **2017**, *13*, 398–411. [CrossRef]
12. Braat, L.C.; De Groot, R. The Ecosystem Services Agenda: Bridging the Worlds of Natural Science and Economics, Conservation and Development, and Public and Private Policy. *Ecosyst. Serv.* **2012**, *1*, 4–15. [CrossRef]
13. Heckwolf, M.J.; Peterson, A.; Jänes, H.; Horne, P.; Künne, J.; Liversage, K.; Sajeva, M.; Reusch, T.B.H.; Kotta, J. From Ecosystems to Socio-Economic Benefits: A Systematic Review of Coastal Ecosystem Services in the Baltic Sea. *Sci. Total Environ.* **2021**, *755*, 142565. [CrossRef] [PubMed]
14. Schröter, M.; Van Der Zanden, E.H.; Van Oudenhoven, A.P.E.; Remme, R.P.; Serna-Chavez, H.M.; De Groot, R.S.; Opdam, P. Ecosystem Services as a Contested Concept: A Synthesis of Critique and Counter-Arguments. *Conserv. Lett.* **2014**, *7*, 514–523. [CrossRef]
15. Van Oudenhoven, A.P.E.; Schröter, M.; Drakou, E.G.; Geijzendorffer, I.R.; Jacobs, S.; Van Bodegom, P.M.; Chazee, L.; Czúcz, B.; Grunewald, K.; Lillebø, A.I.; et al. Key Criteria for Developing Ecosystem Service Indicators to Inform Decision Making. *Ecol. Indic.* **2018**, *95*, 417–426. [CrossRef]
16. Millenium Ecosystem Assessment. *Ecosystems and Human Well-Being: Synthesis*; Island Press: Washington, DC, USA, 2005; ISBN 1-56973-597-2.
17. Tengberg, A.; Fredholm, S.; Eliasson, I.; Knez, I.; Saltzman, K.; Wetterberg, O. Cultural Ecosystem Services Provided by Landscapes: Assessment of Heritage Values and Identity. *Ecosyst. Serv.* **2012**, *2*, 14–26. [CrossRef]
18. Milcu, A.I.; Hanspach, J.; Abson, D.; Fischer, J. Cultural Ecosystem Services: A Literature Review and Prospects for Future Research. *Ecol. Soc.* **2013**, *18*, 44. [CrossRef]
19. Nevzati, F.; Veldi, M.; Külvik, M.; Bell, S. Analysis of Landscape Character Assessment and Cultural Ecosystem Services Evaluation Frameworks for Peri-Urban Landscape Planning: A Case Study of Harku Municipality, Estonia. *Land* **2023**, *12*, 1825. [CrossRef]
20. Schaich, H.; Bieling, C.; Plieninger, T. Linking Ecosystem Services with Cultural Landscape Research. *GAIA—Ecol. Perspect. Sci. Soc.* **2010**, *19*, 269–277. [CrossRef]
21. Elwell, T.L.; López-Carr, D.; Gelcich, S.; Gaines, S.D. The Importance of Cultural Ecosystem Services in Natural Resource-Dependent Communities: Implications for Management. *Ecosyst. Serv.* **2020**, *44*, 101123. [CrossRef]
22. Blicharska, M.; Smithers, R.J.; Hedblom, M.; Hedenås, H.; Mikusiński, G.; Pedersen, E.; Sandström, P.; Svensson, J. Shades of Grey Challenge Practical Application of the Cultural Ecosystem Services Concept. *Ecosyst. Serv.* **2017**, *23*, 55–70. [CrossRef]
23. Drakou, E.G.; Kermagoret, C.; Liquete, C.; Ruiz-Frau, A.; Burkhard, K.; Lillebø, A.I.; Van Oudenhoven, A.P.E.; Ballé-Béganton, J.; Rodrigues, J.G.; Nieminen, E.; et al. Marine and Coastal Ecosystem Services on the Science–Policy–Practice Nexus: Challenges and Opportunities from 11 European Case Studies. *Int. J. Biodivers. Sci. Ecosyst. Serv. Manag.* **2017**, *13*, 51–67. [CrossRef]
24. Cabana, D.; Ryfield, F.; Crowe, T.P.; Brannigan, J. Evaluating and Communicating Cultural Ecosystem Services. *Ecosyst. Serv.* **2020**, *42*, 101085. [CrossRef]
25. Leyshon, C. Finding the Coast: Environmental Governance and the Characterisation of Land and Sea. *Area* **2018**, *50*, 150–158. [CrossRef]
26. Pikner, T.; Piwowarczyk, J.; Ruskule, A.; Printsmann, A.; Veidemane, K.; Zaucha, J.; Vinogradovs, I.; Palang, H. Sociocultural Dimension of Land–Sea Interactions in Maritime Spatial Planning: Three Case Studies in the Baltic Sea Region. *Sustainability* **2022**, *14*, 2194. [CrossRef]
27. Banela, M.; Kitsiou, D. Mapping Cultural Ecosystem Services: A Case Study in Lesvos Island, Greece. *Ocean Coast. Manag.* **2023**, *246*, 106883. [CrossRef]
28. Garcia Rodrigues, J.; Conides, A.; Rivero Rodriguez, S.; Raicevich, S.; Pita, P.; Kleisner, K.; Pita, C.; Lopes, P.; Alonso Roldán, V.; Ramos, S.; et al. Marine and Coastal Cultural Ecosystem Services: Knowledge Gaps and Research Priorities. *One Ecosyst.* **2017**, *2*, e12290. [CrossRef]

29. Kosanic, A.; Petzold, J. A Systematic Review of Cultural Ecosystem Services and Human Wellbeing. *Ecosyst. Serv.* **2020**, *45*, 101168. [CrossRef]
30. Hernández-Morcillo, M.; Plieninger, T.; Bieling, C. An Empirical Review of Cultural Ecosystem Service Indicators. *Ecol. Indic.* **2013**, *29*, 434–444. [CrossRef]
31. Brown, G.; Hausner, V.H. An Empirical Analysis of Cultural Ecosystem Values in Coastal Landscapes. *Ocean Coast. Manag.* **2017**, *142*, 49–60. [CrossRef]
32. Vieira Da Silva, C.; Ortigão, M.; Willaert, T.; Rosa, R.; Nunes, L.C.; Cunha-e-Sá, M.A. Participatory Geographic Information Systems (PGIS): Alternative Approaches to Identify Potential Conflicts and Positional Accuracy in Marine and Coastal Ecosystem Services. *Mar. Policy* **2021**, *131*, 104650. [CrossRef]
33. Brown, G.; Reed, P.; Raymond, C.M. Mapping Place Values: 10 Lessons from Two Decades of Public Participation GIS Empirical Research. *Appl. Geogr.* **2020**, *116*, 102156. [CrossRef]
34. Brown, G.; Montag, J.M.; Lyon, K. Public Participation GIS: A Method for Identifying Ecosystem Services. *Soc. Nat. Resour.* **2012**, *25*, 633–651. [CrossRef]
35. Fagerholm, N.; Raymond, C.M.; Olafsson, A.S.; Brown, G.; Rinne, T.; Hasanzadeh, K.; Broberg, A.; Kyttä, M. A Methodological Framework for Analysis of Participatory Mapping Data in Research, Planning, and Management. *Int. J. Geogr. Inf. Sci.* **2021**, *35*, 1848–1875. [CrossRef]
36. Burdon, D.; Potts, T.; McKinley, E.; Lew, S.; Shilland, R.; Gormley, K.; Thomson, S.; Forster, R. Expanding the Role of Participatory Mapping to Assess Ecosystem Service Provision in Local Coastal Environments. *Ecosyst. Serv.* **2019**, *39*, 101009. [CrossRef]
37. Vihervaara, P.; Viinikka, A.; Brander, L.; Santos-Martín, F.; Poikolainen, L.; Nedkov, S. Methodological Interlinkages for Mapping Ecosystem Services—From Data to Analysis and Decision-Support. *One Ecosyst.* **2019**, *4*, e26368. [CrossRef]
38. Ruskule, A.; Klepers, A.; Veidemane, K. Mapping and Assessment of Cultural Ecosystem Services of Latvian Coastal Areas. *One Ecosyst.* **2018**, *3*, e25499. [CrossRef]
39. McKinley, E.; Acott, T.; Yates, K.L. Marine Social Sciences: Looking towards a Sustainable Future. *Environ. Sci. Policy* **2020**, *108*, 85–92. [CrossRef]
40. Jernberg, S.; Kuosa, H.; Boström, C.; Burdon, D.; Haavisto, F.; Heiskanen, A.-S.; Kiviluoto, S.; Kuningas, S.; Kunnasranta, M.; Uusitalo, L.; et al. Linking Natural Capital Stocks with Ecosystem Services in the Northern Baltic Sea. *Ecosyst. Serv.* **2024**, *65*, 101585. [CrossRef]
41. National Statistical System of Latvia. *Population in Urban (Densely Populated) and Rural (Sparsely Populated) Areas in Regions, Cities, Municipalities, Towns and Rural Territories (after Administrative-Territorial Reform in 2021) 2022–2023*; National Statistical System of Latvia: Riga, Latvia, 2023.
42. Veidemane, K. The Impact of Driving Forces and Protection Policies on Future Coastal Landscapes: A Case Study of Latvia. In *Global Change and Baltic Coastal Zones*; Schernewski, G., Hofstede, J., Neumann, T., Eds.; Coastal Research Library; Springer: Dordrecht, The Netherlands, 2011; Volume 1, pp. 193–210. ISBN 978-94-007-0399-5.
43. Ruskule, A.; Kotta, J.; Saha, C.R.; Arndt, P.; Ustups, D.; Strāķe, S.; Bergström, L. Testing the Concept of Green Infrastructure at the Baltic Sea Scale to Support an Ecosystem-Based Approach to Management of Marine Areas. *Mar. Policy* **2023**, *147*, 105374. [CrossRef]
44. Swanwick, C. *Landscape Character Assessment: Guidance for England and Scotland*; The Countryside Agency & Scottish Natural Heritage: Cheltenham, UK, 2002.
45. Tudor, C. *An Approach to Landscape Character Assessment*; Natural England: London, UK, 2014; ISBN 978-78367-141-0.
46. Anderson, C. *Guidance on Coastal Character Assessment.*; Scottish National Heritage: Inverness, UK, 2017; p. 24.
47. Sowińska-Świerkosz, B.N.; Chmielewski, T.J. A New Approach to the Identification of Landscape Quality Objectives (LQOs) as a Set of Indicators. *J. Environ. Manag.* **2016**, *184*, 596–608. [CrossRef]
48. Hazeu, G.W.; Metzger, M.J.; Mücher, C.A.; Perez-Soba, M.; Renetzeder, C.; Andersen, E. European Environmental Stratifications and Typologies: An Overview. *Agric. Ecosyst. Environ.* **2011**, *142*, 29–39. [CrossRef]
49. Zariņa, A.; Vinogradovs, I.; Vološina, M.; Reķe, A.; Briede, A.; Ušča, M.; Avotiņš, A.; Vītols, J.; Immurs, D.V. Zemgales Reģionālais Ainavas un Zaļās Infrastruktūras Plāns 2020–2027. Gadam. 2019. Available online: https://www.zemgale.lv/lv/media/109/download?attachment (accessed on 31 January 2024).
50. Haines-Young, R.; Potschin, M. Common International Classification of Ecosystem Services (CICES) V5.1: Guidance on the Application of the Revised Structure. Available online: https://cices.eu/content/uploads/sites/8/2018/01/Guidance-V51-01012018.pdf (accessed on 31 January 2024).
51. Morf, A.; Kull, M.; Piwowarczyk, J.; Gee, K. Towards a Ladder of Marine/Maritime Spatial Planning Participation. In *Maritime Spatial Planning*; Zaucha, J., Gee, K., Eds.; Springer International Publishing: Cham, Switzerland, 2019; pp. 219–243. ISBN 978-3-319-98695-1.
52. JASP Team. JASP (Version 0.18.3); 2024. Available online: https://jasp-stats.org/download/ (accessed on 31 January 2024).
53. Plieninger, T.; Van Der Horst, D.; Schleyer, C.; Bieling, C. Sustaining Ecosystem Services in Cultural Landscapes. *Ecol. Soc.* **2014**, *19*, 59. [CrossRef]
54. Medeiros, A.; Fernandes, C.; Gonçalves, J.F.; Farinha-Marques, P. Research Trends on Integrative Landscape Assessment Using Indicators—A Systematic Review. *Ecol. Indic.* **2021**, *129*, 107815. [CrossRef]
55. Bruni, D. Landscape Quality and Sustainability Indicators. *Agric. Agric. Sci. Procedia* **2016**, *8*, 698–705. [CrossRef]

56. Stauskis, G. Identifying Key Criteria for Quality Assessment of Landscape Architecture Projects. *Archit. Urban Plan.* **2020**, *16*, 5–11. [CrossRef]
57. Galler, C.; Von Haaren, C.; Albert, C. Planning Multifunctional Measures for Efficient Landscape Management: Quantifying and Comparing the Added Value of Integrated and Segregated Management Concepts. In *Landscape Ecology for Sustainable Environment and Culture*; Fu, B., Jones, K.B., Eds.; Springer: Dordrecht, The Netherlands, 2013; pp. 249–284. ISBN 978-94-007-6529-0.
58. Maron, M.; Cockfield, G. Managing Trade-Offs in Landscape Restoration and Revegetation Projects. *Ecol. Appl.* **2008**, *18*, 2041–2049. [CrossRef]
59. Gunton, R.M.; Marsh, C.J.; Moulherat, S.; Malchow, A.; Bocedi, G.; Klenke, R.A.; Kunin, W.E. Multicriterion Trade-offs and Synergies for Spatial Conservation Planning. *J. Appl. Ecol.* **2017**, *54*, 903–913. [CrossRef]
60. Wiggering, H.; Müller, K.; Werner, A.; Helming, K. The Concept of Multifunctionality in Sustainable Land Development. In *Sustainable Development of Multifunctional Landscapes*; Helming, K., Wiggering, H., Eds.; Springer: Berlin/Heidelberg, Germany, 2003; pp. 3–18. ISBN 978-3-642-05502-7.
61. Severin, M.I.; Raes, F.; Notebaert, E.; Lambrecht, L.; Everaert, G.; Buysse, A. A Qualitative Study on Emotions Experienced at the Coast and Their Influence on Well-Being. *Front. Psychol.* **2022**, *13*, 902122. [CrossRef] [PubMed]
62. Ashbullby, K.J.; Pahl, S.; Webley, P.; White, M.P. The Beach as a Setting for Families' Health Promotion: A Qualitative Study with Parents and Children Living in Coastal Regions in Southwest England. *Health Place* **2013**, *23*, 138–147. [CrossRef]
63. Clarke, B.; Thet, A.K.; Sandhu, H.; Dittmann, S. Integrating Cultural Ecosystem Services Valuation into Coastal Wetlands Restoration: A Case Study from South Australia. *Environ. Sci. Policy* **2021**, *116*, 220–229. [CrossRef]
64. Bullock, C.; Joyce, D.; Collier, M. An Exploration of the Relationships between Cultural Ecosystem Services, Socio-Cultural Values and Well-Being. *Ecosyst. Serv.* **2018**, *31*, 142–152. [CrossRef]
65. Ahtiainen, H.; Liski, E.; Pouta, E.; Soini, K.; Bertram, C.; Rehdanz, K.; Pakalniete, K.; Meyerhof, J. Cultural Ecosystem Services Provided by the Baltic Sea Marine Environment. *Ambio* **2019**, *48*, 1350–1361. [CrossRef]
66. Dimmock, K. The Importance of Marine Leisure Environments: Investigating Visitor and Local Resident Perceptions and Uses of Marine Environments on Coffs Coast, NSW Australia. *Ann. Leis. Res.* **2012**, *15*, 243–260. [CrossRef]
67. Veidemane, K.; Nikodemus, O. Latvijas pludmales ainavu sniegto rekreācijas pakalpojumu vērtējums. *Latv. Univ. Raksti. Zemes Un Vides Zinātnes* **2013**, *791*, 61–74.
68. Reimann, M.; Ehrlich, Ü.; Tõnisson, H. Regional Differences in Recreational Preferences of Estonian Coastal Landscapes. *J. Coast. Res.* **2014**, *70*, 420–425. [CrossRef]
69. Bell, S.L.; Phoenix, C.; Lovell, R.; Wheeler, B.W. Seeking Everyday Wellbeing: The Coast as a Therapeutic Landscape. *Soc. Sci. Med.* **2015**, *142*, 56–67. [CrossRef]
70. Nieminen, E.; Ahtiainen, H.; Lagerkvist, C.-J.; Oinonen, S. The Economic Benefits of Achieving Good Environmental Status in the Finnish Marine Waters of the Baltic Sea. *Mar. Policy* **2019**, *99*, 181–189. [CrossRef]
71. Maguire, G.S.; Miller, K.K.; Weston, M.A.; Young, K. Being beside the Seaside: Beach Use and Preferences among Coastal Residents of South-Eastern Australia. *Ocean Coast. Manag.* **2011**, *54*, 781–788. [CrossRef]
72. Inácio, M.; Gomes, E.; Bogdzevič, K.; Kalinauskas, M.; Zhao, W.; Pereira, P. Mapping and Assessing Coastal Recreation Cultural Ecosystem Services Supply, Flow, and Demand in Lithuania. *J. Environ. Manag.* **2022**, *323*, 116175. [CrossRef] [PubMed]
73. Fish, R.; Church, A.; Winter, M. Conceptualising Cultural Ecosystem Services: A Novel Framework for Research and Critical Engagement. *Ecosyst. Serv.* **2016**, *21*, 208–217. [CrossRef]
74. Agnew, S.; Kopke, K.; Power, O.-P.; Troya, M.D.C.; Dozier, A. Transdisciplinary Research: Can Citizen Science Support Effective Decision-Making for Coastal Infrastructure Management? *Front. Mar. Sci.* **2022**, *9*, 809284. [CrossRef]
75. Kuhn, T.; Vikström, S.; Suškevičs, M.; Storie, J.T.; Saikkonen, L.; Rees, R.; Håkansson, C.; Oinonen, S.; Burkhard, B. Literature Syntheses to Inform Marine Ecosystem Management: Lessons Learned from Stakeholder Participation. *Ecosyst. People* **2023**, *19*, 2188970. [CrossRef]
76. Longato, D.; Cortinovis, C.; Albert, C.; Geneletti, D. Practical Applications of Ecosystem Services in Spatial Planning: Lessons Learned from a Systematic Literature Review. *Environ. Sci. Policy* **2021**, *119*, 72–84. [CrossRef]
77. Dang, A.N.; Jackson, B.M.; Benavidez, R.; Tomscha, S.A. Review of Ecosystem Service Assessments: Pathways for Policy Integration in Southeast Asia. *Ecosyst. Serv.* **2021**, *49*, 101266. [CrossRef]

land

MDPI

Review

Coastal Cultural Ecosystem Services: A Bridge between the Natural Ecosystem and Social Ecosystem for Sustainable Development

Yuqing Zhao [1], Zenglin Han [2,3], Changren Zhang [1], Yuqiao Wang [4], Jingqiu Zhong [2,3] and Mengfan Gao [1,*]

1 School of Geography, Liaoning Normal University, Dalian 116029, China; zhaoyuq157@163.com (Y.Z.); changren0906@163.com (C.Z.)
2 Key Research Base of Humanities and Social Sciences of the Ministry of Education, Center for Studies of Marine Economy and Sustainable Development, Liaoning Normal University, Dalian 116029, China; hzl@lnnu.edu.cn (Z.H.); zjq@lnnu.edu.cn (J.Z.)
3 University Collaborative Innovation Center of Marine Economy High-Quality Development of Liaoning Province, Liaoning Normal University, Dalian 116029, China
4 College of Humanities and Social Sciences, University of Macau Cultural Industry Management, Macau 999078, China; yqwang0173@163.com
* Correspondence: gao_mengfan@163.com

Abstract: Cultural Ecosystem Services (CESs), as non-material benefits and well-being provided by ecosystems to humans, possess the ability to bridge nature and human society and interpret their complex interrelationships. Coastal areas are regions with concentrated human activities, where coastal zones are often subject to human development, pollution, and degradation. Compared to other ecosystems, coastal ecosystems face greater pressures and threats, and the cultural services they provide are more vulnerable. Research on coastal ecosystem cultural services needs to consider ecosystem vulnerability and find ways to protect and restore ecosystem functions. Therefore, this paper explores the intrinsic logical system and feasibility of guiding natural resource management and enhancing human well-being through coastal CESs, discussing related research data acquisition, method analysis, and perceptual application. Based on this, this paper analyzes the development trends of coastal CESs in natural resource management and enhancing human well-being from aspects such as biodiversity, human–nature interaction processes, cultural heritage conservation, local economic development, and community management. Finally, it proposes advancing the in-depth research of coastal CESs from the perspectives of integrating multi-source data, interdisciplinary development, and incorporating CESs into policy making, providing theoretical support for the systematic study of rational resource utilization and sustainable ecosystem development.

Keywords: coastal CESs; natural resource management; human well-being; policy making

1. Introduction

With the rapid development of the socio-economy, the theory of sustainable development can guide humanity to actively explore solutions to development issues in environmental, economic, and social domains, collectively addressing new global challenges [1–3]. As human activities continue to intensify, issues such as overexploitation of resources, environmental pollution, climate change, and loss of biodiversity have emerged globally, subjecting the atmosphere, water bodies, and land to sustained pressure, gradually expanding the negative impacts on the resilience and sustainability of ecosystems [4]. In recent years, as a bridge and link connecting natural processes with human activities, research on ecosystem services (ESs) has become a frontier and hotspot in international geography, ecology, and related disciplines. ESs, as vehicles for resource allocation and benefit coordination, can be divided into four major categories: provisioning services, regulating services, cultural services, and supporting services, linking ecosystems with human well-being and serving as important tools for achieving land space optimization and guiding planning decisions in line with the concept of ecological civilization [1,5–7]. Among them, cultural ecosystem services (CESs), due to their intangible, subjective, and non-economic characteristics, have been controversial in their development. However, as the basic social and cultural benefits provided by ecosystems to humanity, they embody the profound interrelationships between human spatial perceptions and cultural practices reflected in outdoor recreation, scientific exploration, artistic creation, and other production and consumption activities [8,9]. Effective protection and management of CESs are of significant theoretical and practical importance for achieving rational resource utilization, cultural inheritance and innovation, safeguarding the sustainable development of ecosystems, and building a harmonious coexistence between humans and nature.

Coastal areas constitute one of the most important geographical structures and functional systems on Earth. From an ecological perspective, coastal areas are among the most biodiverse ecosystems on the planet, including important ecosystems such as coastal wetlands, coral reefs, and marine grass beds, which provide habitats and breeding grounds for many species and play a crucial role in maintaining global marine ecological balance and biodiversity [10,11]. From an economic standpoint, coastal areas are often key nodes for international trade, port transportation, and fisheries activities, providing abundant fishery resources and serving as important bases for multiple industries such as tourism and shipping. From a social perspective, coastal areas are typically densely populated regions, home to numerous cities and communities with rich historical and cultural traditions. Social activities, cultural heritage, and community life in coastal areas often present distinctive characteristics compared to inland regions. However, coastal areas also face a series of challenges including environmental pollution, sea level rise caused by climate change, natural disasters, and coastal erosion. These challenges not only affect sustainable development at the local level but also pose threats to the global environment and development [12,13]. Therefore, there is a need to enhance attention to the ecological, economic, and social development of coastal areas, seeking ways to safeguard coastal CESs from the perspective of community governance, protect their unique ecosystems and social cultures, foster public sentiments towards environmental friendliness, and thereby increase understanding of coastal ecosystems. Such efforts play a crucial role in human health and well-being.

Research on CESs covers a range of disciplines including geography, ecology, economics, and psychology. It not only focuses on natural aspects such as biodiversity, soil quality, and water resources but also emphasizes the integration of human society with nature. Centered around the framework of "perception-attitude, values-worldview", CESs research seeks to understand the perceptual differences and behavioral preferences of different actors in the interaction with natural ecosystems, thereby providing theoretical support and guidance for optimizing land space [14–17]. Currently, research on ecosystem services in coastal areas mainly focuses on provisioning, regulating, and supporting services, with relatively limited attention to cultural services. This comprehensive research contributes to the establishment of more comprehensive sustainable development strategies and promotes the coordinated development of society in economic, cultural, and ecological aspects. By deeply understanding the role of CESs in sustainable development, it is possible to provide a scientific basis for ecological protection, cultural inheritance, social stability, and other aspects, thereby offering scientific support for the sustainable development of human society. In this context, research on coastal CESs not only helps scientists explore the relationship between ecosystems and culture but also provides important policy recommendations for decision makers.

Increasing attention has been paid to the development and application of CESs. As a special zone of sea–land interaction, coastal areas are characterized by rich natural resources, fragile ecological environment, and frequent human activities. Coastal CESs not only reflect the natural ecological characteristics of specific geographical environments, such as coastlines, wetlands, and coral reefs, but are also closely related to the lifestyles, cultural traditions, and economic activities of local communities. Additionally, factors such as the level of economic development, population density, urban planning, and policy orientation in coastal areas significantly influence the quality and sustainability of CESs. It is very important and necessary to evaluate coastal CESs on unique coastal ecosystems and coastal communities. However, much of the current research focuses on the quantification of assessments of coastal CESs, and there is a relative lack of in-depth exploration and systematic review of coastal CESs. This has led to an unclear understanding of the research focus and trend of coastal CESs. In order to better utilize and expand the application scenarios of coastal CESs, there is an urgent need for more comprehensive and in-depth research on them to reveal the diversity, complexity and potential value of coastal CESs. Therefore, based on the published coastal CESs literature in the WOS core collection and Scopus database, this study uses knowledge mapping tools, such as VOS viewer 1.6.20, to quantify and visualize the analysis, to systematically analyze and summarize the coastal CESs research, and to put forward the future research outlook accordingly. This study aims to provide useful explorations for coastal CESs to promote the positive interactions between natural ecosystems and social ecosystems.

2. The Interactive Relationship among Coastal CESs, Natural Resource Management, and Human Well-Being

In coastal areas, a close interactive relationship exists between coastal CESs, natural resource management, and human well-being (Figure 1). Coastal ecosystems provide rich CESs such as landscape aesthetics, biodiversity conservation, and cultural heritage, which directly impact human spiritual, cultural, and economic life [18,19].

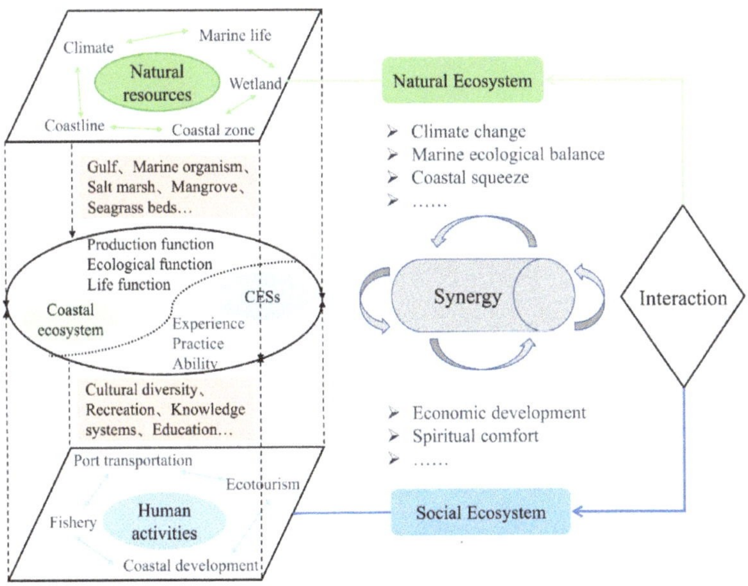

Figure 1. Interaction between natural resources, human activities, and CESs in coastal areas.

Firstly, coastal CESs are a crucial component of natural resource management. Coastal ecosystems, such as salt marshes, mangroves, and seagrass beds, provide rich biodiversity and ecological functions. The goal of natural resource management is to ensure the health and sustainability of these ecosystems, thereby ensuring their continued provision of coastal CESs. For example, by protecting and restoring coastal wetlands, their ecological functions can be enhanced, which in turn supports cultural diversity and recreational activities.

Secondly, coastal CESs have a direct impact on human well-being. They enhance people's mental comfort and social cohesion by providing opportunities for recreation and leisure. Coastal CESs are also closely related to education and cultural experiences; through participation in ecotourism and environmental education activities, individuals can relax, gain knowledge and skills, and improve their awareness and practical abilities regarding environmental issues.

Additionally, coastal CESs are closely connected with economic activities, particularly in the fields of ecotourism and fisheries. These activities not only provide economic income to local communities but also promote the sustainable development of socio-ecological systems. However, human activities, such as coastal development and port transportation, can place stress on coastal ecosystems, leading to what is termed "Coastal squeeze". This necessitates effective natural resource management to balance economic development with ecological protection.

In summary, the interdependence and influence between coastal CESs, natural resource management, and human well-being form a mutually reinforcing system. Effective natural resource management is fundamental to ensuring the sustained provision of coastal CESs, which in turn directly contributes to enhancing human well-being. Concurrently, the improvement in human well-being supports the rational use and protection of natural resources. This synergy underscores the importance of achieving sustainable development in coastal regions.

3. Materials and Methods

3.1. Database Search

This study conducted a search on the Web of Science Core Collection database and Scopus on 5 June 2024. The search covered all years from 2009 to 2024. In order to understand the amount of ongoing research, different terms and combinations have been used, as shown in Table 1. To focus on research related to coastal CESs, proceeding papers, book chapters, letters, and other document types were excluded. Only articles and reviews were selected, and duplicate data were removed. Ultimately, 150 documents were chosen for analysis.

Table 1. Searched terms and results from queries to WOS core collection and SCOPUS.

Searched Terms	Occurrences	
	WOS Core Collection	Scopus
"cultural ecosystem services" AND "coastal"	62	83
"culture ecosystem services" AND "coastal"	3	2
"cultural ecosystem services" AND "ocean"	7	16
"cultural ecosystem services" AND "sea"	19	31
"cultural ecosystem services" AND "marine"	35	48
"cultural ecosystem services" AND "fish"	21	16
"cultural ecosystem services" AND "mangrove"	7	16
"cultural ecosystem services" AND "bay"	7	14
"cultural ecosystem services" AND "shellfish"	3	4
Total	164	230

3.2. Literature Classification

For the identified coastal CESs literature, this study classified, summarized, and concluded the selected literature from the following aspects: (1) geographic distribution; (2) types of coastal ecosystems; (3) subcategories of CESs; (4) research methods for coastal CESs; and (5) scientific issues concerned by coastal CESs research.

3.2.1. Geographic Distribution

In this study, we conducted an in-depth reading of each piece of literature and paid particular attention to the geographic location of the study. During the reading process, we meticulously scrutinized the specific regions covered by each empirical study and made exhaustive records of the countries in which these regions are located. These records not only provided us with the geographic context of each study, but also served as the basis for further analysis of the relationships between different countries and coastal ecosystem types. By gaining insight into these geographic locations, we were able to more accurately explore and understand the diversity of coastal ecosystems and their interactions with human activities.

3.2.2. Identification of Coastal Ecosystem Types

The diversity of ecosystems determines that the CESs they provide also vary. Different researchers will focus on CESs in different ecosystem types, depending on the characteristics of the study area, research interests, and goals. In this study, we synthesized the coastal ecosystem types involved in the study of coastal CESs with reference to the usual classification of coastal ecosystem types [20]. We propose that coastal ecosystem types should include the following ten types: marine organism, estuary, salt marsh, dune, mangroves, lagoon, seagrass beds, estuaries, coral reef, and gulf (represents the shoreline area near the gulf). These types not only cover the natural ecological characteristics of coastal areas, but also reflect their important role in providing various types of ecosystem services. Through this categorization, this study aims to more systematically analyze and understand the contribution of different coastal ecosystems in maintaining ecological balance and supporting human well-being.

3.2.3. Identification of Subcategories of Coastal CESs

The current international classification and definition of CESs is mainly provided by Millennium Ecosystem Assessment (MA), The Economics of Ecosystems and Biodiversity (TEEB), Common International Classification of Ecosystem Services (CICES), Intergovernmental Science-Policy Platform for Biodiversity and Ecosystem Services (IPBES), UK National Ecosystem Assessment (UK NEA), which includes types such as aesthetic value and recreational value. In order to ensure consistency in the understanding of CESs, this paper selects the definitions and categories proposed by the MA reports, which are referenced and emphasized in a number of studies (plus references). These categories were identified as Recreation, Aesthetic, Sense of place, Spiritual and religious, Education, Social relations, Cultural heritage, Cultural diversity, Knowledge system, Inspiration.

3.2.4. Identification of Methods Used in Coastal CESs Research

Based on a systematic review of the literature on coastal CESs, we summarized and concluded the data types and research methods in detail. We took detailed notes on the types of data used by researchers when reading the literature and initially categorized them into three main groups: natural environment data, social data, and perceptual data. For each type of data, we not only recorded its main characteristics, but also looked in detail at the sources, collection, and analysis processes used to obtain these data. In terms of research methodology, we classified the methods into two categories, statistical analysis and spatial visualization analysis, based on the research process and technical support. In addition, the literature systematic review was an iterative process that adjusted the selection of data types and research methods based on new findings and feedback to ensure that systematic generalizations and summaries were obtained.

3.2.5. Identification of Scientific Issues Concerned by Coastal CESs Research

When conducting a systematic review of the literature on coastal CESs, we first familiarized ourselves with the current state and trends in the research field through a literature preview, which in turn identifies key themes. Next, we defined the purpose of this study, around which we formed specific scientific questions designed to guide the review process and ensure research direction. We then refined these questions to ensure that they were both relevant and actionable. Finally, we documented the questions and updated them with new findings and understandings throughout the review process to ensure the currency and depth of the research.

3.3. Bibliometric Analysis Methods

Visual mapping software usually has powerful data mining and analysis functions, which can perform operations such as keyword extraction and topic clustering, reveal the interrelationships and citation networks among the literature through the layout of nodes and links, sort out the overall structure and development of the research field, and understand the core topics and development trends of the research field in a more in-depth manner. Providing researchers with tools for quantitative literature trend analysis, in-depth knowledge structure visualization, and interdisciplinary correlation exploration can effectively meet diverse research needs. Therefore, we utilized the keyword clustering and emergence functions of VOS viewer 1.6.20 software to explore the research and hot frontiers of coastal CESs in recent years.

4. Results

4.1. Results of the Bibliometric Analysis

From the trend of annual publications in coastal CESs research (Figure 2), it can be observed that literature related to coastal CESs first appeared in 2009. Over the following decade, scholars' attention to this field has continuously increased, especially after 2018, when the number of related research documents rapidly increased and continued to grow. In 2015, the United Nations adopted the "2030 Agenda for Sustainable Development", which includes 17 Sustainable Development Goals (SDGs), emphasizing ecosystem conservation, cultural heritage protection, and other aspects. Meanwhile, academia's understanding of CESs has gradually deepened, shifting from a simple focus on resource utilization to a comprehensive perspective that considers ecology, culture, society, and economy. This shift helps strengthen the connection between CESs and coastal area policy formulation. At present, researchers from 32 countries and regions, including the UK, the US, China, and Spain, have carried out special studies on coastal economic benefits. Among them, the UK and the US have the highest number of publications, with 17 and 11 papers, respectively. In terms of research subjects, the diversification of spatial dimensions and value types further reflects the complexity of the relationship between humans and nature in coastal areas.

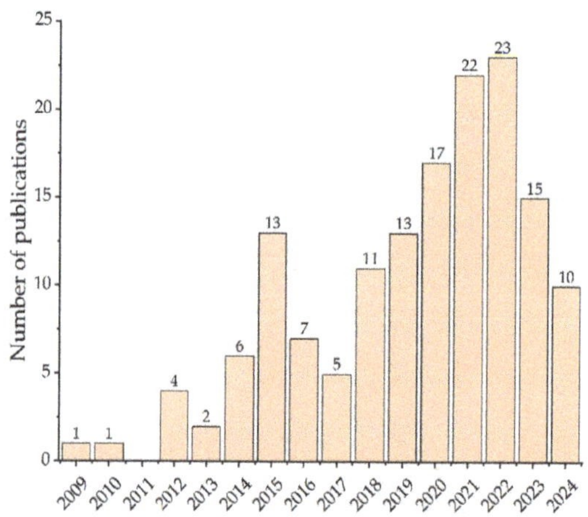

Figure 2. Number of publications per year that addressed coastal CESs.

To better understand the research hotspots in different countries and regions and people's perceptual preferences for different coastal ecosystems, this study further analyzes the relationship between the country of the study area and types of coastal ecosystems, the connection between different types of coastal ecosystems and CESs subtypes (Figure 3). In terms of the location of the study area, the UK, US, Spain, Australia, and China have the highest number of studies of relevant coastal CESs. This is due to the fact that these countries have extensive and diverse coastal ecosystems. The abundance of resources provides a large number of objects and opportunities for research. At the same time, the economies of these countries are highly dependent on coastal resources, especially in fisheries, tourism, port trade, and coastal urban development, all of which have contributed to the increased demand for research on coastal ecosystems and their CESs. Coastal ecosystems refer to natural systems in the ocean composed of biological communities and their interactions with the environment. They mainly include ten typical ecosystem types such as marine organism, estuary, salt marsh, dune, mangroves, lagoon, seagrass beds, estuaries, coral reef and gulf (represents the shoreline area near the gulf). These ecosystems possess natural ecological values and serve various social functions. Currently, CESs research is primarily focused on these types of ecosystems. In terms of individual coastal ecosystem types, the marine organism receives the highest level of attention. This is because the richness of the marine organism plays a crucial role in maintaining ecosystem balance and constitutes a significant component of coastal CESs. Additionally, the marine organism holds special symbolic and cultural value in many societies, closely intertwined with local traditions, religions, and arts. Additionally, gulf, mangroves, salt marshes, and estuary receive considerable attention from researchers. This is because they have significant impacts on local communities' livelihoods, culture, and socio-economic development. They also face threats such as land development, pollution, and sea-level rise. Conducting CESs research on these ecosystems helps understand their vulnerabilities and challenges, aiding in the formulation of appropriate conservation and management strategies.

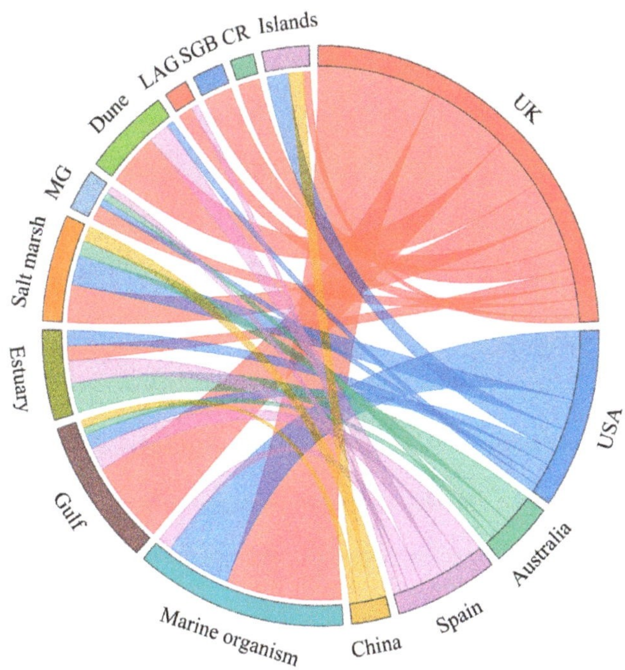

Figure 3. Study of the relationship between countries and coastal ecosystem types. Notes: MG for mangrove; LAG for Lagoon; SGB for Seagrass beds; CR for Coral reef.

From the correspondence between different types of coastal ecosystems and CESs subclasses (Figure 4), it can be observed that the differences in structure and function of different coastal ecosystem types, environmental pressures, ecological vulnerabilities, as well as the needs and cultural traditions of coastal communities, can all influence the CESs provided by coastal ecosystem types. In terms of individual coastal ecosystem types, there is a higher number of studies related to marine organisms and the gulf. This is because marine organisms provide services to humans, including food resources and marine tourism. Additionally, the gulf faces various threats from human activities such as overfishing, pollution, and climate change. Understanding the impacts of these threats on marine organisms and ecosystems, as well as how to effectively protect these biological resources through public participation, is one of the current research directions. In terms of CESs subclasses, there is a higher number of studies related to recreation and cultural heritage. This is because governments and stakeholders need to consider how to balance the development of recreational activities with the need for ecosystem protection, as well as how to effectively protect and manage cultural heritage. Moreover, recreational services and cultural heritage often attract broader public participation, promoting public awareness and protection of ecosystems and cultural heritage.

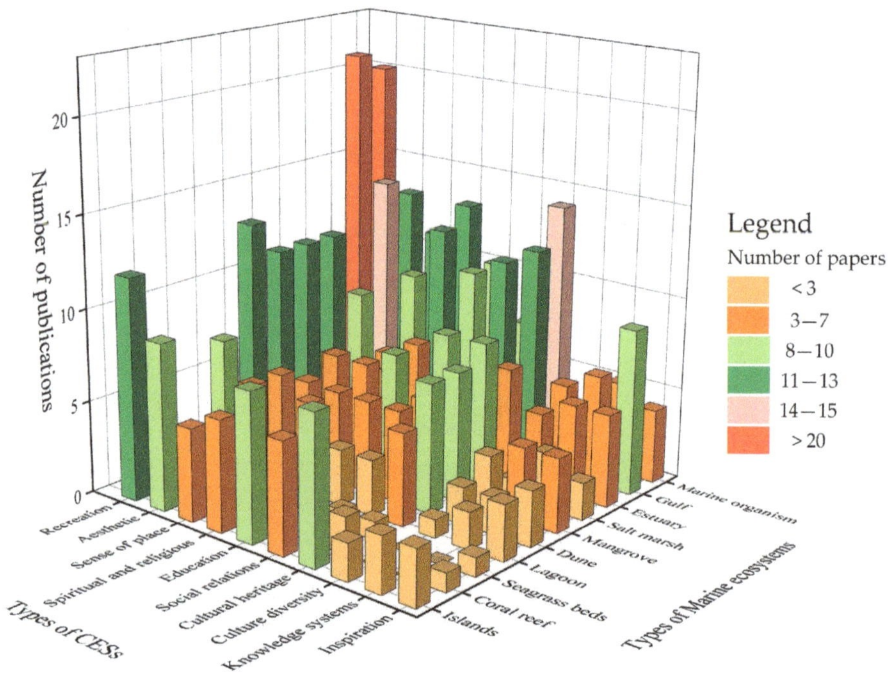

Figure 4. Correspondence between coastal ecosystem types and CESs subcategories in the publications.

In a particular field, research hotspots and the evolutionary context can be explored through keyword mining. With the support of VOS viewer 1.6.20 software, keywords with a co-occurrence frequency exceeding two were subjected to co-occurrence visualization processing (Figure 5). From the figure, it is evident that by excluding terms related to literature search queries such as "cultural ecosystem services", "ecosystem services", and "coastal", keywords like "management" (27), "values" (25), "conservation" (22), and "framework" (19) emerge as hotspots in the field. Additionally, a multi-center research structure has formed, with the search terms as the central point and research hotspots as secondary centers. This structure is further elaborated upon through various entry points such as "benefits" (14), "biodiversity" (12), "social media" (10), "landscape" (6), and "health" (5). In addition to the aforementioned keywords, environmental ethics is also particularly important in the study of coastal ecosystem cultural services. It establishes the value of nature, emphasizes human responsibility towards the environment, advocates for respect for biodiversity and cultural heritage, supports sustainable development, and encourages proactive efforts to protect these cultural services. Research on CESs operates at spatial and local scales, exploring effective pathways for local resource management by analyzing the perceptual preferences and behavioral patterns of different societal actors in various social contexts. As an interdisciplinary research direction, coastal CESs research utilizes theories from disciplines such as geography, ecology, economics, and sociology to address issues across different domains. It provides theoretical support and methodological guidance for sustainable ecosystem management, coastal land use planning, and the high-quality development and spatial layout of related industries.

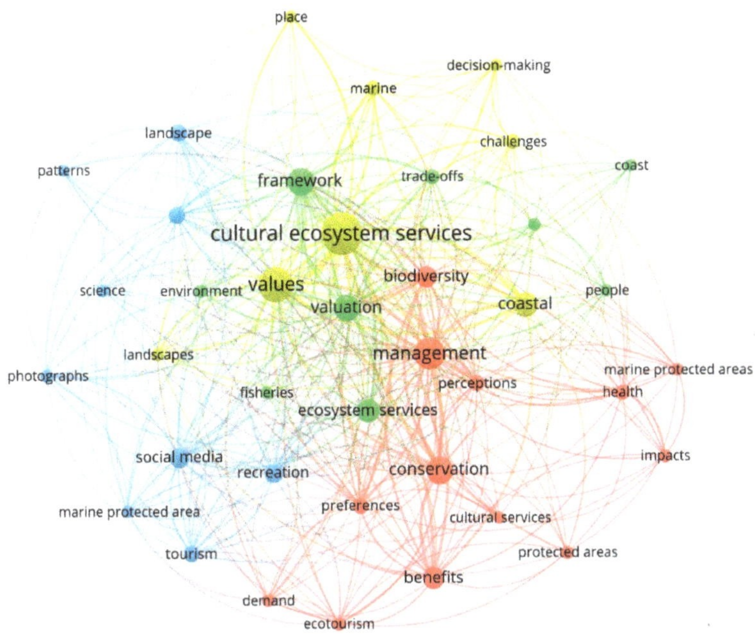

Figure 5. Keywords co-occurrence mapping for coastal CESs research.

The distribution of journals can visualize the distribution of mainstream journals in coastal CESs research, and the distribution of co-cited journals can characterize their influence and relevance to the academic research in this field. From the Figure 6a knowledge graph of coupled distribution of articles, it can be seen that the journals with high volume of publications are *Ecosystem Services*, *Marine Policy*, *Ocean & Coastal Management*, *Ecological Indicators*, *Ecological Economics*, and the fields of the journals are mainly environmental sciences, ecology, and oceanography. As can be seen from the timeline in Figure 6a, the above journals with a high volume of publications focusing on the development of coastal CESs are mainly concentrated in the period of 2018–2021. *Journal of Environmental Management*, *Land*, *Ecosystems and People*'s focus on coastal CESs research has intensified in recent years. As can be seen from the distribution of journal co-citations shown in Figure 6b, the 155 sample documents with a total citation frequency of 200 or more are *Bioscience* (1486), *Ecosystem Services* (672), *Ecological Indicators* (432), *Ecological Complexity* (330), and *Landscape Ecology* (319). Overall, environmental sciences, ecology, and landscape journals continued to pay more attention to coastal CESs research, while oceanography journals paid less attention, suggesting that environmental, ecological, and landscape journals play an important role in supporting and leading coastal CESs research.

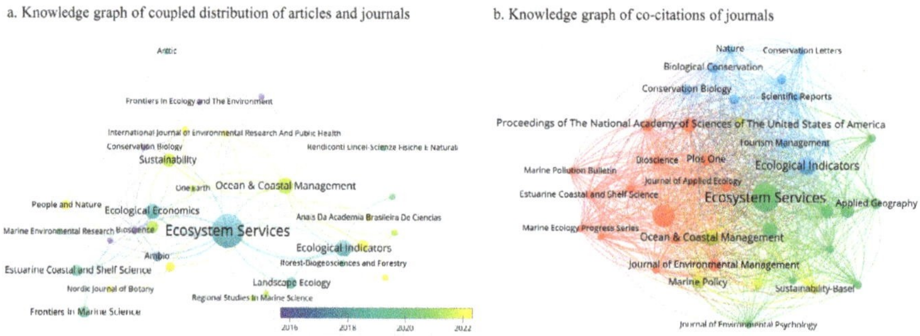

Figure 6. Knowledge map of the distribution of journals carrying articles on coastal CESs research.

4.2. Research Methods for Coastal CESs

Compared to the other three categories of services in coastal ecosystems (supporting, provisioning, and regulating services), research on CESs started relatively late for two main reasons. Firstly, cultural services possess intangible and subjective characteristics, with complex content and thus presenting greater research challenges. Secondly, attention to non-material dimensions of development often builds on material dimensions development, with resulting CESs research having a certain degree of delay. As of now, in terms of data acquisition, coastal CESs have established methods such as remote sensing monitoring, collection of local statistical data, and survey questionnaires to obtain the necessary data on the natural environment, social development, and public perception. Regarding data analysis, approaches predominantly focus on text analysis and spatial visualization, supporting more in-depth practical research in areas such as community-based sustainable management, climate change, and marine protected areas.

4.2.1. Data Acquisition for Coastal CESs

Coastal CESs research typically requires multi-source data including environmental, social, and perceptual data. Environmental data at various scales are mainly obtained through field sampling and remote sensing monitoring methods [21,22]. The identification of CESs hotspots and cold spots, as well as the matching of supply–demand patterns based on environmental data, are equally crucial for landscape sustainable management and biodiversity conservation. Pouso et al. (2018) conducted questionnaire surveys on beach users' behaviors and perceptions, comparing them with environmental time series data (microbial concentrations and water transparency) [23]. The results underscored the significance of water sanitation actions in restoring degraded coastal environments and maintaining ecosystem services. Additionally, remote sensing imagery monitoring of land surface forms and land use types at different historical stages serves as vital reference material for CESs temporal scale analysis. By delineating the timeline of events influencing tidal and freshwater flows in the Tamar River estuary since 1806, Davis and Kidd emphasized the importance of identifying pressure sources [24]. They regarded this as foundational in balancing the potential for reducing various ecosystem services under environmental restoration, particularly between provisioning and cultural ecosystem services.

In terms of social data acquisition, local statistical data serve as the primary source. Uddin et al. utilized tourism revenue data collected by the Forest Department to assess the cultural services provided by the Sundarbans Reserve Forest to the local population. They argued for a comprehensive economic assessment of the total stock and potential of all ecosystem services, and the determination of sustainable yield limits for services under different socio-economic and climate change scenarios to enhance the forest's sustainable management [25]. Apart from local statistical data, geolocated data, which record human activities associated with geographic entities and their interrelationships with geographic locations, offer new data sources for studying human–nature interactions and regional development. Their application in coastal CESs research is increasingly widespread [26]. Martinez Pastur et al. employed geotagged digital images posted by residents and tourists on the Panoramio web platform to identify hotspots of four cultural ecosystem services. They correlated these hotspots with social and biophysical landscape features, demonstrating the utility of this method for assessing CESs at the regional scale, particularly in areas with low data availability and on-site accessibility [27].

In terms of perception data acquisition, early CESs research often relied on a combination of survey questionnaires as the primary method and semi-structured interviews as a supplementary method to collect perceptions and behavioral preferences of different stakeholders regarding environmental spaces [28]. Urquhart and Acott, through semi-structured interviews with stakeholders in fishing communities, explored the contribution of local perceptions to understanding the relationship between fisheries and coastal CESs. They argued that while ensuring the sustainability of fish resources, the contribution of fisheries to the social and cultural well-being of coastal communities should not be overlooked [29]. To adapt to emerging trends, the application of social media big data in CESs research has become increasingly popular. Moreover, the application of multi-source social media data has successfully overcome the limitations of geographical administrative boundaries, supporting researchers in conducting large-scale assessments of CESs when field surveys are not feasible [30]. To understand the visitation patterns and benefits of different types of natural and cultural heritage tourists, Ghermandi et al. combined social media photos with high-resolution land cover maps to identify different cultural services and their associations with specific ecosystem and land cover types. The results can provide valuable information for decision makers to improve tourism management and conservation strategies [31].

4.2.2. Methodology for Analyzing Coastal CESs

Evaluation of coastal CESs primarily focuses on non-material values that are difficult to quantify, such as spiritual, religious, and local perceptions. The data composition typically includes various types such as textual data, point data, and spatial data. Therefore, analysis of the results often involves statistical analysis and spatial visualization analysis to meet the subsequent applications in natural resource management and enhancement of human well-being.

Coastal CESs research typically involves a large amount of survey questionnaire data, interview data, etc. Statistical analysis can help researchers extract useful information from these data and reveal hidden patterns and trends in the research. Firstly, based on the collection of relevant data such as tourism revenue and cultural heritage, statistical analysis can better understand their contributions to communities and economies. De la Lama et al. identified and described the historical cultural benefits of small-scale fisheries in Peru and their transformative role in early social development, emphasizing the contribution and historical significance of small-scale fisheries to society [32]. Secondly, in coastal CESs research, researchers often find it difficult to conduct comprehensive surveys of entire populations or ecosystems. Therefore, they often need to use statistical analysis for inference and generalization, such as understanding community residents' and stakeholders' perceptions, attitudes, and involvement in coastal CESs, analyzing tourists' behavior patterns and preferences in ecosystems, and determining their needs, concerns, and willingness to participate, thus promoting community involvement and co-management, guiding the planning and management of ecotourism. Based on stakeholder mapping, semi-structured interviews, observation, and surveys of socio-cultural preferences, Malinauskaite et al used grounded theory induction to qualitatively analyze data, examining the CESs derived from whales by people residing in Skjálfandi Bay, North Iceland, and its surroundings, to incorporate CESs related to whale resources into marine resource management in Iceland [33]. Finally, statistical analysis can also assess the impact of human behavioral preferences on ecosystems, including their effects on biodiversity, ecosystem functions, and ecosystem health, thereby guiding ecosystem conservation and sustainable management. Based on identifying trends in the consumption and collection of wild plants, Reyes-García et al. analyzed the correlation between CESs and this trend, finding that CESs are an important factor in explaining different development trends among different plant species [34].

Through spatial visualization analysis, researchers can display the spatial distribution of ecosystems, the supply and demand of coastal CESs, and related data on maps, helping to reveal the spatial correlations between landscapes, human activities, and coastal CESs. This provides intuitive spatial information support for decision making. By conducting spatial visualization analysis of factors such as ecosystem characteristics, distribution of cultural landmarks, and population density in different regions, researchers can understand the spatial distribution characteristics of coastal CESs and identify important ecological landscapes and cultural landmarks to explore the spatial correlations between human activities and ecosystems. Zhao et al. used a combination of social media data and the SolVES model to simulate the spatial distribution patterns of tidal wetland CESs from different stakeholders' perspectives, thus reconciling the contradiction between agricultural infrastructure construction and coastal wetland ecosystem management, and providing planning recommendations for the sustainable development of coastal wetland agriculture [35]. Additionally, by mapping the spatial relationships between the supply and demand of coastal CESs in different regions, the spatial accessibility and sustainability of coastal CESs can be clarified, helping policymakers and planners assess whether the supply of coastal CESs in different areas meets demand and identify areas with supply–demand imbalances, providing scientific evidence for ecosystem protection and optimizing user experience. Santana-Santana et al. analyzed and discussed the preference statistical relationships between two user types: those with disabilities and those without disabilities. They also studied whether environmental management and land use in the study area met users' expectations, providing a reference for optimizing user experience [36].

4.3. Trends in Research on Coastal CESs

The trend of coastal CESs research can be determined based on the intensity and timing of keyword emergence in the keyword burst diagram generated by VOSviewer 1.6.20 software (Figure 7). It can be observed that keywords focusing on natural resource management, such as marine spatial planning, biodiversity, and environment management, mostly appeared during the initial stage of coastal CESs research. As the research progresses, the frequency of keywords such as decision making, tourism, health, and perceptions increases, indicating that researchers are no longer solely focused on natural resource management and ecosystem protection, but are also paying attention to the dynamic process of CESs from perception to application. This highlights the greater role played by coastal CESs research in improving human–environment relationships and enhancing human well-being [37–40].

Research on coastal CESs is advancing towards a new interdisciplinary and integrative era. With the integration of modern technologies such as social media and big data, researchers can more accurately capture public perceptions and attitudes towards coastal ecosystems. This, in turn, provides scientific evidence for fields closely related to ecosystems, such as tourism and health, while also assessing the impact of human behavior on ecosystems and formulating effective environmental education and conservation strategies. Additionally, the quantitative assessment of coastal CESs, adaptive management in response to climate change, the provision of scientific foundations for policy and governance, as well as the enhancement of community engagement and local management, are all critical directions in the development of coastal CESs research. These efforts aim to achieve long-term ecosystem sustainability and the maximization of human well-being.

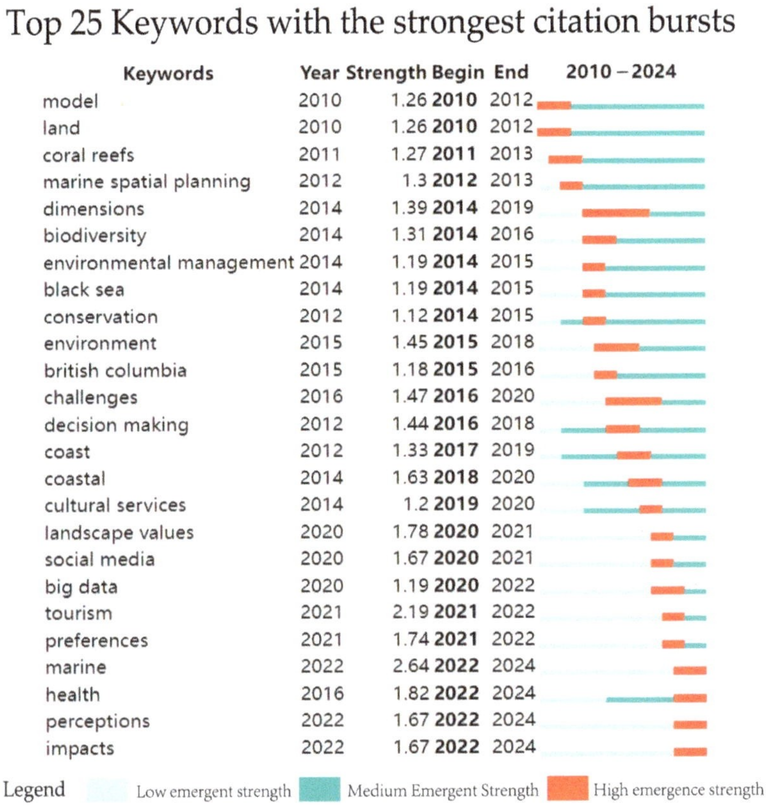

Figure 7. Coastal CESs study high mutability keywords.

4.3.1. Coastal CESs and Natural Resource Management

Coastal areas harbor rich biodiversity and essential ecological functions, including marine fisheries resources, coastal wetlands, and coral reefs. Effective natural resource management can ensure the sustainable utilization of these resources, preventing overfishing and development, thereby protecting the health and ecological balance of marine ecosystems.

Due to the vast array of services and resources provided by the ocean, it is evident that marine biodiversity is crucial for both human and planetary health [41,42]. Through cultural activities, ecotourism, the preservation and promotion of local marine culture, and the traditions of fishermen, people can gain deeper insights into the diversity and fragility of coastal ecosystems. This promotes respect and reliance on the natural environment, reducing overexploitation and damage to natural resources, and enhances awareness and actions towards biodiversity conservation [43–45]. Furthermore, the establishment of marine protected areas, ecological reserves, and similar regions to protect the integrity and stability of rare species and critical ecosystems can effectively maintain the biodiversity of coastal ecosystems [46,47]. Kochalski et al. (2022) surveyed in France, Germany, Sweden, and Norway to clarify the public's perceptions of CESs provided by threatened species such as Salmon salar. The results indicate that improving ecological conditions and raising public awareness can promote the development of perceptible ecosystem services and the value of a species in the public eye [48]. Coastal CESs positively impact biodiversity conservation by increasing awareness and appreciation of biodiversity, fostering harmonious relationships between humans and nature, providing protective measures, and promoting community participation and cooperation. This contributes to achieving the sustainable development and protection of coastal ecosystems.

Furthermore, coastal CESs play a significant role in the interaction between humans and the natural environment, carrying people's cognition, emotions, and values towards the natural environment, thereby playing an important guiding and supportive role in natural resource management [49,50]. Coastal CESs are not only reflected in the aesthetic enjoyment of natural landscapes but also encompass the protection and transmission of traditional fisheries, marine lifestyles, and local cultural heritage [51]. Through shaping people's attitudes and behaviors toward the natural environment, coastal CESs influence the development, utilization, and protection of natural resources [52]. At the same time, the research framework of CESs helps enhance respect and attention towards natural resources, promotes community participation in natural resource management, and drives the formulation and implementation of relevant policies [53–55]. Gajardo et al. quantified the CESs provided by Taytay Bay, identifying human activities that threaten CESs. The results indicate that CESs contribute to promoting human well-being at the community level, and integrating cultural and non-monetary environmental values into ecosystem assessments and on-site policy-making is crucial [56]. Therefore, coastal CESs complement natural resource management, and understanding coastal CESs can enhance the sustainability of natural resource management.

4.3.2. Coastal CESs and Human Well-Being

The unique natural landscapes and biodiversity of coastal areas can provide coastal CESs to meet the growing spiritual needs of humanity. The lifestyles of coastal residents, fisheries culture, and cultural heritage of coastal communities also serve similar functions. The well-being of contemporary and future human societies depends on whether there are ecologically sustainable and socially just ways of life worldwide, which in turn directly impact the functionality, services, and landscape patterns of ecosystems.

The natural environment and cultural landscapes of coastal areas are often closely intertwined, forming unique cultural landscapes and traditional ways of life. These cultural heritages carry rich history, traditions, and cultural memories, representing valuable assets of local communities [57]. By exploring and preserving local cultural heritage, and promoting the development and production of cultural products and traditional craftsmanship, we can not only meet market demands but also enhance the added value of local fishery products. Additionally, by engaging in activities such as ecotourism, marine cultural events, and traditional fishing experiences, people gain a deeper understanding and appreciation of local cultural traditions, which can strengthen awareness and protection of cultural heritage [58,59]. Cabana et al. used the coastal site of Dublin Bay as an example, and proposed a comprehensive evaluation and communication strategy for CESs, integrating research from humanities, social sciences, and environmental sciences. The study suggests that different values and services depend on specific coastal habitats, and integrating CESs into cultural heritage conservation frameworks can enhance contributions from multiple stakeholders and the public [60]. By exploring how coastal ecosystems meet human spiritual needs through CESs, promoting awareness, inheritance, and protection of cultural heritage, CESs contribute significantly to the conservation of cultural heritage in coastal areas.

The natural landscapes and ecological environments of coastal areas are often essential factors attracting tourists [47]. By offering a diverse range of coastal leisure activities such as beach vacations, marine cultural experiences, and traditional fishing tours, coastal areas can attract a large number of tourists, thereby driving the development of the tourism industry. This not only generates tourism revenue for the local community but also stimulates the prosperity of related industries such as catering, accommodation, and fishery product sales, promoting industry upgrading and brand building [61,62]. Through cultural activities, cultural tourism projects, and the provision of employment and entrepreneurship platforms, residents' entrepreneurial and innovative awareness is stimulated, contributing to talent mobility and optimizing resource allocation, thus making a significant contribution to the diversification of the local economy. Ondiek et al. assessed the CESs provided by natural wetlands and rice fields in Kenya, and the results showed that although rice fields could provide higher economic value, natural wetlands contributed more to meeting residents' spiritual needs. Ensuring the sustainable use of both ecological types is equally important for maintaining residents' livelihoods [63]. Coastal CESs, through promoting tourism development, fostering the growth of the cultural and creative industries, and encouraging community employment and entrepreneurship, provide vital support for the growth and diversification of the local economy, thereby enhancing the competitiveness and sustainability of regional economic development.

Understanding CESs also contributes to enhancing community cohesion and social stability, fostering communication and cooperation among community members, strengthening the inheritance and development of social culture, enhancing community cohesion and identity, and promoting harmonious development within the community [64–66]. At the same time, understanding CESs can inspire community members to pay attention to and participate in environmental protection and ecological civilization construction. By organizing activities such as ecotourism, marine cultural experiences, and environmental volunteer work, community members gain a better understanding and respect for the local ecological environment, enhance their sense of responsibility and awareness of environmental protection, and promote the implementation of community environmental management and ecological protection efforts. Treviño used a local perspective to identify the CESs provided by mangroves to clarify how they support local well-being. The results indicate that users of mangroves attribute profound cultural importance and personal attachment to the mangrove ecosystem [67]. Therefore, coastal CESs, through pathways such as enhancing spiritual enjoyment, promoting economic development, and strengthening social cohesion, play an undeniable positive role in enhancing human well-being.

5. Discussions

5.1. Applications and Challenges of Multi-Source Data in the Study of Coastal CESs

The coordinated integration of "natural ecosystem-socio-economic-public perception" data is the key to conducting coastal CESs research. Among them, public perception data play a vital role, directly reflecting the public's cognition, attitude, and demand for coastal ecosystem cultural services. These data can help researchers understand the public's views on ecosystem protection, utilization, and development, thereby providing a basis for policy making and resource management.

In the research of coastal CESs, the collection and analysis of perceptual data are crucial for understanding human cognition and attitudes toward CESs. Traditionally, perceptual data are primarily obtained through questionnaires or semi-structured interviews. In these cases, researchers can tailor questions according to their research objectives to deeply explore respondents' views and perceptions of coastal ecosystem cultural services. This approach helps clarify public perceptions of CESs categories [68], CESs responses to environmental changes [69], and the threats faced by CESs supply [70]. However, traditional data collection methods also have certain limitations. Field surveys often require high costs and are limited in terms of the selected research areas, which may affect the generalizability and representativeness of the research results.

With the rapid development of information technology, social media data have provided new perspectives and data support for coastal CESs research. Social media platforms such as Flickr and Instagram, due to their real-time nature, large data volume, and rich content interpretation potential, have gradually become favored data sources for scholars. These data can reflect the public's immediate reactions and attitudes toward coastal CESs [55], reveal cognitive differences and value perceptions among different social groups [31], and quickly identify the impact of environmental disasters on CESs at large spatial scales [71]. However, there are limitations in terms of data noise and user bias. As Cao et al. noted, sentiments related to coastal CESs vary based on social media characteristics [62]. Additionally, social media data involve user privacy, and researchers must follow relevant data protection laws and regulations, such as the EU's General Data Protection Regulation (GDPR) or privacy protection laws in other regions, to ensure the legality of data collection and processing.

While social media provides a rich foundation for coastal CESs research, the diversity and vast volume of data also present challenges for data collection and processing. In current research, geotagged photos are widely used and are often manually inspected and classified according to their content to correspond to CESs categories. This manual classification can lead to subjectivity and inconsistency in the results. Identifying CESs through photos can only consider CESs categories that reflect direct, in-site, and outdoor interactions depending on actual presence in the environmental setting [31]. In the context of the explosive growth of social media data, applying machine learning for intelligent classification of photo content can not only reduce subjective bias in the classification process and ensure the objectivity and consistency of the results but also significantly improve data processing efficiency, providing a new solution for coastal CESs data analysis.

In terms of data integration, coastal CESs research needs to involve various types of data, including ecological environment, socio-economic, cultural heritage, and others, typically sourced from different domains and origins, including government agencies, research institutions, community organizations, etc. The quality and reliability of data directly influence the accuracy and credibility of research results. Issues such as low data quality, unclear data sources, and missing data may lead to biases in analysis results. Therefore, ensuring data consistency and reliability is a challenge. In summary, coastal CESs research, leveraging digital technology and data analysis tools, can enhance the quality and benefits of coastal CESs, promoting sustainable development and ecological protection of communities.

5.2. Promoting Interdisciplinary Integration to Enhance a Comprehensive Understanding of Coastal CESs

The complexity and diversity of coastal CESs necessitate the comprehensive consideration of ecological, economic, and social factors to fully understand their essence and value. Such integrated analysis requires the establishment of effective communication and collaboration mechanisms between ecology, culture, sociology, and economics to advance in-depth research on coastal CESs. Coastal areas face unique natural disaster risks such as sea-level rise and coastal squeeze due to their specific geographic location and environmental conditions, which are exacerbated by climate change. Ecology provides an in-depth understanding and analysis of coastal ecosystems, including knowledge of marine biodiversity and ecosystem functions. Researchers in this field explore changes in coastal CESs against the backdrop of sea-level rise and climate change [22,71], as well as the contributions of typical ecosystems and marine organisms to local coastal CESs [41,63,72,73]. Cultural studies focus on aspects such as maritime culture, traditional customs, and cultural heritage, using CESs as tools to deepen the understanding of marine cultural heritage and emphasizing the impact of cultural differences on CESs [58,74,75]. Sociology examines social factors such as community organization, social relationships, and cultural identity, delving into changes in human emotions, which helps to highlight the importance of fisheries in the social and cultural well-being of coastal communities [29]. Economics addresses issues related to the value of CESs, ecotourism, and industrial development, typically exploring the impact of economic activities such as tourism and aquaculture on CESs and seeking to optimize resource allocation to meet the needs of human society [76].

The composition and functions of coastal CESs are highly complex, involving multiple dimensions such as biodiversity, geographical distribution, and human activities. Coastal regions often face the trade-off between ecological protection and economic development. Coastal area management policies need to comprehensively consider various factors such as ecological protection, economic development, and social well-being. Only by conducting research on coastal CESs from a multidisciplinary perspective can we provide a scientific basis for policy formulation and ensure the comprehensiveness and effectiveness of policy implementation. For instance, ecology provides a fundamental understanding of ecosystem structures and functions, while cultural studies reveal how human values and traditional customs influence ecosystem services. Sociology analyzes how social structures and behavioral patterns are interconnected with CESs, and economics assesses the economic value of CESs and their contributions to human well-being. The interaction of these disciplines enriches the multidimensional understanding of CESs. Through interdisciplinary integration, knowledge and methodologies from different academic fields can be organically combined, expanding the application scenarios of coastal CESs and providing theoretical support and practical guidance for achieving the dual objectives of marine environmental protection and social development.

5.3. *Integrating Public Perception and Coastal CESs into Sustainable Policymaking*

Incorporating public perception into the policymaking process is crucial for ensuring that policies are inclusive, effective, and sustainable. Public perception offers a diverse range of perspectives that can reflect the needs and expectations of different groups. This process helps to raise public awareness of environmental issues, thereby encouraging various sectors of society to participate in the maintenance of biodiversity and ecological balance. This two-way interactive policymaking model is an important pathway to building a sustainable society. Coastal CESs, as intangible benefits arising from the interaction between humans and natural ecosystems, can play an effective role in public perception and policymaking.

Coastal CESs have a significant guiding role in enhancing human well-being and promoting the development of coastal communities. By assessing the contributions of coastal CESs to residents' well-being, policymakers can find a balance between development and conservation in coastal areas, especially in communities that are dependent on natural resources. As Elwell et al. (2020) suggested, for communities more reliant on natural resources, small-scale tourism and conservation interventions may be reconsidered as potential means to increase well-being [77]. Similarly, Ghermandi et al. (2020) consider ecotourism a priority [31]. Although the development of ecotourism can balance the ecological and socio-economic benefits of coastal areas to some extent, Malinauskaite et al. (2021) revealed some coastal CESs management challenges originating from intensified tourism [33]. Ecotourism should be developed in such a way as to ensure that local natural resources are not destroyed. Whether a region prioritizes environmental protection or economic development will influence the strategies for the conservation and utilization of coastal CESs. Boeri et al. points out that a multi-dimensional conceptualization of biodiversity needs to be considered when planning and funding conservation actions [45]. And Ruiz-Frau et al. mentioned that the integration of spatially explicit socioeconomic data for various uses of marine biodiversity contributes to the development of comprehensive marine spatial plans [73].

Encouraging stakeholder participation is crucial for maintaining coastal CESs, which are fundamental to fairness, health, and well-being. Identifying and measuring intangible values can be successful only when those with stakes in the decision-making context participate as collaborators throughout [78]. Stakeholder participation can collect diverse perspectives and information, helping to create an inclusive decision-making environment. For example, Tajima et al. investigated the differences in the perception of CESs between residents and tourists, suggesting that the interests of stakeholders might conflict during marine development [28]. Santana-Santana et al. (2022) explored the impact of disabilities on the perception of CESs, finding that disability itself is not a statistically significant conditional factor, but the specific type of disability is [36]. Unlike these findings, which showed differences in the perception of coastal CESs among various stakeholders, Barrena et al. (2013) indicated that coastal CESs can be equally important for both local and distant populations [51].

Coastal CESs not only contribute to the development of local communities but also promote public awareness, ecological restoration, and biodiversity maintenance. Many authors have used coastal CESs as a research perspective to explore topics such as the views of mangrove users [67], the impact of human factors on CESs [79], and the shared benefits of human and coastal restoration projects [80,81]. These research findings can provide valuable information for local marine resource management. Additionally, clarifying public preferences and environmental education can promote ecological protection and restoration efforts to a certain extent, which is beneficial for maintaining biodiversity and ecological balance [45,82]. Furthermore, Quevedo et al. (2024) suggest that there is an opportunity for CESs to be integrated into restoration and climate mitigation strategies. Such research can help develop management strategies that are environmentally and socially just and promote environmental decision making [83].

6. Conclusions

This paper adopts a systematic approach to provide a new perspective for understanding the complexity and dynamics of coastal ecosystems. Through this research, we gain a more comprehensive understanding of the multifaceted value of coastal CESs in maintaining ecological balance, supporting human well-being, and promoting economic activities. Specifically, starting from the interaction between coastal CESs and natural resource management and human well-being, this paper systematically reviews published literature in terms of publication quantity, research content, research methods, and research trends. The results of this study show the following:

(1) Coastal CESs began to be studied by scholars in 2009, and the number of publications was in a fluctuating state in the early period, and has steadily increased since 2018. Among the corresponding coastal ecosystem types, marine organism has received the most attention from researchers. Among the corresponding subcategories of CESs, recreation received the most attention, and correspondingly, culture diversity received the least attention. Management and values were the most frequent keywords, and the largest number of related studies were published in *Ecosystem Services*.

(2) Multi-source data for coastal CESs typically include natural environmental data obtained from field sampling, social data obtained from local statistics, and perceptual data obtained from questionnaires or social media big data. The analysis method also gradually shifted from early statistical analysis to spatial visualization analysis. In terms of coastal CESs research trends, researchers' attention has gradually shifted from natural resource management itself to the dynamic process of CESs from perception to application.

(3) Effective coastal CESs management requires integrating ecological, socio-economic, and public perception data through interdisciplinary collaboration. Addressing challenges in data collection, including cost, representativeness, and social media's biases and privacy issues, is vital. Stakeholder engagement ensures inclusive policy making that balances ecological protection with socio-economic well-being.

In order to include more literature on relevant topics, we conducted an extensive literature search from the WOS Core Collection Database and Scopus Database, which are currently recognized databases in the academic community, and used as many subject word combinations as possible to locate relevant studies. Building on this foundation, the visualization of knowledge maps provided by VOS viewer 1.6.20 software is utilized to assist in identifying research hotspots. However, this has certain limitations. This search method may not cover all studies that are substantially related to coastal CESs. In particular, those works that do not directly use the term "CESs" but whose research content is substantially related to CESs may not be included in this review. Additionally, VOS viewer 1.6.20 may present potential biases and interpretive limitations when handling large datasets. Therefore, in future studies, in addition to major academic databases, professional databases and gray literature should also be considered to capture a wider range of research results. At the same time, more attention should be paid to the expansion and refinement of the CES conceptual framework to better identify and integrate those studies that are not directly labeled but are substantially related. In terms of research methodology, it is essential to explore and integrate various comprehensive approaches to enhance the depth and breadth of the review, ensuring that the analysis results are both comprehensive and multidimensional.

Author Contributions: Conceptualization, Z.H. and J.Z.; methodology, Y.Z. and C.Z.; software, Y.Z., Y.W. and M.G.; formal analysis, Y.Z. and C.Z.; writing—original draft preparation, Y.Z.; writing—review and editing, Z.H. and J.Z.; project administration, J.Z.; funding acquisition, Z.H. All authors have read and agreed to the published version of the manuscript.

Funding: This research was funded by the Natural Science Foundation of China, grant number 42476241, 42101113 and 41976206; the Ministry of Education Humanities and Social Sciences Research Youth Fund Project, grant number 21YJCZH193; the Ministry of Education Key Research Base of Humanities and Social Sciences Major Projects, grant number 22JJD790029 and 22JJD790031; Commissioned Projects for Young Top Talents in Philosophy and Social Sciences in Liaoning Province, grant number 2024lslqnbjrckt-14.

Data Availability Statement: Data are contained within the article.

Conflicts of Interest: The authors declare no conflicts of interest. The funders had no role in the design of this study; in the collection, analyses, or interpretation of data; in the writing of the manuscript, or in the decision to publish the results.

References

1. Xu, W.; Xiao, Y.; Zhang, J.; Yang, W.; Zhang, L.; Hull, V.; Wang, Z.; Zheng, H.; Liu, J.; Polasky, S.; et al. Strengthening Protected Areas for Biodiversity and Ecosystem Services in China. *Proc. Natl. Acad. Sci. USA* **2017**, *114*, 1601–1606. [CrossRef] [PubMed]
2. Sharma, R.; Malaviya, P. Ecosystem Services and Climate Action from a Circular Bioeconomy Perspective. *Renew. Sustain. Energy Rev.* **2023**, *175*, 113164. [CrossRef]
3. Fu, B.; Liu, Y.; Meadows, M.E. Ecological Restoration for Sustainable Development in China. *Natl. Sci. Rev.* **2023**, *10*, nwad033. [CrossRef] [PubMed]
4. Smale, D.A.; Wernberg, T.; Oliver, E.C.J.; Thomsen, M.; Harvey, B.P.; Straub, S.C.; Burrows, M.T.; Alexander, L.; Benthuysen, J.A.; Donat, M.G.; et al. Marine Heatwaves Threaten Global Biodiversity and the Provision of Ecosystem Services. *Nat. Clim. Change* **2019**, *9*, 306–312. [CrossRef]
5. Sandifer, P.A.; Sutton-Grier, A.E.; Ward, B.P. Exploring Connections among Nature, Biodiversity, Ecosystem Services, and Human Health and Well-Being: Opportunities to Enhance Health and Biodiversity Conservation. *Ecosyst. Serv.* **2015**, *12*, 1–15. [CrossRef]
6. Bratman, G.N.; Anderson, C.B.; Berman, M.G.; Cochran, B.; de Vries, S.; Flanders, J.; Folke, C.; Frumkin, H.; Gross, J.J.; Hartig, T.; et al. Nature and Mental Health: An Ecosystem Service Perspective. *Sci. Adv.* **2019**, *5*, eaax0903. [CrossRef] [PubMed]
7. Zhang, Z.; Shen, Z.; Liu, L.; Zhang, Y.; Yu, C.; Cui, L.; Gao, Y. Integrating Ecosystem Services Conservation into the Optimization of Urban Planning Policies in Eco-Fragile Areas: A Scenario-Based Case Study. *Cities* **2023**, *134*, 104200. [CrossRef]
8. Fish, R.; Church, A.; Winter, M. Conceptualising Cultural Ecosystem Services: A Novel Framework for Research and Critical Engagement. *Ecosyst. Serv.* **2016**, *21*, 208–217. [CrossRef]
9. Huynh, L.T.M.; Gasparatos, A.; Su, J.; Lam, R.D.; Grant, E.; Fukushi, K. Linking the Nonmaterial Dimensions of Human-Nature Relations and Human Well-Being through Cultural Ecosystem Services. *Sci. Adv.* **2022**, *8*, eabn8042. [CrossRef]
10. Jiao, M.; Ren, L.; Wang, Y.; Ding, C.; Li, T.; Cao, S.; Li, R.; Wang, Y. Mangrove Forest: An Important Coastal Ecosystem to Intercept River Microplastics. *Environ. Res.* **2022**, *210*, 112939. [CrossRef]
11. Massei, K.; Souza, M.C.S.; Silva, R.M.D.; Neumann, V.H.D.M.L.; Manso, V.D.A.V.; Vianna, P.C.G.; Ferreira Junior, A.V.; Moura, C.M.D.S.; Lavôr, L.F.D.; Araújo, R.D.C.; et al. Multi-Proxy Assessment of Coral Reef Formation and Biotic-Abiotic Diversity in an Urban Coastal Reef Ecosystem in Northeastern Brazil. *Sci. Total Environ.* **2023**, *891*, 164487. [CrossRef] [PubMed]
12. Kirwan, M.L.; Megonigal, J.P. Tidal Wetland Stability in the Face of Human Impacts and Sea-Level Rise. *Nature* **2013**, *504*, 53–60. [CrossRef] [PubMed]
13. Fluet-Chouinard, E.; Stocker, B.D.; Zhang, Z.; Malhotra, A.; Melton, J.R.; Poulter, B.; Kaplan, J.O.; Goldewijk, K.K.; Siebert, S.; Minayeva, T.; et al. Extensive Global Wetland Loss over the Past Three Centuries. *Nature* **2023**, *614*, 281–286. [CrossRef]
14. Plieninger, T.; Dijks, S.; Oteros-Rozas, E.; Bieling, C. Assessing, Mapping, and Quantifying Cultural Ecosystem Services at Community Level. *Land Use Policy* **2013**, *33*, 118–129. [CrossRef]
15. Gómez-Baggethun, E.; Barton, D.N. Classifying and Valuing Ecosystem Services for Urban Planning. *Ecol. Econ.* **2013**, *86*, 235–245. [CrossRef]
16. Zheng, S.; Yang, S.; Ma, M.; Dong, J.; Han, B.; Wang, J. Linking Cultural Ecosystem Service and Urban Ecological-Space Planning for a Sustainable City: Case Study of the Core Areas of Beijing under the Context of Urban Relieving and Renewal. *Sustain. Cities Soc.* **2023**, *89*, 104292. [CrossRef]

17. Noor, M.I.M.; Alagappar, P.N.; Then, A.Y.-H.; Justine, E.V.; Lim, V.-C.; Goh, H.C. Perspectives of Youths on Cultural Ecosystem Services Provided by Tun Mustapha Park, Malaysia through a Participatory Approach. *Environ. Educ. Res.* **2023**, *29*, 63–80. [CrossRef]
18. Chaigneau, T.; Brown, K.; Coulthard, S.; Daw, T.M.; Szaboova, L. Money, Use and Experience: Identifying the Mechanisms through Which Ecosystem Services Contribute to Wellbeing in Coastal Kenya and Mozambique. *Ecosyst. Serv.* **2019**, *38*, 100957. [CrossRef]
19. Inácio, M.; Gomes, E.; Bogdzevič, K.; Kalinauskas, M.; Zhao, W.; Pereira, P. Mapping and Assessing Coastal Recreation Cultural Ecosystem Services Supply, Flow, and Demand in Lithuania. *J. Environ. Manag.* **2022**, *323*, 116175. [CrossRef]
20. Federal Geographic Data Committee (FGDC). Coastal and Marine Ecological Classification Standard. Available online: https://www.fgdc.gov/standards/projects/cmecs-folder/CMECS_Version_06-2012_FINAL.pdf (accessed on 15 July 2024).
21. Carranza, M.L.; Drius, M.; Marzialetti, F.; Malavasi, M.; de Francesco, M.C.; Acosta, A.T.R.; Stanisci, A. Urban Expansion Depletes Cultural Ecosystem Services: An Insight into a Mediterranean Coastline. *Rend. Lincei Sci. Fis. e Nat.* **2020**, *31*, 103–111. [CrossRef]
22. Smart, L.S.; Vukomanovic, J.; Sills, E.O.; Sanchez, G. Cultural Ecosystem Services Caught in a 'Coastal Squeeze' between Sea Level Rise and Urban Expansion. *Glob. Environ. Change* **2021**, *66*, 102209. [CrossRef]
23. Pouso, S.; Uyarra, M.C.; Borja, A. The Recovery of Estuarine Quality and the Perceived Increase of Cultural Ecosystem Services by Beach Users: A Case Study from Northern Spain. *J. Environ. Manag.* **2018**, *212*, 450–461. [CrossRef] [PubMed]
24. Davis, J.; Kidd, I.M. Identifying Major Stressors: The Essential Precursor to Restoring Cultural Ecosystem Services in a Degraded Estuary. *Estuaries Coasts* **2012**, *35*, 1007–1017. [CrossRef]
25. Uddin, M.S.; De Ruyter Van Steveninck, E.; Stuip, M.; Shah, M.A.R. Economic Valuation of Provisioning and Cultural Services of a Protected Mangrove Ecosystem: A Case Study on Sundarbans Reserve Forest, Bangladesh. *Ecosyst. Serv.* **2013**, *5*, 88–93. [CrossRef]
26. Depietri, Y.; Ghermandi, A.; Campisi-Pinto, S.; Orenstein, D.E. Public Participation GIS versus Geolocated Social Media Data to Assess Urban Cultural Ecosystem Services: Instances of Complementarity. *Ecosyst. Serv.* **2021**, *50*, 101277. [CrossRef]
27. Martinez Pastur, G.; Peri, P.L.; Lencinas, M.V.; Garcia-Llorente, M.; Martin-Lopez, B. Spatial Patterns of Cultural Ecosystem Services Provision in Southern Patagonia. *Landsc. Ecol.* **2016**, *31*, 383–399. [CrossRef]
28. Tajima, Y.; Hashimoto, S.; Dasgupta, R.; Takahashi, Y. Spatial Characterization of Cultural Ecosystem Services in the Ishigaki Island of Japan: A Comparison between Residents and Tourists. *Ecosyst. Serv.* **2023**, *60*, 101520. [CrossRef]
29. Urquhart, J.; Acott, T. A Sense of Place in Cultural Ecosystem Services: The Case of Cornish Fishing Communities. *Soc. Nat. Resour.* **2014**, *27*, 3–19. [CrossRef]
30. Retka, J.; Jepson, P.; Ladle, R.J.; Malhado, A.C.M.; Vieira, F.A.S.; Normande, I.C.; Souza, C.N.; Bragagnolo, C.; Correia, R.A. Assessing Cultural Ecosystem Services of a Large Marine Protected Area through Social Media Photographs. *OCEAN Coast. Manag.* **2019**, *176*, 40–48. [CrossRef]
31. Ghermandi, A.; Camacho-Valdez, V.; Trejo-Espinosa, H. Social Media-Based Analysis of Cultural Ecosystem Services and Heritage Tourism in a Coastal Region of Mexico. *Tour. Manag.* **2020**, *77*, 104002. [CrossRef]
32. De la Lama, R.L.; Puente, S.; Carlos Sueiro, J.; Chan, K.M.A. Reconnecting with the Past and Anticipating the Future: A Review of Fisheries-Derived Cultural Ecosystem Services in Pre-Hispanic Peru. *People Nat.* **2021**, *3*, 129–147. [CrossRef]
33. Malinauskaite, L.; Cook, D.; Davidsdottir, B.; Ogmundardottir, H. Socio-Cultural Valuation of Whale Ecosystem Services in Skjalfandi Bay, Iceland. *Ecol. Econ.* **2021**, *180*, 106867. [CrossRef]
34. Reyes-García, V.; Menendez-Baceta, G.; Aceituno-Mata, L.; Acosta-Naranjo, R.; Calvet-Mir, L.; Domínguez, P.; Garnatje, T.; Gómez-Baggethun, E.; Molina-Bustamante, M.; Molina, M.; et al. From Famine Foods to Delicatessen: Interpreting Trends in the Use of Wild Edible Plants through Cultural Ecosystem Services. *Ecol. Econ.* **2015**, *120*, 303–311. [CrossRef]
35. Zhao, Q.; Chen, Y.; Gone, K.P.; Wells, E.; Margeson, K.; Sherren, K. Modelling Cultural Ecosystem Services in Agricultural Dykelands and Tidal Wetlands to Inform Coastal Infrastructure Decisions: A Social Media Data Approach. *Mar. Policy* **2023**, *150*, 105533. [CrossRef]
36. Santana-Santana, S.B.; Marrero-Rodríguez, N.; García-Romero, L.; Peña-Alonso, C.; Pérez-Chacón Espino, E. Is Disability a Conditioning Factor to Perceive Cultural Ecosystem Services? Assessing Social Perception in a Coastal Protected Dunefield. *Ocean Coast. Manag.* **2022**, *228*, 106298. [CrossRef]
37. Wang, W.; Wu, C.; Fang, Q.; Harrison, O.I. Cultural Ecosystem Services Evaluation in a Coastal City of China Using Social Media Data. *Ocean Coast. Manag.* **2023**, *242*, 106693. [CrossRef]
38. Martin, C.L.; Momtaz, S.; Gaston, T.; Moltschaniwskyj, N.A. A Systematic Quantitative Review of Coastal and Marine Cultural Ecosystem Services: Current Status and Future Research. *Mar. Policy* **2016**, *74*, 25–32. [CrossRef]
39. Lakshmi, A. Coastal Ecosystem Services & Human Wellbeing. *Indian J. Med. Res.* **2021**, *153*, 382–387. [CrossRef]
40. Cabana, D.; Pinna, S.; Farina, S.; Grech, D.; Barbieri, N.; Guala, I. Coastal Cultural Ecosystem Services and Adolescents' Subjective Well-Being. *Ambio* **2024**. [CrossRef] [PubMed]
41. Baulcomb, C.; Fletcher, R.; Lewis, A.; Akoglu, E.; Robinson, L.; Von Almen, A.; Hussain, S.; Glenk, K. A Pathway to Identifying and Valuing Cultural Ecosystem Services: An Application to Marine Food Webs. *Ecosyst. Serv.* **2015**, *11*, 128–139. [CrossRef]
42. Jo, T.; Sato, M.; Minamoto, T.; Ushimaru, A. Valuing the Cultural Services from Urban Blue-Space Ecosystems in Japanese Megacities during the COVID-19 Pandemic. *People Nat.* **2022**, *4*, 1176–1189. [CrossRef]

43. Jackson-Bué, M.; Brito, A.C.; Cabral, S.; Carss, D.N.; Carvalho, F.; Chainho, P.; Ciutat, A.; Couñago Sanchez, E.; De Montaudouin, X.; Fernández Otero, R.M.; et al. Inter-country Differences in the Cultural Ecosystem Services Provided by Cockles. *People Nat.* **2022**, *4*, 71–87. [CrossRef]
44. Michaelis, A.K.; Walton, W.C.; Webster, D.W.; Shaffer, L.J. Cultural Ecosystem Services Enabled through Work with Shellfish. *Mar. Policy* **2021**, *132*, 104689. [CrossRef]
45. Boeri, M.; Stojanovic, T.A.; Wright, L.J.; Burton, N.H.K.; Hockley, N.; Bradbury, R.B. Public Preferences for Multiple Dimensions of Bird Biodiversity at the Coast: Insights for the Cultural Ecosystem Services Framework. *Estuar. Coast. Shelf Sci.* **2020**, *235*, 106571. [CrossRef]
46. Erskine, E.; Baillie, R.; Lusseau, D. Marine Protected Areas Provide More Cultural Ecosystem Services than Other Adjacent Coastal Areas. *One Earth* **2021**, *4*, 1175–1185. [CrossRef]
47. Jobstvogt, N.; Watson, V.; Kenter, J.O. Looking below the Surface: The Cultural Ecosystem Service Values of UK Marine Protected Areas (MPAs). *Ecosyst. Serv.* **2014**, *10*, 97–110. [CrossRef]
48. Kochalski, S.; Riepe, C.; Arlinghaus, R. Perceived Socio-Cultural Ecosystem Services Provided by Wild Atlantic Salmon Populations in Four European Countries. *Aquat. Ecosyst. Health Manag.* **2022**, *25*, 12–21. [CrossRef]
49. Martin, C.L.; Momtaz, S.; Gaston, T.; Moltschaniwskyj, N.A. Mapping the Intangibles: Cultural Ecosystem Services Derived from Lake Macquarie Estuary, New South Wales, Australia. *Estuar. Coast. Shelf Sci.* **2020**, *243*, 106885. [CrossRef]
50. Richardson, R.B.; Nicholls, S. Characterizing the Cultural Ecosystem Services of Coastal Sand Dunes. *J. Gt. Lakes Res.* **2021**, *47*, 546–551. [CrossRef]
51. Barrena, J.; Nahuelhual, L.; Baez, A.; Schiappacasse, I.; Cerda, C. Valuing Cultural Ecosystem Services: Agricultural Heritage in Chiloe Island, Southern Chile. *Ecosyst. Serv.* **2014**, *7*, 66–75. [CrossRef]
52. Kaltenborn, B.P.; Linnell, J.D.C.; Baggethun, E.G.; Lindhjem, H.; Thomassen, J.; Chan, K.M. Ecosystem Services and Cultural Values as Building Blocks for 'The Good Life'. A Case Study in the Community of Røst, Lofoten Islands, Norway. *Ecol. Econ.* **2017**, *140*, 166–176. [CrossRef]
53. Pleasant, M.M.; Gray, S.A.; Lepczyk, C.; Fernandes, A.; Hunter, N.; Ford, D. Managing Cultural Ecosystem Services. *Ecosyst. Serv.* **2014**, *8*, 141–147. [CrossRef]
54. Gould, R.K.; Klain, S.C.; Ardoin, N.M.; Satterfield, T.; Woodside, U.; Hannahs, N.; Daily, G.C.; Chan, K.M. A Protocol for Eliciting Nonmaterial Values through a Cultural Ecosystem Services Frame. *Conserv. Biol.* **2015**, *29*, 575–586. [CrossRef]
55. Mouttaki, I.; Bagdanavičiūtė, I.; Maanan, M.; Erraiss, M.; Rhinane, H.; Maanan, M. Classifying and Mapping Cultural Ecosystem Services Using Artificial Intelligence and Social Media Data. *Wetlands* **2022**, *42*, 86. [CrossRef]
56. Gajardo, L.J.; Sumeldan, J.; Sajorne, R.; Madarcos, J.R.; Goh, H.C.; Culhane, F.; Langmead, O.; Creencia, L. Cultural Values of Ecosystem Services from Coastal Marine Areas: Case of Taytay Bay, Palawan, Philippines. *Environ. Sci. Policy* **2023**, *142*, 12–20. [CrossRef]
57. Brannigan, J.; Ryfield, F.; Crowe, T.; Cabana, D. "The Languo of Flows" Ecosystem Services, Cultural Value, and the Nuclear Legacy in the Irish Sea. *Environ. Humanit.* **2019**, *11*, 280–301. [CrossRef]
58. Ichumbaki, E.B.; Pollard, E. Valuing the Swahili Cultural Heritage: A Maritime Cultural Ecosystem Services Study from Kilwa, Tanzania. *Conserv. Manag. Archaeol. Sites* **2019**, *21*, 230–255. [CrossRef]
59. Dou, Y.; Liu, M.; Bakker, M.; Yu, X.; Carsjens, G.J.; De Groot, R.; Liu, J. Influence of Human Interventions on Local Perceptions of Cultural Ecosystem Services Provided by Coastal Landscapes: Case Study of the Huiwen Wetland, Southern China. *Ecosyst. Serv.* **2021**, *50*, 101311. [CrossRef]
60. Cabana, D.; Ryfield, F.; Crowe, T.P.; Brannigan, J. Evaluating and Communicating Cultural Ecosystem Services. *Ecosyst. Serv.* **2020**, *42*, 101085. [CrossRef]
61. Zhao, N.; Wang, H.; Zhong, J.; Sun, D. Assessment of Recreational and Cultural Ecosystem Services Value of Islands. *Land* **2022**, *11*, 205. [CrossRef]
62. Cao, H.; Wang, M.; Su, S.; Kang, M. Explicit Quantification of Coastal Cultural Ecosystem Services: A Novel Approach Based on the Content and Sentimental Analysis of Social Media. *Ecol. Indic.* **2022**, *137*, 108756. [CrossRef]
63. Ondiek, R.A.; Kitaka, N.; Omondi Oduor, S. Assessment of Provisioning and Cultural Ecosystem Services in Natural Wetlands and Rice Fields in Kano Floodplain, Kenya. *Ecosyst. Serv.* **2016**, *21*, 166–173. [CrossRef]
64. Oleson, K.L.L.; Barnes, M.; Brander, L.M.; Oliver, T.A.; Van Beek, I.; Zafindrasilivonona, B.; Van Beukering, P. Cultural Bequest Values for Ecosystem Service Flows among Indigenous Fishers: A Discrete Choice Experiment Validated with Mixed Methods. *Ecol. Econ.* **2015**, *114*, 104–116. [CrossRef]
65. Reyes-Arroyo, N.; Camacho-Valdez, V.; Saenz-Arroyo, A.; Infante-Mata, D. Socio-Cultural Analysis of Ecosystem Services Provided by Mangroves in La Encrucijada Biosphere Reserve, Southeastern Mexico. *Local Environ.* **2021**, *26*, 86–109. [CrossRef]
66. Ahtiainen, H.; Liski, E.; Pouta, E.; Soini, K.; Bertram, C.; Rehdanz, K.; Pakalniete, K.; Meyerhof, J. Cultural Ecosystem Services Provided by the Baltic Sea Marine Environment. *Ambio* **2019**, *48*, 1350–1361. [CrossRef]
67. Treviño, M. "The Mangrove Is Like a Friend": Local Perspectives of Mangrove Cultural Ecosystem Services Among Mangrove Users in Northern Ecuador. *Hum. Ecol.* **2022**, *50*, 863–878. [CrossRef]
68. Fletcher, R.; Baulcomb, C.; Hall, C.; Hussain, S. Revealing Marine Cultural Ecosystem Services in the Black Sea. *Mar. Policy* **2014**, *50*, 151–161. [CrossRef]

69. Pouso, S.; Uyarra, M.C.; Borja, Á. Recreational Fishers' Perceptions and Behaviour towards Cultural Ecosystem Services in Response to the Nerbioi Estuary Ecosystem Restoration. *Estuar. Coast. Shelf Sci.* **2018**, *208*, 96–106. [CrossRef]
70. Martin, C.L.; Momtaz, S.; Gaston, T.; Moltschaniwskyj, N.A. Estuarine Cultural Ecosystem Services Valued by Local People in New South Wales, Australia, and Attributes Important for Continued Supply. *Ocean Coast. Manag.* **2020**, *190*, 105160. [CrossRef]
71. Azevedo, A.K.; Vieira, F.A.S.; Guedes-Santos, J.; Gaia, J.A.; Pinheiro, B.R.; Bragagnolo, C.; Correia, R.A.; Ladle, R.J.; Malhado, A.C.M. A Big Data Approach to Identify the Loss of Coastal Cultural Ecosystem Services Caused by the 2019 Brazilian Oil Spill Disaster. *An. Acad. Bras. Ciênc.* **2022**, *94*, e20210397. [CrossRef]
72. Chen, Y.; Caesemaecker, C.; Rahman, H.T.; Sherren, K. Comparing Cultural Ecosystem Service Delivery in Dykelands and Marshes Using Instagram: A Case of the Cornwallis (Jijuktu'kwejk) River, Nova Scotia, Canada. *Ocean Coast. Manag.* **2020**, *193*, 105254. [CrossRef]
73. Ruiz-Frau, A.; Hinz, H.; Edwards-Jones, G.; Kaiser, M.J. Spatially Explicit Economic Assessment of Cultural Ecosystem Services: Non-Extractive Recreational Uses of the Coastal Environment Related to Marine Biodiversity. *Mar. Policy* **2013**, *38*, 90–98. [CrossRef]
74. Strand, M.; Rivers, N.; Snow, B. The Complexity of Evaluating, Categorising and Quantifying Marine Cultural Heritage. *Mar. Policy* **2023**, *148*, 105449. [CrossRef]
75. Hynes, S.; Ghermandi, A.; Norton, D.; Williams, H. Marine Recreational Ecosystem Service Value Estimation: A Meta-Analysis with Cultural Considerations. *Ecosyst. Serv.* **2018**, *31*, 410–419. [CrossRef]
76. Banarsyadhimi, U.R.A.M.F.; Dargusch, P.; Kurniawan, F. Assessing the Impact of Marine Tourism and Protection on Cultural Ecosystem Services Using Integrated Approach: A Case Study of Gili Matra Islands. *Int. J. Environ. Res. Public Health* **2022**, *19*, 12078. [CrossRef] [PubMed]
77. Elwell, T.L.; López-Carr, D.; Gelcich, S.; Gaines, S.D. The Importance of Cultural Ecosystem Services in Natural Resource-Dependent Communities: Implications for Management. *Ecosyst. Serv.* **2020**, *44*, 101123. [CrossRef]
78. Chan, K.M.A.; Guerry, A.D.; Balvanera, P.; Klain, S.; Satterfield, T.; Basurto, X.; Bostrom, A.; Chuenpagdee, R.; Gould, R.; Halpern, B.S.; et al. Where Are Cultural and Social in Ecosystem Services? A Framework for Constructive Engagement. *BioScience* **2012**, *62*, 744–756. [CrossRef]
79. Ahmed, A.; Mahmud, H.; Sohel, M.S.I. DPSIR Framework to Analyze Anthropogenic Factors Influence on Provisioning and Cultural Ecosystem Services of Sundarbans East Reserve Forest, Bangladesh. Reg. *Stud. Mar. Sci.* **2021**, *48*, 102042. [CrossRef]
80. Clarke, B.; Thet, A.K.; Sandhu, H.; Dittmann, S. Integrating Cultural Ecosystem Services Valuation into Coastal Wetlands Restoration: A Case Study from South Australia. *Environ. Sci. Policy* **2021**, *116*, 220–229. [CrossRef]
81. Saeki, I.; Li, Y. Restoration Contributes to Maintain Ecosystem Services and Bio-Cultural Linkages Between Wetlands and Local Communities: A Case from a Botanical Diversity Hotspot in Japan. *Wetlands* **2022**, *42*, 117. [CrossRef]
82. Hutcheson, W.; Hoagland, P.; Jin, D. Valuing Environmental Education as a Cultural Ecosystem Service at Hudson River Park. *Ecosyst. Serv.* **2018**, *31*, 387–394. [CrossRef]
83. Quevedo, J.M.D.; Kohsaka, R. A Systematic Review of Cultural Ecosystem Services of Blue Carbon Ecosystems: Trends, Gaps, and Challenges in Asia and Beyond. *Mar. Policy* **2024**, *159*, 105898. [CrossRef]

land

MDPI

Article

Mapping an Indicator Species of Sea-Level Rise along the Forest–Marsh Ecotone

Bryanna Norlin [1], Andrew E. Scholl [1], Andrea L. Case [2] and Timothy J. Assal [1,3,*]

[1] Department of Geography, Kent State University, Kent, OH 44242, USA; bnorlin@kent.edu (B.N.); ascholl1@kent.edu (A.E.S.)
[2] Department of Plant Biology, Michigan State University, East Lansing, MI 48824, USA; caseand4@msu.edu
[3] Bureau of Land Management, National Operations Center, Denver, CO 80225, USA
* Correspondence: tassal@kent.edu

Abstract: Atlantic White Cedar (*Chamaecyparis thyoides*) (AWC) anchors a globally threatened ecosystem that is being impacted by climate change, as these trees are vulnerable to hurricane events, sea-level rises, and increasing salinity at the forest–marsh ecotone. In this study, we determined the current amount and distribution of AWC in an area that is experiencing sea-level rises that are higher than the global average rate. We used a combination of a field investigation and aerial photo interpretation to identify known locations of AWC, then integrated Sentinel-1 and 2A satellite data with abiotic variables into a species distribution model. We developed a spectral signature of AWC to aid in our understanding of phenology differences from nearby species groups. The selected model had an out-of-bag error of 7.2%, and 8 of the 11 variables retained in the final model were derived from remotely sensed data, highlighting the importance of including temporal data to exploit divergent phenology. Model predictions were strong in live AWC stands and, accurately, did not predict live AWC in stands that experienced high levels of mortality after Hurricane Sandy. The model presented in this study provides high utility for AWC management and tracking mortality dynamics within stands after disturbances such as hurricanes.

Keywords: species distribution modeling; *Chamaecyparis thyoides*; sea-level rise; ecotone; ghost forests

1. Introduction

The outcomes of climate change (e.g., sea-level rises and increasing storm frequency and intensity) threaten numerous coastal species [1–4]. These threats are especially detrimental to plant species that exist in a narrow range and are sensitive to environmental change. Atlantic White Cedar (referred to as AWC), *Chamaecyparis thyoides* (L.) Britton, Sterns & Poggenb. (Cupressaceae), anchors a globally threatened ecosystem that is being impacted by climate change. These trees are vulnerable to hurricane events, wind-driven water-level fluctuations and salinity, sea-level rises, and drainage ditches designed to lower the water table for agriculture [5]. Collectively, these changes, driven in large part by the climate, are shifting the marsh–forest ecotone [1]. Therefore, in order to better manage and conserve AWC, a species sensitive to sea-level dynamics, more accurate information is needed on the location and condition of live stands. An understanding of ecotone dynamics is critical to understanding the effects of climate change; yet it is these areas of the landscape that pose challenges to species mapping efforts as AWC is interspersed with other types of similar vegetation.

1.1. Coastal Forests and Climate Change

Coastal forests and wetlands provide many ecosystem services, including wildlife habitat, coastal erosion protection, water purification and regulation, nutrient cycling, and recreation [6]. However, coastal forests and wetlands are currently threatened by long-term sea-level rises and increased storm intensity and frequency. The spatial distribution of forested wetland areas appears to be closely related to saltwater intrusion and the occurrence of extreme events that control short-term water levels [7]. Forested wetlands are often dominated by tree communities that are not well adapted to increasing saline conditions [7], causing entire forests in the Mid-Atlantic region (of the United States) to shift in favor of more salt-tolerant species. It can be difficult to estimate the projected changes in these coastal forests because the systems are dynamic [8]. However, ecotones, including the marsh–forest boundary, provide a critical location to study the impact of climate change and extreme events on species residing in ecotones because they are often sensitive to change [2].

Globally, wetlands and coastal forests are at risk from sea-level rises, but wetlands in the Mid-Atlantic region are especially at risk. New Jersey has experienced increased rates of sea-level rises over the past 40 years, with a rate of 0.2 inches/year, which is higher than the global average of 0.1 inch/year [9]. Coastal wetlands will need to migrate in order to compete with sea-level rises; however, inland migration could be blocked by bluffs, development, and shoreline protection structures [10]. Trees and other vegetation within the coastal forests can be killed by tidal flooding and storm surges if they are not flood-tolerant or tolerant to the salinity of the flood waters [3]. The exposure to saltwater can create "ghost forests", areas of standing dead trees [4], which represent the conversion of forest into marsh as those areas become inundated, leaving a patch of dead trees behind [1]. This is driven by sea-level rises as well as disturbance events [7]. Therefore, ghost forests provide evidence of a shifting marsh–forest ecotone and can be used to track the movement of the ecotone boundary over time [11].

Hurricanes and severe storms are a threat to the wetlands and coastal forests in the Mid-Atlantic region of the United States. An estimated 400,000 acres of wetlands were inundated as a result of Hurricane Sandy, a category 3 hurricane that struck in late October 2012 [12]. Hurricane Sandy inundated forested areas by creating breaches in the natural barriers that protect these areas from saltwater intrusion. However, it is challenging to understand the impacts of Hurricane Sandy alone on these forests because there are multiple stressors [13]. After Hurricane Sandy, it was reported that severe degradation occurred in 41.4% of wetlands, and 51.1% of the damage from all types of degradation (including severe, moderate, and low) was long-term damage [6]. Some plant species will be more tolerant to the increased salinity that can be the result of a disturbance event; however, those others unable to adapt will need to migrate or could face widespread mortality due to these increasing threats.

1.2. Atlantic White Cedar Ecosystem Biogeography and Vulnerability

Atlantic White Cedar is a coniferous, evergreen species that is found in the Eastern United States, from Maine to Mississippi, in a narrow band [5], rarely more than 200 km from the coast [14]. AWC generally occurs in narrow belts along streams and can persist seaward to tidal areas where it anchors the forest–marsh ecotone [15]. AWC influences soil characteristics through peat building and provides a habitat for rare and endangered species [5,16,17].

AWC is vulnerable to anthropogenic climate change, which is causing increased storm frequency, intensity, and coastal flooding. The species is salt-intolerant and can experience reduced growth and increased mortality in response to inundation. AWC ghost forests are prevalent in coastal areas of the Mid-Atlantic regions of the US due to increased pressure from sea-level rises [4]. The increase in ghost forests at the marsh–forest ecotone is effectively tracking hotspots of increased salinity and inundation [7,11]. There have been ghost forests further inland in response to periods of inundation [4], and AWC ghost forest formation was reported after Hurricane Sandy in New Jersey [5].

1.3. Predictive Habitat Distribution Modelling

Species distribution models (SDMs) relate species distribution data to the environmental characteristics of an area and have been utilized in diverse environments for a wide variety of species (e.g., Assal et al., 2015, 2021; Hailu et al., 2017; He et al., 2019) [18–21]. SDMs can be used to estimate the potential distribution of a species, then to predict the extent of suitable habitat over a given area [22,23]. SDMs are used to forecast forest species ranges that may be the most impacted by climate change or extreme events, providing valuable knowledge for potential management or mitigation practices [24]. Atlantic White Cedar wetlands are difficult to access due to the surrounding environmental conditions, making it more difficult to monitor these stands [25]. Therefore, an AWC SDM would be beneficial for informing the current and potential areas, as well as identifying areas of the landscape that may be more susceptible to impacts from climate change.

SDMs use environmental data, including both abiotic and biotic variables, to map and predict species occurrence across a landscape. Incorporating remote sensing data into the process, known as predictive habitat distribution modelling [19,26], allows for a more dynamic model because it captures vegetation phenology [27]. Advances in multispectral satellite capabilities provide the opportunity to identify a more complete spectral signature, or response, of a species. A spectral signature is the unique reflectance of a specific material or object over a range of wavelengths [28,29] and aids in discriminating vegetation types and plant stress [30,31]. More recent satellite sensors such as Sentinel-1 (Synthetic Aperture Radar) and Sentinel-2 (multispectral imagery) have been used for vegetation classification and habitat monitoring [32]. The inclusion of both Sentinel satellites in forest mapping has led to higher model accuracy [33]. The combination of remotely sensed and abiotic variables to map and predict species distributions across spatiotemporal scales is a technique that is applicable to many scales and locations due to increases in data availability and computational power.

1.4. Research Questions and Objectives

Understanding AWC distribution will help inform management and mitigation loss strategies associated with climate change impacts. A species distribution model will enable efficient mapping, as AWC grows in wetlands that can be challenging to access. The purpose of this study is to determine the distribution of Atlantic White Cedar throughout southern New Jersey, an area that experienced a severe disturbance, Hurricane Sandy, in 2012. Our primary goal is to create a transferable framework to map the distribution of Atlantic White Cedar and provide managers with the current extent of the range. The main objective of this study is to (A) determine the distribution of AWC throughout southern New Jersey. To help address the primary research goal, there were two smaller research questions addressed: (B) What are the driving predictors of AWC presence across the landscape? (C) What is the spectral signature of Atlantic White Cedar? This research will address a data gap about the distribution of Atlantic White Cedar, provide insight on the vegetation dynamics within these stands of trees, and lead to a better understanding of where AWC is able to persist on the landscape.

2. Materials and Methods

2.1. Study Area

The study domain comprises 36 HUC-12 watersheds (Hydrologic Unit Code (HUC) 12) [34] throughout southern New Jersey that contained known stands of Atlantic White Cedar (Figure 1). This area, 3,430 square kilometers, contains part of the New Jersey Pinelands, which support a wide variety of plant and animal communities that are threatened by water-quality changes and degradation of watersheds from development and agriculture throughout the southern part of the state [35]. Average annual precipitation range is 106.7–116.8 cm, and the average annual temperature is 0–2.2 °C in winter and between 22.2 and 25 °C in the summer months [36]. The pinelands are composed of many different tree species, including pitch pine, red maple, oak tree varieties, and Atlantic White Cedar, which dominates the wetlands areas [36,37].

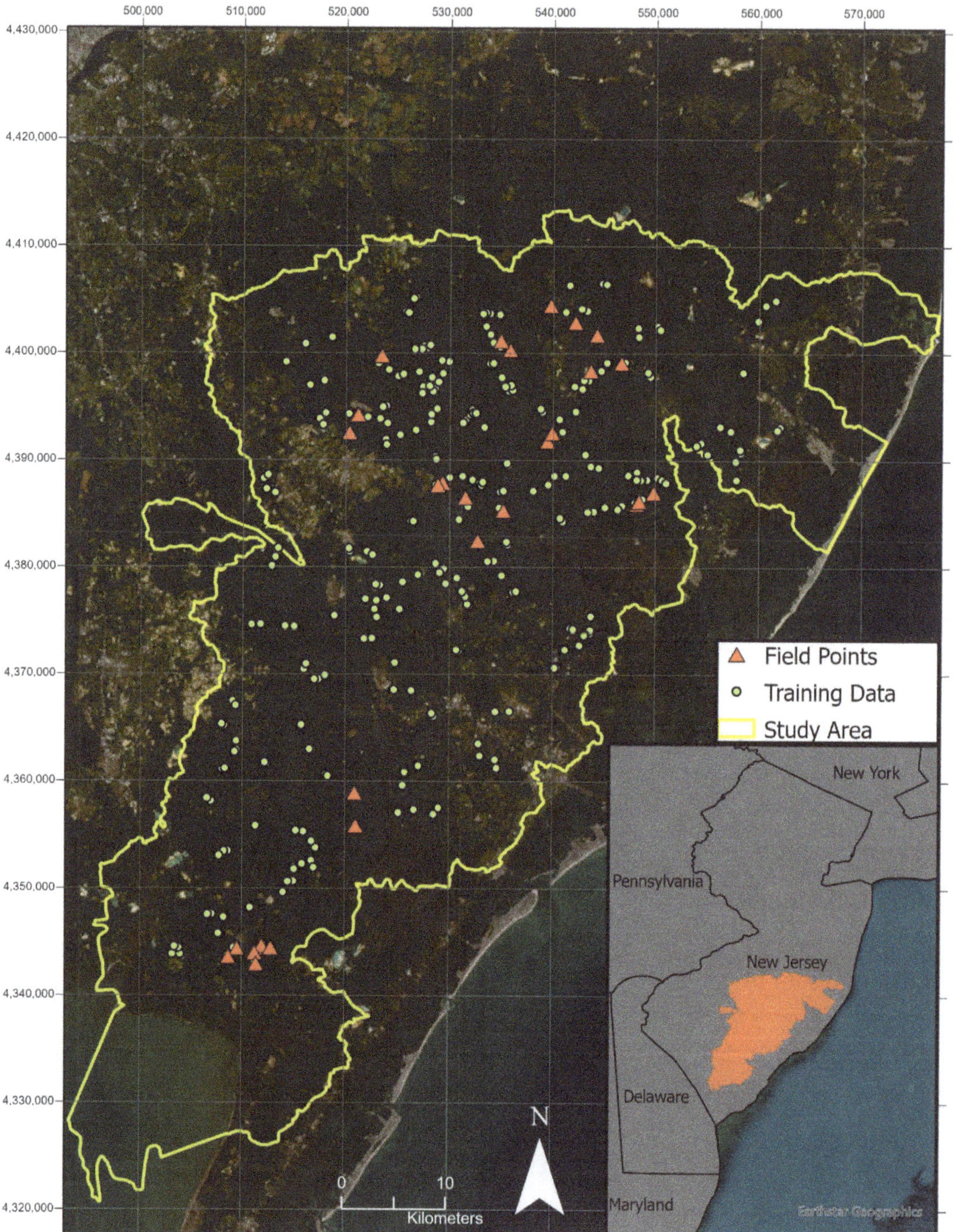

Figure 1. Location of the study area in southern New Jersey. Background: ESRI Base Maps (2023).

2.2. Sample Data Collection

Live and dead AWC stands were documented using a GPS unit (Geode GNSS Receiver; Juniper Systems; Logan, UT, USA) and photographs during a field campaign in August 2022 (Figure 2). Information on the stand condition and setting was documented, including photographs, to provide reference for the position of the stand of AWC during the aerial photograph interpretation process. A total of 30 field sites were visited. The number of stands able to be reached was limited due to the relative inaccessibility of AWC stands on public land (we accessed private lands when permission from the landowner was received) for aerial photograph characterization of AWC. Additional AWC presence points were created through aerial photo interpretation in inaccessible areas.

Figure 2. Multiple perspectives of Atlantic White Cedar stands. Panel (**A**)—Field photograph of an AWC stand taken during the 2022 field season. Panel (**B**)—Natural color NAIP imagery showing an AWC stand that was visited during the 2022 field season. Panel (**C**)—Same AWC stand as panel B, shown in color infrared with a contrast stretch applied. AWC is distinguished by its unique texture, tone (the blueish-gray tone), and shape as these stands are typically linear.

Images from the National Agriculture Imagery Program (NAIP) were obtained from the USDA data gateway [38] for the southern New Jersey area. The most recent NAIP photos at the time of analysis were used (26 July–29 September 2019) and had a spatial resolution is 0.6 m in four spectral bands (blue, green, red, and near-infrared (NIR)). The NIR was critical to delineate vegetation from the surrounding environments and distinguish differences in vegetation types (Figure 2). The stands in the field sites served as reference areas for photo interpretation of AWC stands to then develop additional training points (n = 362, 134 AWC presence points, 228 absence points). Contrast stretches were applied to the NAIP imagery to highlight the unique texture and pattern of AWC compared to the surrounding vegetation. This photo interpretation technique has been used previously to develop presence/absence data in SDMs [18–21].

We derived a dataset of mapped AWC polygons from a 2012 Land Use/Land Change (LULC) dataset [39] that had been iteratively updated from the original 1986 dataset created through photo interpretation using a minimum mapping unit of 2.5 acres. This is the only known dataset of mapped AWC in this area, and it was last generated in 2012, just prior to Hurricane Sandy (late October 2012). This afforded an opportunity to compare the pre-hurricane AWC with our updated model and document any change in AWC stands since that time. The LULC data also informed the training dataset using a modified stratified random technique to ensure adequate coverage of vegetation types that may occur near AWC stands. Training points were deterministically placed in six LULC classes: AWC, other coniferous species, water, built/urban, deciduous, and agriculture. Later, additional absence points were added to assist with model training: 10.7% of the training points were water, 12.4% deciduous, 3.0% built environment, 4.4% agriculture, and 18.5% other conifers/pine. More points were placed in areas that may have had more similar characteristics to AWC, such as deciduous and other conifers/pine. After points were classified, the classification scheme was later collapsed to simply AWC presence (1) and absence (0) for use in the random forests model.

2.3. Satellite Data

Remotely sensed variables were derived from Sentinel-1 and Sentinel-2 imagery, as both provide information about stand structure, age, vegetation biophysical parameters, chlorophyll content, leaf area index, aboveground biomass, and more [33,40,41]. We utilized a suite of multispectral bands from the Sentinel-2 surface reflectance (Level 2A) product ("COPERNICUS/S2_SR") in Google Earth Engine (GEE), as well as two derived vegetation indices (Table 1). We acquired imagery during a leaf-on period (15 June–15 September) and a leaf-off period (1 January–15 February and 1–31 December), all during 2021 for consistency (a portion of the study area experienced a wildfire in 2022). We enforced a minimum observation threshold of four clear-sky observations during the leaf-on period for a given pixel to be considered in the analysis; the leaf-off period only required one observation. If a pixel failed to meet the minimum valid observations, it was dropped from the analysis as it could influence the composite image [18]. AWC is a coniferous species that retains needles during the late fall and winter seasons, whereas deciduous vegetation does not retain leaves during that time. We use two time periods as an opportunity to exploit the divergent phenology of AWC from nearby deciduous trees [18,19]. After the minimum observation filter was applied, the mean composite images were exported from GEE for each band and time period (Table 1). We also chose to incorporate the Sentinel-1 C-Band Synthetic Aperture Radar (SAR) ("COPERNICUS/S1_GRD") to account for potential distinct polarization characteristics that coniferous AWC might exhibit over the growth form of deciduous species. We compiled mean composites for the VV and VH band from the same time periods from ascending orbit of the satellite. No minimum observation threshold was applied as SAR is not affected by clouds [33]. Sentinel 2-A bands 5–7, 8A, and 11–12 were resampled from 20 m to 10 m using bilinear interpolation to match the spatial resolution of Sentinel-1 and Sentinel-2-A bands (2–4 and 8) in GEE. All composite bands (Table 1) were exported for further processing in the R statistical software (version R-4.2.2) [42] using the raster package [43].

Table 1. Description of explanatory variables considered in the analysis.

Variable	
	Biotic
VV Band	C-Band SAR vertical transmission/reception
VH Band	C-Band SAR vertical transmission and horizontal reception
Band 2	Blue band
Band 3	Green band
Band 4	Red band
Band 5	Red Edge band 1
Band 6	Red Edge band 2
Band 7	Red Edge band 3
Band 8	Visible and near infrared
Band 8A	Visible and near infrared
Band 11	Shortwave infrared (SWIR)
Band 12	Shortwave infrared (SWIR)
NDVI (Normalized Difference Vegetation Index)	NDVI = (B3 refletance − B2 reflectance)/(B3 reflectance + B2 reflectace) [44]
NDRE (Normalized Difference Red Edge)	NDRE = (B8 reflectance − B6 reflectance)/(B8 reflectance + B6 reflectance) [45]

Table 1. *Cont.*

Variable	
	Abiotic
Percent sand in soil	Percent sand found in soil at a depth of 30–60 cm
Saturated soil water content	Saturated soil water content at a depth of 30–60 cm
Elevation	Derived from DEM
Slope	Derived from DEM
Aspect	Derived from DEM
Topographic Position Index	A measure of slope position and landform type with respect to adjacent grid cells [46]
Topographix Wetness Index	A measure of topographic control on hydrologic processes [47]
Topographic Ruggedness Index	A measure of the elevational difference between a cell and adjacent cells [48]
Northness	Cosine transformation of aspect
Eastness	Sine transformation of aspect
Distance to water	Distance from each cell center to the closest water source
Distance to head of tide	Distance from each cell center to the closest point inland that the tide reaches
Latitude	Latitude at the cell center
Longitude	Longitude at the cell center

Spectral Signatures and Phenology

Spectral signatures of multiple vegetation types were created to determine if the spectral signature of AWC was similar to other vegetation types within the study area, as the inclusion of phenology in SDMs is often overlooked [49], and, to our knowledge, the spectral signature of AWC measured by Sentinel-2A has never been documented. Therefore, we investigated the spectral signature of AWC and other nearby vegetation during the leaf-off and leaf-on periods to aid our understanding of satellite-derived reflectance changes throughout the year in these vegetation types. Spectral responses are influenced by plant physiological traits such as leaf type, chlorophyll content, and water content [31,50]. Coniferous species are known to be more absorptive than deciduous forests and tend to have low values in the near-infrared region [51]. Some vegetation types emit similar spectral responses, and to determine if any of the surrounding vegetation had similar reflectance values in any of the Sentinel-2A bands, the responses for 8 different vegetation types were determined for both the leaf-on and leaf-off periods.

2.4. *Topographic and Abiotic Data*

We derived a suite of topographic variables from a 10 m digital elevation model (DEM) accessed in GEE ("USGS/3DEP/10m") to be used as predictor variables (Table 1). A new dataset was derived to determine the distance to water using an NJ Coastline dataset [52] and an NHDPlus dataset [34] that provided mapped tributaries, canals, and drainage ditches. The two datasets were intersected in a GIS to create a single dataset representing the coastline, river and tributary banks, and the centerline of smaller streams. Finally, a proximity raster was created that represented the Euclidian distance for each cell to the nearest water source. The distance to the head of tide (HOT), the furthest point inland on any tributary where the tide reaches, was created by calculating the Euclidian distance of each raster cell to the HOT [53]. Given AWC's sensitivity to changing salinities, this was an important variable to consider as vegetation within the tidal zones may be more prone to mortality in the future with an increase in the sea level. We obtained both percent sand and saturated soil water content (at depth of 30 to 60 cm) from the Polaris dataset [54], available through GEE. This depth was chosen based on the root characteristics of AWC, which are shallow and only reach about 0.3–0.6 m [25]. The 30 m data were resampled to 10 m to match the scale of all other covariates using bilinear interpolation.

2.5. Predictive Habitat Distribution Modelling

To relate the presence and absence points to the underlying environmental conditions to be able to predict AWC presence, a random forests (RF) machine learning approach was used [55]. RF has a long history in ecological studies and is shown to have a high classification accuracy when used in species distribution modelling [56]. Multicollinearity is common among remotely sensed variables and derived indices [18,57] and can lead to uncertainty within a model [58]. Multicollinearity between the predictor variables was addressed by calculating the variable inflation factors (VIFs) and the maximum linear correlation of each variable using the usdm package in R (version 4.2.2) [59]. Variables with high correlation (VIF > 0.8) were removed from consideration, resulting in 22 variables considered for model testing [18,60,61].

The selected model with optimal out-of-bag error and class error was then used to predict the probability of AWC occurrence throughout the study area. We converted the continuous prediction to a binary presence/absence map using the threshold where the sensitivity was equal to the specificity. The receiver-operator characteristic curve was calculated, and the area under the curve (AUC) was used to assess model accuracy. Finally, the model was spatially constrained, a posteriori, to only include the zones delineated in the LULC dataset (based on the Anderson Classification System) that are consistent with the ecological characteristics of AWC.

3. Results

3.1. Spectral Signatures and Phenology

The spectral signatures of the selected vegetation types and zones showed that AWC had a similar signature to that of other coniferous species within the study area (Figure 3). The deciduous and coniferous species exhibited clear differences during the leaf-on period in the red edge 2 and 3 (bands 6 and 7), and the NIR, and the NIR narrow regions (bands 8–8a). In the short-wave infrared bands (bands 11 and 12), there is little separation during the leaf-on period. During the leaf-off period, there is greater separation between the vegetation types in the short-wave infrared region. The spectral signatures of AWC and pitch pine (*Pinus rigida*) are extremely similar. However, there is subtle separation between these two species in the red edge, NIR, and NIR narrow regions during the leaf-on period (Figure 3), confirming that phenology considerations are important when considering remotely sensed data.

3.2. Species Distribution Model Evaluation

Atlantic White Cedar presence was best predicted by a combination of spectral and topographic variables selected during the model fitting process. The RF model used to predict AWC presence had an out-of-bag error of 7.2% and a maximized occurrence threshold of 0.32. The probability of AWC class presence and absence error was 11.9% and 4.4%, respectively, and the model predicted AWC with high accuracy, as indicated by an AUC value of 0.97. There were 11 variables included in the final model, and NDVI during the leaf-off period was the most important variable for predicting Atlantic White Cedar presence (Figure 4). Distance to water was the third most important variable in the model predictions, accounting for AWC preference to grow near streams and tidal areas. The NIR band (band 8) for both the leaf-on and leaf-off periods was retained in the model and potentially important to discriminate between AWC and pitch pine (Figure 3). The leaf-off VH band, a proxy for vegetation structure retained in the final model, was likely important in delineating AWC from deciduous and potentially other coniferous species with a different growth form.

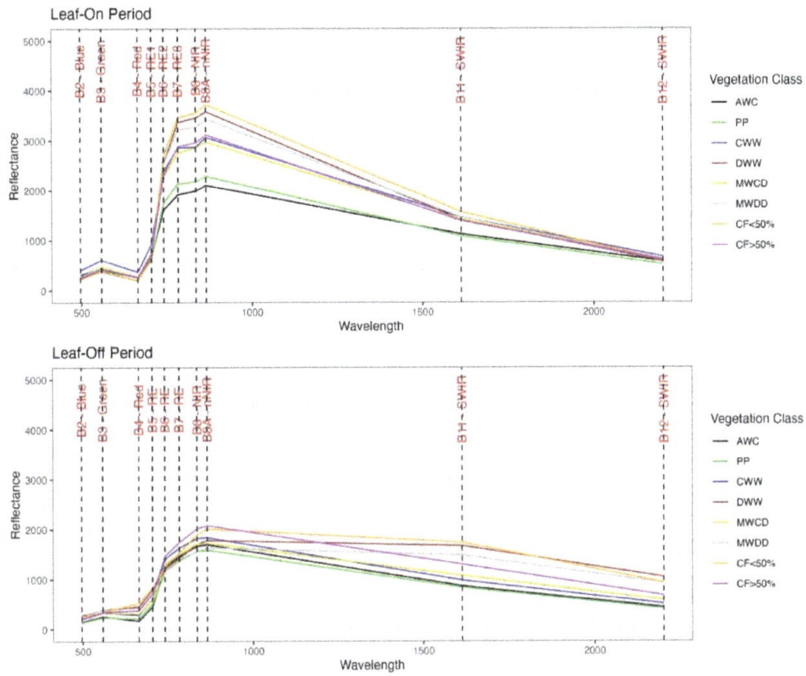

Figure 3. Spectral signatures of chosen classes to compare to AWC during the leaf-on period (top panel) and leaf-off period (bottom panel). AWC = Atlantic White Cedar, PP = Pitch Pine, CWW = Coniferous Wooded Wetland, CF < 50% = Coniferous Forest (10–50% crown closure), CF > 50% = Coniferous Forest (>50% crown closure), MWCD = Mixed Wooded Wetlands Coniferous Dominated, DWW = Deciduous Wooded Wetlands, MWDD = Mixed Wooded Wetlands Deciduous Dominated.

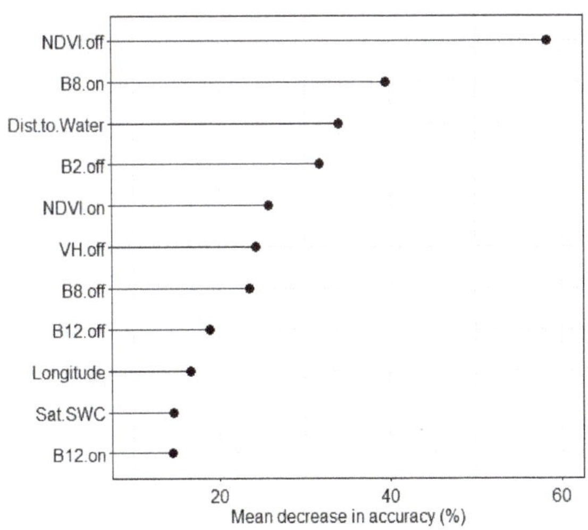

Figure 4. Variable importance plot for predictor variables used in the classification model. Higher values in mean decrease in accuracy on the x-axis indicate more important variables in the classification. Note: see Table 1 for predictor variable names; on refers to leaf-on periods, and off refers to leaf-off periods.

3.3. Atlantic White Cedar Mapping

The model output was used to create a spatially explicit map of AWC presence (Figure 5). The model was then spatially constricted to wetland zones and predicted 149 km^2 of AWC within those zones. The Atlantic White Cedar Wetlands class from the LULC map had the highest amount of the total predicted AWC (42.6%), followed by coniferous forest with greater than 50% crown closure (25.7%). The mean probability of occurrence mirrored the area of predicted AWC for each zone (Table 2). The predicted AWC showed strong agreement with aerial photographs at the stand scale (Figures 6 and 7). At the watershed scale (HUC-10), the vast majority of the predicted AWC occurs in the Wading River (39.4%) and Mullica River (33.8%) watersheds, which contain the majority of the protected areas in a number of state forests (Table 3).

Table 2. The percentage of total predicted AWC within the spatially constrained area of RF predictions by zone.

Zone Type	Percentage	Mean Probability
AWC Wetlands	42.6%	0.67
Coniferous Forest (>50% crown closure)	25.7%	0.46
Coniferous Wooded Wetlands	18.2%	0.5
Mixed Wooded Wetlands (Coniferous Dom.)	5.9%	0.48
Coniferous Forest (10–50% crown closure)	4.8%	0.44
Mixed Wooded Wetlands (Deciduous Dom.)	1.8%	0.46
Deciduous Wooded Wetlands	1.0%	0.43

Table 3. Amount of predicted AWC within each HUC-10 watershed. Percentages are the amount of the total predicted AWC within each watershed.

Name	Amount of Predicted AWC in Each Watershed (km^2)
Wading River	58.8 (39.4%)
Mullica River	50.4 (33.8%)
Lower Great Egg Harbor River	7.4 (5.0%)
Dennis Creek-Delaware Bay	7.1 (4.8%)
Manahawkin Bay-Little Egg Harbor	6.2 (4.2%)
Upper Great Egg Harbor River	6.0 (4.0%)
Lower Maurice River	5.2 (3.5%)
Barnegat Bay	4.8 (3.2%)
Tuckahoe River	1.8 (1.2%)
Great Egg Harbor Bay-Barrier Islands	1.3 (0.9%)

While the model was able to distinguish live AWC stands, it likely overpredicted in some of the upland areas, where pitch pine was more abundant. The model predicted 149 km^2 across the study area in the spatially constrained zones compared with the NJDEP 2012 dataset which identified 116 km^2 of AWC. Although the model may have overpredicted, the mapping effort could have missed AWC stands that were less conspicuous to aerial photography interpreters. However, the model only identifies 64 km^2 of live AWC inside the 2012 mapped AWC polygons, indicating a decrease of 51 km^2 (44%) of AWC between 2012 and 2021.

AWC mortality was documented both in the field and in aerial photographs throughout the study area. The model performed well to identify live AWC and distinguish it from dead AWC (Figures 8 and 9). The model did not predict live AWC in areas at the edge of the forest–marsh ecotone that were mapped as AWC prior to Hurricane Sandy (Figure 8). Aerial photo interpretation identified these areas as dead stands of AWC. Low-lying areas between the Bass and Wading Rivers experienced heavy flooding in the weeks after Hurricane Sandy and experienced large amounts of mortality confirmed in the field. The model was able to distinguish between the live and dead AWC (ghost forest) and correctly mapped the healthy, live stands (Figures 8 and 9).

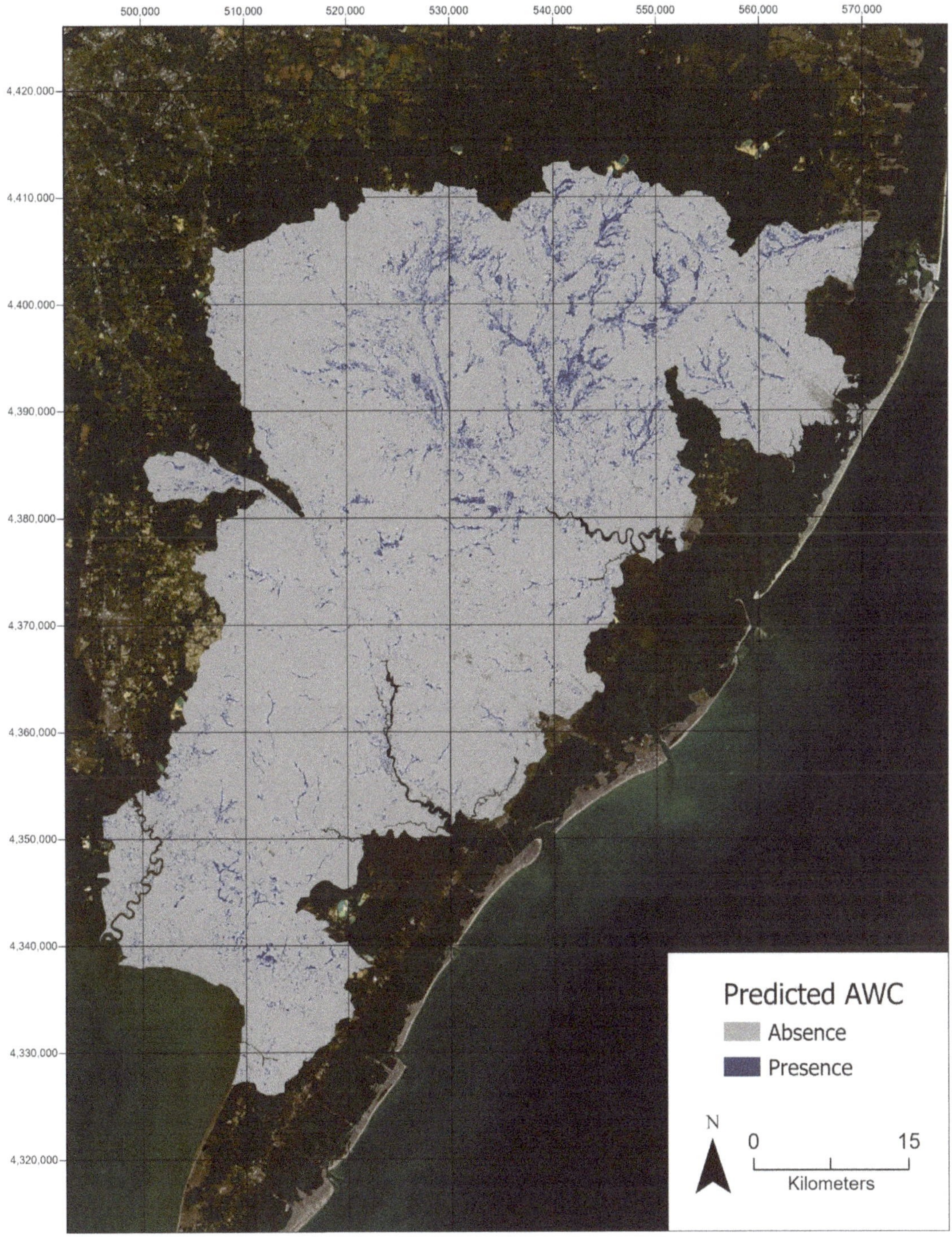

Figure 5. Spatially explicit prediction of AWC developed from the random forest model. The data are classified into AWC presence and absence based on a probability threshold of 0.32.

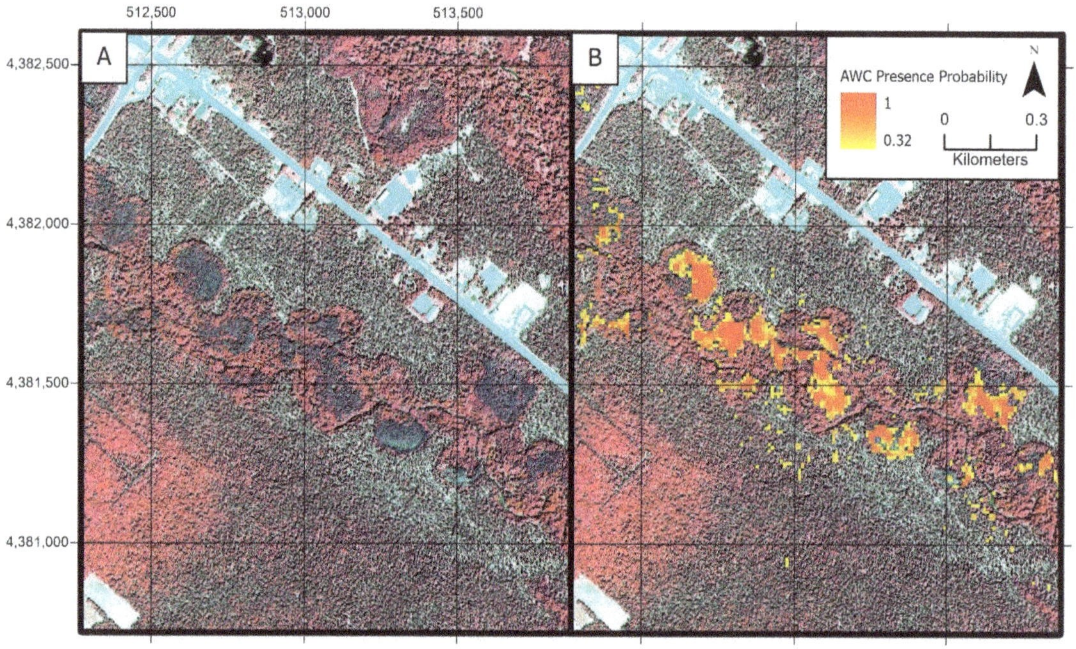

Figure 6. Panel (**A**)—AWC visible as areas of dark gray in NAIP color infrared imagery. Panel (**B**)—Model output of predicted probability of AWC presence. The model does not predict in adjacent areas of deciduous vegetation (appears red in color infrared).

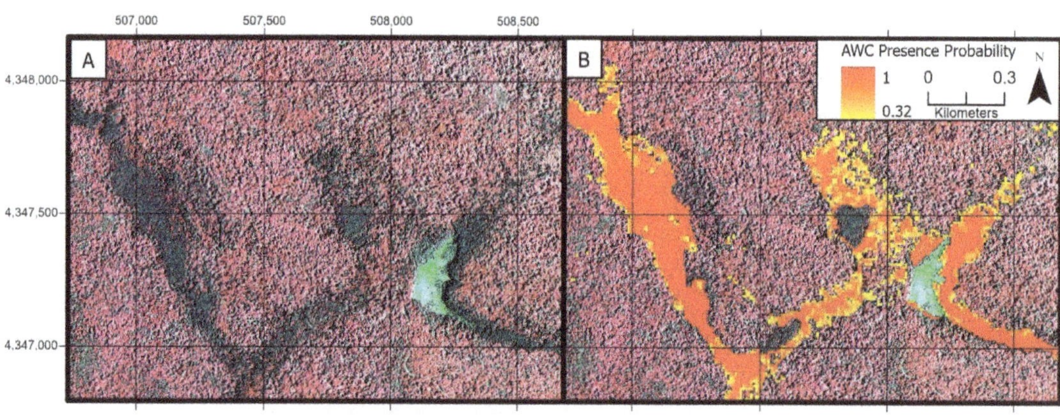

Figure 7. Panel (**A**)—Color infrared NAIP imagery of an area with AWC presence. Panel (**B**)—Model output of AWC presence closely follows the narrow bands of AWC on the landscape.

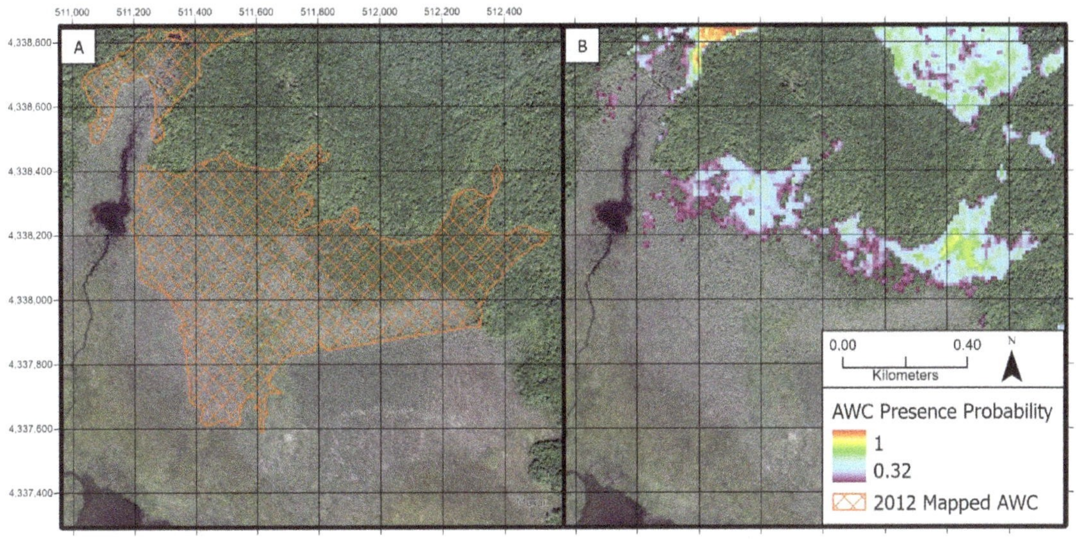

Figure 8. Panel (**A**)—Mapped AWC polygons from NJDEP (vintage 2012) overlaid on 2019 NAIP imagery. Areas of dead AWC are visible as large gray patches on the landscape inside the polygons. Panel (**B**)—The model only predicts where live patches of AWC exist and does not predict in locations of dead AWC stands.

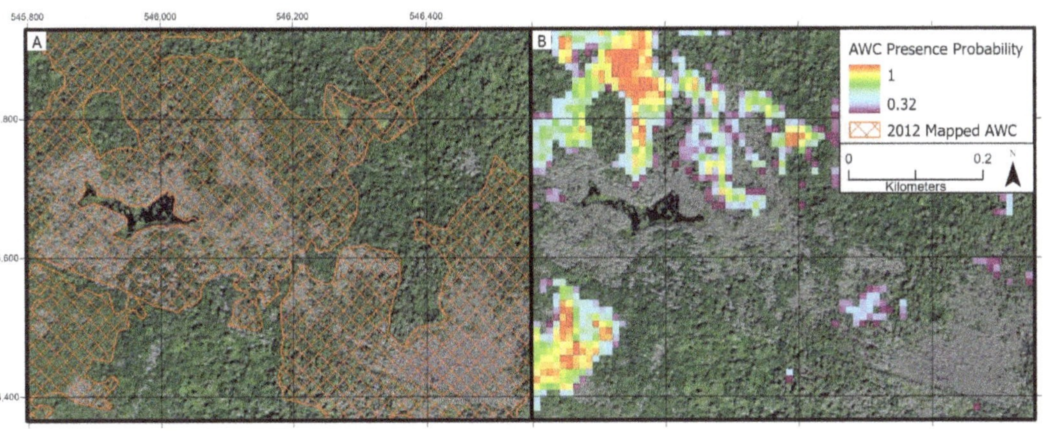

Figure 9. Panel (**A**)—Mapped AWC polygons from NJDEP (vintage 2012) overlaid on 2019 NAIP imagery. Areas of dead AWC are visible as large gray patches on the landscape that was confirmed with a field visit in 2022. Panel (**B**)—The model only predicts where live patches of AWC exist and does not predict AWC presence in the distinct areas of mortality.

4. Discussion

This model fills a needed data gap on the interaction of landscape position, soils, and hydrology on Atlantic White Cedar occurrence [62] and the driving variables of AWC presence in this region. Using an SDM is an improvement upon more traditional species distribution data types such as polygons or occurrence points because it uses fewer data to create distribution estimations, relies on open-source data, can be recreated for future time periods, and balances omission and commission errors, which are typically higher in distributions that are based solely on species occurrences or range maps [63]. The

2012 dataset, while useful and providing a good reference for this project, was created by updating a previous dataset using aerial photograph interpretation. It was limited by the minimum mapping unit and human interpretation. Furthermore, a severe disturbance event (Hurricane Sandy) occurred after that product was created, resulting in unmapped mortality of AWC. The RF model created here provides a transferable framework for the mapping of AWC throughout its range and is an improvement on previously created distribution maps.

4.1. Remote Sensing Is Critical in Mapping Atlantic White Cedar

It is not uncommon to have under- or overprediction within SDMs, which can either make the distribution of a species appear to be smaller or larger than it truly is. While SDMs may not always capture the true distribution of a species, they can be used as a hypothesis of where a species is located and how the species fares in different locations [64]. Although we were unable to include species interactions or dispersal constraints in our SDM [63,64], we were successful in identifying the current distribution of live, healthy AWC stands (Figure 5). Furthermore, the model can be utilized as a tool for tracking changes in the distribution of this species across the landscape. The model was developed using fine-scale (~10 m) data, which enabled small stands of AWC to be mapped across this landscape. The framework relies on open-source data and is, therefore, transferrable to model other areas within AWC's range. Consideration of a large suite of potential predictor variables allowed the modelling process to identify those most important for predicting AWC presence (Figure 4). Eight of the eleven predictor variables were from satellite data, highlighting the value of including remotely sensed data and phenology insight into species distribution models.

Our results show that there has been a loss of AWC in some areas that were previously mapped as live AWC, as the model is able to distinguish between the live and dead AWC (Figures 8 and 9). We identified a 44% loss of AWC between 2012 and 2021 in southern New Jersey. There is clear aerial photographic evidence that some of this stand loss occurred since 2012. However, a qualitative assessment of aerial photos from the 2000s identified small pockets of mortality within 2012 mapped stands, suggesting that the AWC dataset had not been updated during that period, thus making direct comparisons difficult. Moving forward, this model could be used to quantify and track AWC loss across a landscape due to climate change or other pressures in the future. Sea-level rises, inundation, and more frequent and intense storms all present a threat to AWC, as this species is sensitive to changes in the water level and salinity. These issues are especially pressing in the Mid-Atlantic region, which has the highest rates of SLR globally [9] and will continue to press AWC further inland in areas where the migration is not blocked by urbanization. The conversion of AWC to ghost forest has been documented by other researchers as well [4,7]. In this study we extend that knowledge and document AWC conversion to ghost forest in a new region. Furthermore, the model presented here can be extended to future time periods to track those changes across the landscape and determine which stands are experiencing increased stress or mortality.

4.2. Ecological and Technological Considerations

The model had a low error component with reasonable predictions of AWC presence; however, there were some areas of overprediction (Figure 8). Overprediction within an SDM occurs when the model predicts the species to have a broader distribution than reality due to the inadequate capture of biotic interactions or dispersal constraints on the species [63]. We believe there are a few reasons for overprediction in the model. First, the available data might be too coarse to capture fine scale variability across the landscape. SDMs should be created at a scale relevant to biological interactions, and for many sessile species (such as plants), it is necessary to use smaller scales to produce models with high predictive capabilities [65]. For all the datasets to match in resolution, we chose to use more coarse-scale abiotic variables (10 m) than were available to match the scale of the spectral variables. Using finer-scale abiotic variables would have required scaling up datasets, which can cause issues with fine-scale heterogeneity, interactions, nonlinearity, and feedbacks [66]. To maintain the integrity of the data, coarser-scale abiotic variables were used to best match the resolution of the Sentinel-1 and Sentinel-2A bands. The coarser spatial scale of the datasets likely contributed to the overprediction in the model due to the generalization of the underlying environmental conditions. Mapping narrow, linear vegetation patterns on the landscape similar to AWC presents a challenge with satellite data [18]. Furthermore, AWC is influenced by changes in microtopography [14], and the spatial resolution of 10 m may not have been adequate to detect those fine-scale changes in topography.

While RF models are known for their utility within ecology and their high classification accuracy within SDMs, there are some limitations when creating an SDM utilizing a machine learning approach, such as overfitting in some cases [24]. Random forests provide more accurate predictions across a landscape than other methods, but if RF is used to extrapolate and project those predictions to an area outside of the placement of the training data provided, there can be some resulting issues with both over- and underprediction [67].

Another potential reason for the overprediction in the model may have been the placement of the presence and absence points used in the training data (Figure 1). There may have been AWC in locations that were not truly suitable for species to live long term, such as in relict populations, and there may have been absence points placed in areas that are truly suitable for this species, but no AWC was present due to dispersal limitations, biotic interactions, or environmental stochasticity [64]. It is advised to avoid placing absence points in an area near presence points [68]. However, the small study area made the placement of points an appropriate distance apart from each other more challenging. The placement of the presence and absence points may not have represented the true niche of this species, and the distance between the presence and absence points may not have been large enough to provide the necessary variability in the data to help the model distinguish between AWC and the other nearby classes.

Overprediction in SDMs is a common issue, and many studies do not find clear support for their model predictions [64]. In this study, we attempted to correct for the overprediction by adding variables that we had not previously considered, as well as spatially constraining the model to only predict within areas that are ecologically relevant to the study. The most frequent model overprediction correction approach is the use of geographical barriers [63], which was performed here by determining ecologically appropriate spatial zones, then constraining the model predictions to those areas. Spatially constraining the model allowed us to identify and remove predictions from the model output that fell within zones that clearly did not contain AWC and those where it was unlikely that AWC was present. It can be challenging to identify if an area is truly overpredicted without quantitative field data; our assumptions of overpredicted areas are based on our interpretation of the aerial imagery and known AWC locations. Even in light of these considerations, the SDM presented here correctly identified live pixels of AWC in a matrix of live and dead AWC (Figure 8), as well as distinguished AWC pixels from the surrounding deciduous vegetation (Figure 6).

There are a number of other coniferous species present, such as pitch pine, which is the most common pine throughout the New Jersey Pinelands. Pitch pine can be codominant within AWC stands in New Jersey [25] and has a very similar spectral response to AWC (Figure 3), which could potentially be contributing to the overprediction in the model. The spectral response of a given pixel is an amalgamation of all elements, including other conifer species, present in the pixel [69,70]. Spectral variables were important predictors in the SDM (Figure 4); therefore, if multiple species had similar responses, it is likely that this would cause the model to predict those areas as AWC, especially if it was a species that can exist in the same underlying environmental conditions, such as pitch pine. We believe it is advantageous to document the spectral response of AWC using the expanded spectral capabilities of Sentinel-2A. Advances in future sensors will likely provide more spectral information that could be used to build upon the work presented here.

4.3. Management Implications for Atlantic White Cedar

Atlantic White Cedar is facing increasing mortality along the coast in the face of anthropogenic climate change. Species within the marsh–forest ecotone will respond differently to these environmental changes [2], and the ghost forests provide the opportunity to track the migration of species within the marsh–forest ecotone. The marsh–forest ecotone is shifting in response to increasing storms and sea-level rises [3], and AWC is one of the most conspicuous species that is experiencing loss from these disturbances. After Hurricane Sandy in 2012, there were notable AWC ghost forests in southern New Jersey [5], and many of these areas did not recover post-hurricane. The model was able to distinguish between live and dead AWC (Figures 8 and 9), which allows for the tracking of AWC loss at the stand level to be monitored over time. Knowledge of where some potential areas for active management (e.g., planting new trees) or maintaining the live, healthy stands is important to have to help best protect and preserve this species in this region.

We created a transferable model that documents the baseline amount of Atlantic White Cedar throughout southern New Jersey. The use of random forests in species distribution models is beneficial for inventory and monitoring purposes as these models are run without a priori assumptions about the ecological influences of variables on the species, allowing them to better predict a species across space than some traditional SDM methods [71]. In areas where there is limited information about underlying processes, species interactions, dispersal limits, and nonlinear relationships, the use of RF models can help create a model with the best predictive capabilities without knowing any of the information about which processes may be influencing a species distribution across a landscape. The information provided from SDMs can be used to monitor species and help create management and loss mitigation strategies to conserve species in the future. In coastal regions, it is very challenging to predict where and when storm surges will occur [72]; however, SDMs provide information that could help managers locate stands of trees that may be in an area that is more susceptible to these threats. The model outputs capture the loss of AWC on the landscape in recent years, as the model was able to distinguish areas of mortality (Figures 8 and 9). This work provides much needed information to managers about AWC stands and helps us better understand stand dynamics (Table 3). Illuminating which environmental drivers were the most important to AWC within the model provides insight into what drives the distribution of the species across the landscape. Distance to water was the third most important predictor of AWC presence (Figure 4), which was not surprising given this species thrives only within 200 km of the coast [14] and tends to grow along streams and rivers [15] (Figure 6). These areas are more susceptible to inundation from sea-level rises or storm surges and may be more at risk of experiencing loss of AWC and stands within these areas should be carefully monitored.

Our analysis demonstrates that Atlantic White Cedar stands are dynamic and have experienced change since Hurricane Sandy (2012). The use of predictive habitat distribution modelling, using both remotely sensed and abiotic variables, is an effective method for the mapping of species distributions for use in management and loss mitigation. This work demonstrates the importance of including remotely sensed variables in species distribution models [27] (Figure 4). The use of SDMs to create spatially explicit maps of species distributions that are difficult to map using field methods is an important data gap to address due to the pressures of climate change on this important ecotone.

5. Conclusions

In this study, we aimed to create a spatially explicit map of Atlantic White Cedar presence throughout southern New Jersey. To help meet this goal, we determined the driving variables of AWC presence as well as the spectral signature of AWC compared to surrounding vegetation types to better understand the dynamics of this important, sensitive species. A combination of field methods and aerial photograph interpretation allowed for the creation of a training dataset that was used to create a random forests model. The results showed that both remotely sensed and abiotic variables were important in predicting AWC presence and that there has been some AWC mortality experienced on the landscape. The model presented in this study provides high utility for AWC management and tracking mortality dynamics within stands after disturbances such as hurricanes. This study established a baseline on the amount and locations of AWC occurrence throughout southern New Jersey. Our work can be extended across the full distribution of the species and applied to future time periods to track those changes across the landscape to assess changing environmental conditions. Additional research can build on the work presented here by utilizing new sensors, such as those mounted on unmanned aerial vehicles. These would provide imagery with finer resolution and incorporate additional field data to understand if and why overprediction occurs in the model.

Author Contributions: Conceptualization, B.N. and T.J.A.; methodology, B.N., A.E.S., A.L.C. and T.J.A.; formal analysis, B.N. and T.J.A.; data curation, B.N.; writing—original draft preparation, B.N. and T.J.A.; writing—review and editing, B.N., A.E.S., A.L.C. and T.J.A.; visualization, B.N. and T.J.A.; supervision, T.J.A. All authors have read and agreed to the published version of the manuscript.

Funding: This research received no external funding.

Data Availability Statement: All satellite data used in this study are publicly available through Google Earth Engine. Field-collected data and unique abiotic variables (e.g., distance to water) created for this research have been archived in the Knowledge Network for Biocomplexity [73]. R code for analysis is available from the corresponding author upon request.

Acknowledgments: We thank Stephanie Burkey and Kathryn Burns for field assistance. We gratefully acknowledge financial support from the Department of Geography at Kent State University.

Conflicts of Interest: The authors declare no conflicts of interest.

References

1. Schieder, N.W.; Kirwan, M.L. Sea-Level Driven Acceleration in Coastal Forest Retreat. *Geology* **2019**, *47*, 1151–1155. [CrossRef]
2. Jobe, J.G.D.; Gedan, K. Species-Specific Responses of a Marsh-Forest Ecotone Plant Community Responding to Climate Change. *Ecology* **2021**, *102*, e03296. [CrossRef]
3. Fagherazzi, S.; Nordio, G.; Munz, K.; Catucci, D.; Kearney, W.S. Variations in Persistence and Regenerative Zones in Coastal Forests Triggered by Sea Level Rise and Storms. *Remote Sens.* **2019**, *11*, 2019. [CrossRef]
4. Able, K.W. Special Issue: Concepts and Controversies in Tidal Marsh Ecology Revisited From Cedar Cemeteries to Marsh Lakes: A Case Study of Sea-Level Rise and Habitat Change in a Northeastern US Salt Marsh. *Estuaries Coasts* **2021**, *44*, 1649–1657. [CrossRef]
5. Doyle, J.M.; Earley, K.E.; Atkinson, R.B. An Analysis of Atlantic White Cedar (*Chaemaecyparis thyoides* (L.) B.S.P.) Tree Rings as Indicators of Ghost Forest Development in a Globally Threatened Ecosystem. *Forests* **2021**, *12*, 973. [CrossRef]
6. Hauser, S.; Meixler, M.S.; Laba, M. Quantification of Impacts and Ecosystem Services Loss in New Jersey Coastal Wetlands Due to Hurricane Sandy Storm Surge. *Wetlands* **2015**, *35*, 1137–1148. [CrossRef]
7. Ury, E.A.; Yang, X.; Wright, J.P.; Bernhardt, E.S. Rapid Deforestation of a Coastal Landscape Driven by Sea-Level Rise and Extreme Events. *Ecol. Appl.* **2021**, *31*, e02339. [CrossRef]
8. Woods, N.N.; Swall, J.L.; Zinnert, J.C. Soil Salinity Impacts Future Community Composition of Coastal Forests. *Wetlands* **2020**, *40*, 1495–1503. [CrossRef]
9. Kopp, R.E.; Andrews, C.; Broccoli, A.; Garner, A.; Kreeger, D.; Leichenko, R.; Lin, N.; Little, C.; Miller, J.A.; Miller, J.K.; et al. *New Jersey's Rising Seas and Changing Coastal Storms: Report of the 2019 Science and Technical Advisory Panel*; The State University of New Jersey: New Brunswick, NJ, USA, 2019.
10. Cooper, M.J.P.; Beevers, M.D.; Oppenheimer, M. The Potential Impacts of Sea Level Rise on the Coastal Region of New Jersey, USA. *Clim. Chang.* **2008**, *90*, 475–492. [CrossRef]
11. Smith, A.J.; Goetz, E.M. Climate Change Drives Increased Directional Movement of Landscape Ecotones. *Landsc. Ecol.* **2021**, *36*, 3105–3116. [CrossRef]
12. Hasse, J.E.; Lathrop, R.G.; Bognar, J.A. *Changing Landscapes in the Garden State: Land Use Change in NJ 1986 Thru 2012*; Rutgers University: New Brunswick, NJ, USA; Rowan University: Glassboro, NJ, USA, 2016.
13. The American Littoral Society; National Fish and Wildlife Foundation. Assessing the Impacts of Hurricane Sandy. 2012. Available online: https://rucore.libraries.rutgers.edu/rutgers-lib/43631/ (accessed on 19 September 2024).
14. Hnatkovich, A.M.; Yorks, T.E. Current Status and Future Development of the Only *Chamaecyparis thyoides* (Atlantic White-Cedar) Population in Pennsylvania. *Nat. Areas J.* **2009**, *29*, 216–223. [CrossRef]

15. Little, S.J. Ecology and Silviculture of Whitecedar and Associated Hardwoods in Southern New Jersey. *Yale Sch. For. Environ. Stud. Bull. Ser.* **1950**, *54*, 134.
16. Crawford, E.R.; Day, F.P.; Atkinson, R.B. Influence of Environment and Substrate Quality on Root Decomposition in Naturally Regenerating and Restored Atlantic White Cedar Wetlands. *Wetlands* **2007**, *27*, 1–11. [CrossRef]
17. Laing, J.M.; Shear, T.H.; Blazich, F.A. How Management Strategies Have Affected Atlantic White-Cedar Forest Recovery after Massive Wind Damage in the Great Dismal Swamp. *For. Ecol. Manag.* **2011**, *262*, 1337–1344. [CrossRef]
18. Assal, T.J.; Steen, V.A.; Caltrider, T.; Cundy, T.; Stewart, C.; Manning, N.; Anderson, P.J. Monitoring Long-Term Riparian Vegetation Trends to Inform Local Habitat Management in a Mountainous Environment. *Ecol. Indic.* **2021**, *127*, 107807. [CrossRef]
19. Assal, T.J.; Anderson, P.J.; Sibold, J. Mapping Forest Functional Type in a Forest- Shrubland Ecotone Using SPOT Imagery and Predictive Habitat Distribution Modelling. *Remote Sens. Lett.* **2015**, *6*, 755–764. [CrossRef]
20. Hailu, B.T.; Siljander, M.; Maeda, E.E.; Pellikka, P. Assessing Spatial Distribution of *Coffea arabica* L. in Ethiopia's Highlands Using Species Distribution Models and Geospatial Analysis Methods. *Ecol. Inform.* **2017**, *42*, 79–89. [CrossRef]
21. He, Y.; Chen, G.; Potter, C.; Meentemeyer, R.K. Integrating Multi-Sensor Remote Sensing and Species Distribution Modeling to Map the Spread of Emerging Forest Disease and Tree Mortality. *Remote Sens. Environ.* **2019**, *231*, 111238. [CrossRef]
22. Smith, A.B.; Murphy, S.J.; Henderson, D.; Erickson, K.D. Including Imprecisely Georeferenced Specimens Improves Accuracy of Species Distribution Models and Estimates of Niche Breadth. *Glob. Ecol. Biogeogr.* **2023**, *32*, 342–355. [CrossRef]
23. Elith, J.; Leathwick, J.R. Species Distribution Models: Ecological Explanation and Prediction across Space and Time. *Annu. Rev. Ecol. Evol. Syst.* **2009**, *40*, 677–697. [CrossRef]
24. Pecchi, M.; Marchi, M.; Burton, V.; Giannetti, F.; Moriondo, M.; Bernetti, I.; Bindi, M.; Chirici, G. Species Distribution Modelling to Support Forest Management. A Literature Review. *Ecol. Model.* **2019**, *411*, 108817. [CrossRef]
25. Laderman, A.D. *The Ecology of Atlantic White Cedar Wetlands: A Community Profile*; U.S. Department of Interior, Fish and Wildlife Service, National Wetlands Research Center: Washington, DC, USA, 1989.
26. Zimmermann, N.E.; Edwards, T.C.J.; Moisen, G.G.; Frescino, T.S.; Blackard, J.A. Remote Sensing-Based Predictors Improve Distribution Models of Rare, Early Successional and Broadleaf Tree Species in Utah. *J. Appl. Ecol.* **2007**, *44*, 1057–1067. [CrossRef]
27. Cord, A.; Rödder, D. Inclusion of Habitat Availability in Species Distribution Models through Multi-Temporal Remote-Sensing Data? *Ecol. Appl.* **2011**, *21*, 3285–3298. [CrossRef]
28. Jones, H.G.; Vaughan, R.H. *Remote Sensing of Vegetation: Principles, Techniques, and Applications*; OUP Oxford: Oxford, UK, 2010.
29. Price, J.C. How Unique Are Spectral Signatures? *Remote Sens. Environ.* **1994**, *49*, 181–186. [CrossRef]
30. Rajakumari, S.; Mahesh, R.; Sarunjith, K.J.; Ramesh, R. Building Spectral Catalogue for Salt Marsh Vegetation, Hyperspectral and Multispectral Remote Sensing. *Reg. Stud. Mar. Sci.* **2022**, *53*, 102435. [CrossRef]
31. Estrada, F.; Flexas, J.; Araus, J.L.; Mora-Poblete, F.; Gonzalez-Talice, J.; Castillo, D.; Matus, I.A.; Méndez-Espinoza, A.M.; Garriga, M.; Araya-Riquelme, C.; et al. Exploring Plant Responses to Abiotic Stress by Contrasting Spectral Signature Changes. *Front. Plant Sci.* **2023**, *13*, 1026323. [CrossRef]
32. Bhattarai, R.; Rahimzadeh-Bajgiran, P.; Weiskittel, A.; Meneghini, A.; MacLean, D.A. Spruce Budworm Tree Host Species Distribution and Abundance Mapping Using Multi-Temporal Sentinel-1 and Sentinel-2 Satellite Imagery. *ISPRS J. Photogramm. Remote Sens.* **2021**, *172*, 28–40. [CrossRef]
33. Akbari, V.; Solberg, S.; Puliti, S. Multitemporal Sentinel-1 and Sentinel-2 Images for Characterization and Discrimination of Young Forest Stands under Regeneration in Norway. *IEEE J. Sel. Top. Appl. Earth Obs. Remote Sens.* **2021**, *14*, 5049–5063. [CrossRef]
34. United States Geological Survey, U.S.G.S. National Hydrography Dataset Plus High Resolution. Available online: https://www.sciencebase.gov/catalog/item/5d30c268e4b01d82ce84a923 (accessed on 9 September 2022).
35. Bunnell, J.F.; Zampella, R.A.; Lathrop, R.G.; Bognar, J.A. Landscape Changes in the Mullica River Basin of the Pinelands National Reserve, New Jersey, USA. *Environ. Manag.* **2003**, *31*, 696–708. [CrossRef]
36. Miao, Z.; Lathrop, R.G.; Xu, M.; La Puma, I.P.; Clark, K.L.; Hom, J.; Skowronski, N.; Van Tuyl, S. Simulation and Sensitivity Analysis of Carbon Storage and Fluxes in the New Jersey Pinelands. *Environ. Model. Softw.* **2011**, *26*, 1112–1122. [CrossRef]
37. Zampella, R.A.; Lathrop, R.G. Landscape Changes in Atlantic White Cedar (*Chamaecyparis thyoides*) Wetlands of the New Jersey Pinelands. *Landsc. Ecol.* **1997**, *12*, 397–408. [CrossRef]
38. United States Department of Agriculture. Ortho NAIP. Available online: https://nrcs.app.box.com/v/naip/folder/95484293573 (accessed on 21 February 2022).
39. NJDEP Bureau of GIS. Land Use/Land Cover of New Jersey 2012. Available online: https://www.nj.gov/dep/gis/digidownload/zips/OpenData/Land_lu_2012.zip (accessed on 25 February 2022).
40. Fernández-Guisuraga, J.M.; Suárez-Seoane, S.; Calvo, L. Radar and Multispectral Remote Sensing Data Accurately Estimate Vegetation Vertical Structure Diversity as a Fire Resilience Indicator. *Remote Sens. Ecol. Conserv.* **2023**, *9*, 117–132. [CrossRef]
41. Koley, S.; Chockalingam, J. Sentinel 1 and Sentinel 2 for Cropland Mapping with Special Emphasis on the Usability of Textural and Vegetation Indices. *Adv. Sp. Res.* **2022**, *69*, 1768–1785. [CrossRef]
42. R Core Team. *R: A Language and Environment for Statistical Computing 2022*; R Core Team: Vienna, Austria, 2022.
43. Hijmans, R.J. Raster: Geographic Data Analysis and Modeling. 2023. Available online: https://cran.r-project.org/web/packages/raster/raster.pdf (accessed on 19 September 2024).

44. Rouse, J.W.; Haas, R.H.; Schell, J.A.; Deering, D.W.; Harlan, J.C. *Monitoring the Vernal Advancements and Retrogradation of Natural Vegetation*; Final Report; NASA/GSFC: Greenbelt, MD, USA, 1974; pp. 1–137.
45. Sims, D.A.; Gamon, J.A. Relationships between Leaf Pigment Content and Spectral Reflectance across a Wide Range of Species, Leaf Structures and Developmental Stages. *Remote Sens. Environ.* **2002**, *81*, 337–354. [CrossRef]
46. De Reu, J.; Bourgeois, J.; Bats, M.; Zwertvaegher, A.; Gelorini, V.; De Smedt, P.; Chu, W.; Antrop, M.; De Maeyer, P.; Finke, P.; et al. Application of the Topographic Position Index to Heterogeneous Landscapes. *Geomorphology* **2013**, *186*, 39–49. [CrossRef]
47. Sørensen, R.; Zinko, U.; Seibert, J. On the Calculation of the Topographic Wetness Index: Evaluation of Different Methods Based on Field Observations. *Hydrol. Earth Syst. Sci.* **2006**, *10*, 101–112. [CrossRef]
48. Riley, S.J.; DeGloria, S.D.; Elliot, R. Terrain_Ruggedness_Index.Pdf. *Intermt. J. Sci.* **1999**, *5*, 23–27.
49. Ponti, R.; Sannolo, M. The Importance of Including Phenology When Modelling Species Ecological Niche. *Ecography* **2023**, *2023*, e06143. [CrossRef]
50. Roberts, D.A.; Ustin, S.L.; Ogunjemiyo, S.; Greenberg, J.; Bobrowski, S.Z.; Chen, J.; Hinckley, T.M. Spectral and Structural Measures of Northwest Forest Vegetation at Leaf to Landscape Scales. *Ecosystems* **2004**, *7*, 545–562. [CrossRef]
51. Williams, D.L. A Comparison of Spectral Reflectance Properties at the Needle, Branch, and Canopy Level for Selected Conifer Species. *Remote Sens. Environ.* **1991**, *35*, 79–93. [CrossRef]
52. NJDEP Bureau of GIS. Coastline (2012) of New Jersey. Available online: https://gisdata-njdep.opendata.arcgis.com/datasets/njdep::coastline-2012-of-new-jersey/about (accessed on 11 June 2022).
53. NJDEP Bureau of GIS. Head of Tide for New Jersey Watercourses. Available online: https://gisdata-njdep.opendata.arcgis.com/datasets/njdep::head-of-tide-hot-for-new-jersey-watercourses/about (accessed on 14 June 2022).
54. Chaney, N.W.; Minasny, B.; Herman, J.D.; Nauman, T.W.; Brungard, C.W.; Morgan, C.L.S.; McBratney, A.B.; Wood, E.F.; Yimam, Y. POLARIS Soil Properties: 30-m Probabilistic Maps of Soil Properties Over the Contiguous United States. *Water Resour. Res.* **2019**, *55*, 2916–2938. [CrossRef]
55. Liaw, A.; Wiener, M. Classification and Regression by RandomForest. *R J.* **2002**, *2*, 18–22.
56. Cutler, D.R.; Edwards, T.C.; Beard, K.H.; Cutler, A.; Hess, K.T.; Gibson, J.; Lawler, J.J. Random Forests for Classification in Ecology. *Ecology* **2007**, *88*, 2783–2792. [CrossRef] [PubMed]
57. Engler, R.; Waser, L.T.; Zimmermann, N.E.; Schaub, M.; Berdos, S.; Ginzler, C.; Psomas, A. Combining Ensemble Modeling and Remote Sensing for Mapping Individual Tree Species at High Spatial Resolution. *For. Ecol. Manag.* **2013**, *310*, 64–73. [CrossRef]
58. De Marco Junior, P.; Nóbrega, C.C. Evaluating Collinearity Effects on Species Distribution Models: An Approach Based on Virtual Species Simulation. *PLoS ONE* **2018**, *13*, 0202403. [CrossRef]
59. Naimi, B.; Hamm, N.A.S.; Groen, T.A.; Skidmore, A.K.; Toxopeus, A.G. Where Is Positional Uncertainty a Problem for Species Distribution Modelling. *Ecography* **2014**, *37*, 191–203. [CrossRef]
60. Jarnevich, C.S.; Esaias, W.E.; Ma, P.L.A.; Morisette, J.T.; Nickeson, J.E.; Stohlgren, T.J.; Holcombe, T.R.; Nightingale, J.M.; Wolfe, R.E.; Tan, B. Regional Distribution Models with Lack of Proximate Predictors: Africanized Honeybees Expanding North. *Divers. Distrib.* **2014**, *20*, 193–201. [CrossRef]
61. Werkowska, W.; Márquez, A.L.; Real, R.; Acevedo, P. A Practical Overview of Transferability in Species Distribution Modeling. *Environ. Rev.* **2017**, *25*, 127–133. [CrossRef]
62. Keeland, B.D.; McCoy, J.W. Plant Community Composition of a Tidally Influenced, Remnant Atlantic White Cedar Stand in Mississippi. In *Ecology of Tidal Freshwater Forested Wetlands of the Southeastern United States*; Springer: Berlin/Heidelberg, Germany, 2007; pp. 89–112, ISBN 9781402050947.
63. Velazco, S.J.E.; Ribeiro, B.R.; Laureto, L.M.O.; De Marco Júnior, P. Overprediction of Species Distribution Models in Conservation Planning: A Still Neglected Issue with Strong Effects. *Biol. Conserv.* **2020**, *252*, 108822. [CrossRef]
64. Lee-Yaw, J.A.; McCune, J.L.; Pironon, S.; Sheth, S.N. Species Distribution Models Rarely Predict the Biology of Real Populations. *Ecography* **2022**, *2022*, e05877. [CrossRef]
65. Chauvier, Y.; Descombes, P.; Guéguen, M.; Boulangeat, L.; Thuiller, W.; Zimmermann, N.E. Resolution in Species Distribution Models Shapes Spatial Patterns of Plant Multifaceted Diversity. *Ecography* **2022**, *2022*, e05973. [CrossRef]
66. Fritsch, M.; Lischke, H.; Meyer, K.M. Scaling Methods in Ecological Modelling. *Methods Ecol. Evol.* **2020**, *11*, 1368–1378. [CrossRef]
67. Heikkinen, R.K.; Marmion, M.; Luoto, M. Does the Interpolation Accuracy of Species Distribution Models Come at the Expense of Transferability? *Ecography* **2012**, *35*, 276–288. [CrossRef]
68. Lobo, J.M.; Jiménez-Valverde, A.; Hortal, J. The Uncertain Nature of Absences and Their Importance in Species Distribution Modelling. *Ecography* **2010**, *33*, 103–114. [CrossRef]
69. Assal, T.J.; Sibold, J.; Reich, R. Modeling a Historical Mountain Pine Beetle Outbreak Using Landsat MSS and Multiple Lines of Evidence. *Remote Sens. Environ.* **2014**, *155*, 275–288. [CrossRef]
70. Lefsky, M.A.; Cohen, W.B. Selection of Remotely Sensed Data. In *Remote Sensing of Forest Environments: Concepts and Case Studies*; Wulder, M.A., Franklin, S.E., Eds.; Springer: Boston, MA, USA, 2003; pp. 13–46, ISBN 1402074050.
71. Magness, D.R.; Huettmann, F.; Morton, J.M. Using Random Forests to Provide Predicted Species Distribution Maps as a Metric for Ecological Inventory & Monitoring Programs. *Stud. Comput. Intell.* **2008**, *122*, 209–229. [CrossRef]

72. Hanley, M.E.; Bouma, T.J.; Mossman, H.L. The Gathering Storm: Optimizing Management of Coastal Ecosystems in the Face of a Climate-Driven Threat. *Ann. Bot.* **2020**, *125*, 197–212. [CrossRef]
73. Norlin, B.; Assal, T. Data from: Mapping an Indicator Species of Sea Level Rise along the Forest-Marsh Ecotone. Available online: https://knb.ecoinformatics.org/submit/urn:uuid:bd1f89e8-3b73-4945-8d84-c17b5e7f10e3 (accessed on 23 July 2023).

land

MDPI

Article

Ecosystem Services Supporting Environmental Impact Assessments (EIAs): Assessments of Navigation Waterways Deepening Based on Data, Experts, and a 3D Ecosystem Model

Gerald Schernewski [1,2,*], Mara Jekat [1], Frank Kösters [3], Thomas Neumann [4], Swantje Steffen [5] and Miriam von Thenen [1]

1 Research Unit—Coastal Seas and Society, Leibniz-Institute for Baltic Sea Research, Seestraße 15, D-18119 Rostock, Germany; mara.jekat@gmx.de (M.J.)
2 Marine Research Institute, Klaipeda University, Universiteto Ave. 17, LT-92294 Klaipeda, Lithuania
3 Federal Waterway Engineering and Research Institute, Hydraulic Engineering in Coastal Areas, Wedeler Landstraße 157, D-22559 Hamburg, Germany; frank.koesters@baw.de
4 Department of Physical Oceanography and Instrumentation, Leibniz-Institute for Baltic Sea Research, Seestraße 15, D-18119 Rostock, Germany; thomas.neumann@io-warnemuende.de
5 Heinrich Hirdes GmbH, Alter Hafen Nord 210, D-18069 Rostock, Germany
* Correspondence: gerald.schernewski@io-warnemuende.de; Tel.: +49-381-5197207

Abstract: The navigation waterways to the harbors of Rostock (Warnow Estuary), Germany, and Szczecin (Oder/Szczecin Lagoon), Poland, were recently deepened. Both activities required Environmental Impact Assessments. We conducted expert- and data-based ecosystem service assessments for both case studies. Additionally, we performed 3D-ecosystem model simulations. For the Oder Lagoon, the model results show that the waterway deepening increased the burial in sediments by 807 t N/a, 112 t P/a and 4661 T C/a. However, altogether, the impacts of the deepening to 12.5 m draught on the lagoon ecosystem are minor and a model application is not necessary, but the results improve the data basis for ecosystem service assessments. Our expert-based ecosystem service approach is adaptable to the needs of coastal engineering and hydraulic projects and is both easy and quick to apply and transferable. The assessment results highlight the relevance of cultural services and can complement EIAs. Our approach can especially support the early scoping stage of an EIA. It has the potential to enhance cooperation and communication with and between stakeholders, reduce conflicts, and save time. Additionally, it could improve the compilation and addressing of stakeholder concerns, potentially reducing costs associated with unnecessary studies.

Keywords: shipping channel; harbor; ecosystem service assessment; sediments; nutrient burial; Baltic Sea; stakeholder; Environmental Impact Assessment; Natura 2000; coastal waters

1. Introduction

Maritime transport plays a key role in the European economy, transporting about 75% of its external trade and approximately 31% of its internal trade [1]. During the next decades, the European Union's Sustainable and Smart and Mobility Strategy aims at a significant increase in maritime transport, with required total investments until 2050 amounting to about 9.5 billion Euros [2]. This includes the deepening of existing navigation channels to enable larger ships.

The Baltic Sea hosts over 500 ports. It is one of the seas with the highest ship density and traffic volume worldwide. The 21 largest commercial Baltic sea ports belong to the trans-European transport network. In 2020, these 21 harbors recorded passenger traffic of 53.2 million persons, about one third of EU maritime passenger traffic, as well as 523.2 million tons of cargo traffic. Between 2010 and 2020, the 21 harbors showed an annual cargo growth rate of 0.4%. In parallel, between 2000 and the early 2020s, the capacity of the largest container ships increased from 8000 twenty-foot shipping container equivalent units (TEUs) to 24,000 TEUs [3]. As a consequence, ships have increasing draughts. This causes an ongoing need for deeper navigation channels as well as increased sediment maintenance dredging.

The Baltic Sea has a surface area of 404,364 km^2, but with an average depth of only 53 m, it is very shallow. It is connected to the North Sea by the narrow and shallow waters of the sound and the Belt Sea and consists of a several sub-basins, which are separated by shallow sills [4]. The long and ragged Baltic Sea shoreline has a total length of about 8000 km and is characterized by numerous bays, lagoons and estuaries. The southern half of the Baltic Sea consists of soft, usually plain sandy sediments. One consequence of these special features is that storm-induced sediment erosion, re-allocation and accretion are common, and the maintenance dredging of navigational waterways has to take place frequently.

All these aspects mean that the dredging and deepening of navigational waterways play an important role in the coastal management of the Baltic Sea. These activities interfere with the natural dynamic, can cause the degradation of the marine environment and can potentially reduce the ecosystem's ability to provide ecosystem services [5].

The aims of the European Union's marine policies, such as the Water Framework Directive (WFD, 2000/60/EC) and the Marine Strategy Framework Directive (MSFD, 2008/56/EC), are to achieve and maintain a good ecological and environmental status. This includes sea-floor integrity, including the preservation of structure and functions of benthic ecosystems [6]. As a consequence, activities that potentially cause a risk for the environment require an Environmental Impact Assessment (EIA, 2011/92/EU). The EIA assesses the direct and indirect environmental impact of an activity based on a wide range of factors [7].

Ecosystem services are commonly defined as benefits that humans derive from ecosystems [8] and play an important role in European coastal and marine policies [9]. Ecosystem services take an anthropocentric perspective on ecosystems and are considered to support a comprehensive understanding of the structures and (inter-)dependencies between humans and the environment. Integrating ecosystem service assessments into Environmental Impact Assessments has received a lot of attention. In 2013, Karjalainen et al. [10] stated an experimental stage, and Baker et al., 2013 [11], named potential benefits but at the same time pointed out weaknesses and asked for pragmatic approaches. General methodologies, guidelines and case studies are available, but with a focus on terrestrial systems [12–15]. In 2013, Liquete et al. [16] pointed out the deficits with respect to marine ecosystem services assessments, and deficits are still visible, e.g., [17]. But a recent review, with focus on Marine Protected Areas (MPAs), indicates that ecosystem services in general gained importance in marine systems [18].

There are several reasons for the limited application of ecosystem service assessments in coastal waters, including weaker data availability, the highly dynamic nature of aquatic systems, and the limited transferability of existing methodologies and guidelines from land to sea. However, several case studies demonstrate the use of ecosystem service assessments in coastal waters. For instance, Inacio et al. [19] compare various Baltic estuaries, bays, and lagoons using an ecosystem service assessment tool based on compiled data, though it remains largely scientific in its approach. A more practical study compares the historical and potential future ecological states of two estuaries to support the implementation of the Water Framework Directive [20]. Since future ecosystem states are hypothetical and lack field data, an expert-based ecosystem service assessment is conducted.

Other examples include expert-driven assessments that compare different coastal protection scenarios [21] or evaluate the impacts of climate change and sea-level rise on a coastal region [22]. However, to our knowledge, there are no examples of ecosystem service assessments applied to concrete coastal or marine infrastructure projects and their associated EIAs.

Key characteristics of the EIA process include a focus on providing information, public consultations, participation, and the assessment of alternative measures [23]. Recent studies support the idea that expert-based ecosystem service assessments can be particularly valuable in coastal and marine policy implementation [24]. They help compare and prioritize various environmental measures, conservation, and restoration strategies while also raising awareness of environmental challenges, potential solutions, and their consequences [20–22]. These assessments are considered effective in capturing local perspectives, increasing awareness of the multifaceted impacts of measures, and fostering a more structured participatory dialogue with local communities and stakeholders [21,22]. Moreover, they can facilitate communication, broaden the perspectives and knowledge of stakeholders, and support social learning processes [20].

Our overarching questions is whether expert-based ecosystem service assessments can play a role in coastal and marine Environmental Impact Assessments as well as what can be learned from complementing ecosystem model applications. The objectives are to (a) compile a set of ecosystem services that enables a comprehensive assessment of the environmental and societal consequences of waterway deepening; (b) test this approach in two contrasting Baltic Sea case studies using different assessment methods and different scenarios; (c) compare the expert assessments with the results of a 3D ecosystem model to assess the reliability of expert based assessments; and (d) critically assess the results in the EIA context, as well as with respect to transferability and general applicability of the ecosystem services assessment tool for navigational waterway deepening.

Our two contrasting case studies can be regarded as representative of the southern Baltic Sea coasts, and our approach is generally transferable.

2. Materials and Methods

2.1. The Harbors Rostock, Germany, and Swinoujscie/Szczecin, Poland

In 2016, the harbors Rostock, Germany, and Swinoujscie/Szczecin, Poland, recorded 26.9 and 24.7 million tons of cargo traffic, respectively [2]. Only a few harbors in the Baltic Sea have significantly higher cargo traffic.

For Rostock, projections for ship traffic from 2015 to 2030 suggest a slight increase from 4801 to 5094 ship movements (only tankers and cargo ships), but a substantial increase in ship capacities [25,26]. Passenger ship traffic is expected to rise from 11,994 to 17,042 passengers during this period. Notably, the number of large passenger ships (over 10,000 tons deadweight) is projected to increase from 188 in 2015 to 1322 in 2030 [25,26].

Currently, Rostock's waterway in the Warnow Estuary allows a maximum shipping depth (draught) of 13 m up to the industrial harbor. Given future projections, the port authority recognized the need to increase the shipping depth to 15 m (with waterway depths ranging from 16.1 to 16.8 m and bottom widths from 112 to 135 m) (Figure 1).

This requires dredging approximately 4.8 million cubic meters of sediment over a length of 15 km. The dredged material comprises 47% clay, glacial till, and silt and about 30% sand, gravel, and bottom mud. Approximately 0.4 million cubic meters of sediment with higher pollution levels were in advance of the project deposited in a land-based sink field, while unpolluted sediments will be deposited in a nearby offshore dumping area. It is estimated that an annual maintenance dredging of 60,000 cubic meters will be necessary to maintain the water depth over the entire length and a waterway width between 120 and 230 m [25,26]. The dredging began in the summer of 2022 and is expected to be completed by 2025 [27].

Between 2017 and 2022, the port of Szczecin-Swinoujscie experienced a significant increase in turnover, rising from 25.5 million tons to 36.8 million tons [28]. During the same period, regular passenger traffic increased slightly to 1.1 million passengers. To accommodate larger ships in the future, the 62 km waterway across the Szczecin Lagoon was deepened from 10.5 m to 12.5 m and widened to 100 m between 2018 and 2023 (Figure 1). This allowed for a doubling of ship capacity to about 20,000 tons of cargo per vessel [29]. The dredged material was deposited on two newly created islands with areas of approximately 250 hectares and 123 hectares, respectively [30]. The total project costs amounted to 337 million Euros [31].

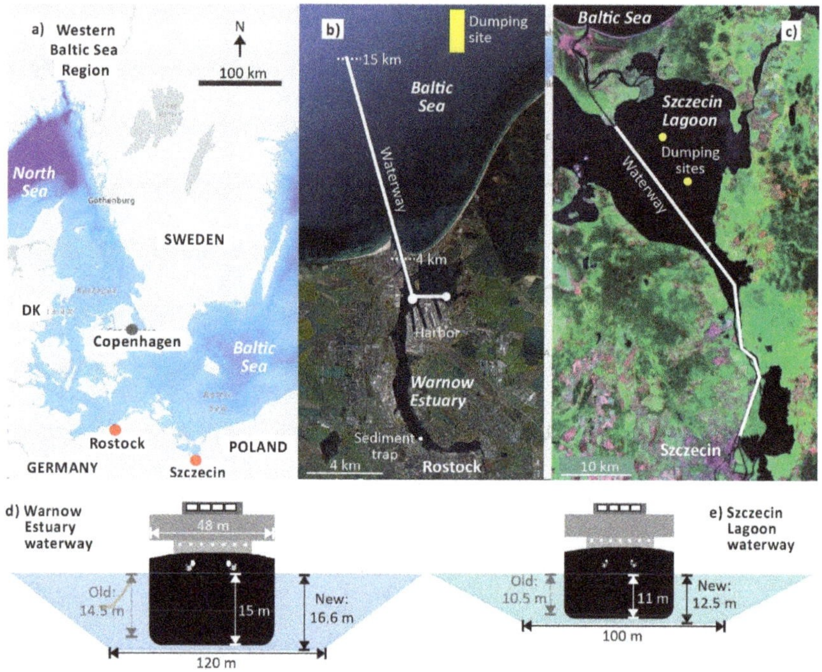

Figure 1. (**a**) Location of Rostock and Szczecin harbors; (**b**) location of the waterway/channel through the Warnow Estuary to Rostock harbor and (**c**) across the Szczecin Lagoon to Szczecin harbor. (**d**,**e**) provide details about the state in the 2010's and the recent navigational channel deepening projects.

The Szczecin Lagoon is designated as European Union Natura 2000 site [32]. These sites are protected areas under the EU Birds (2009/147/EC) and Habitats Directives (92/43/EEC), which implement the EU's biodiversity policy. In such sites, port development and dredging operations are restricted and require an appropriate assessment to complement the EIA.

Conversely, the EU Strategy for the Baltic Sea Region (EUSBSR) [33] has three objectives: saving the sea, connecting the region, and increasing prosperity. The strategy explicitly promotes waterway deepening under the priority area "Transport—Improving internal and external transport links".

2.2. Environmental Impact Assessments: Process and Results

The following outlines the conceptual background and regional results of Environmental Impact Assessments (EIAs), providing a foundation for evaluating the complementary value of ecosystem services assessments.

Major building and development projects in the EU have to be assessed for their environmental impacts under the Environmental Impact Assessment Directive (2011/92/EU as amended by 2014/52/EU). The direct and indirect significant impacts on population and human health; biodiversity; land, soil, water, air, and climate; material assets, cultural heritage and the landscape; and the interactions between these factors, are assessed in an EIA [34]. The EIA procedure starts with a screening to determine if the planned project indeed needs an EIA (Figure 2) [35]. Subsequently, the competent authority determines the content, scope, depth of detail, and methods that the developers shall use in their EIA reporting. In this scoping phase, it is also determined if the project is likely to have cross-border impacts. In that case the ESPOO convention applies, this requires the party of origin to notify the affected party of the expected environmental impacts and to provide the EIA report so that the affected party may comment [36].

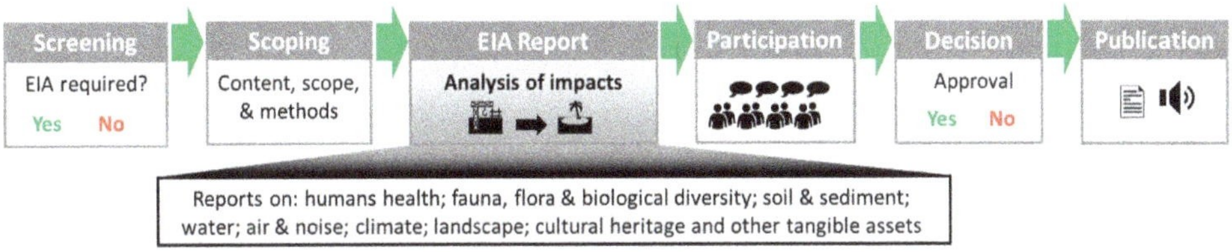

Figure 2. The Environmental Impact Assessments (EIA) process according to European Union Directive (2011/92/EU as amended by 2014/52/EU) and the thematic assessment reports provided for the waterway deepenings in Rostock and in Swinoujscie/Szczecin.

The EIA report provides information on the planned project (e.g., location, type, technical design, size); the potential significant environmental impacts, a description of reasonable alternatives that have been examined by the developer; and a description of the characteristics of the project, the location, and planned measures to exclude, reduce, compensate or replace significant adverse environmental impacts [34]. The report needs to be made publicly available so that the public and public agencies (specialist authorities) may provide comments and objections, which the competent authority needs to address in its approval decision that is also to be made publicly available.

Both waterway deepenings, in Rostock and in Swinoujscie/Szczecin, were subject to an EIA. All documents for the EIA procedure in Rostock are available at the website of Federal Waterway and Shipping Administration [26]. An Espoo report on the environmental impacts of the Swinoujscie/Szczecin waterway deepening was provided to the German authorities upon request and is available at the website of the Ministry for economy, infrastructure, tourism and labor [37]. The report provides a summary of the impacts and some extracts of the original EIA report, but the original EIA report is, to our knowledge, not publicly available (in Germany). The protected assets that were subject to the environmental impact study in both cases are in line with what is outlined in the EIA Directive (Figure 2). For both projects, additional studies are also available in the Appendix A (Table A1).

The environmental impact study describes the current status/situation for each protected good (e.g., the degree of already existing impacts), the sensitivity of the protected good, and the expected impacts. For each protected good, a specific study area or spatial extent, in which the impacts are assessed, is determined. In the case of Rostock, these areas ranged from 100 m around the waterway to 3000 m (Table A2). The expected impacts are assessed in terms of their spatial scale and the duration of the impact. The methodological framework followed is that of an ecological risk analysis. The EIA report thereby provides a verbal-argumentative description of the assessment and references the specialist reports that were commissioned beforehand. While the focus is generally on the significant adverse environmental impacts, beneficial impacts are also reported, following a nine-level scale from very high negative to very high positive impacts (Table A3). The method followed in Rostock complied with the "Guidelines for the Environmental Impact Assessment of Federal Waterways" [38]. The assessment method for the EIA in Poland was not specified in the Espoo report. A number of guidelines exist at the EU level [34]; however, it is not clear to what extent they were followed either in Rostock or in Swinoujscie/Szczecin. In the case of Rostock, the entire process, from the application to deepen the waterway to the final approval, took 12 years (Table A4). The EIA reports from Rostock and Swinoujscie/Szczecin both came to the conclusion that no (significantly) adverse impacts were expected. Only for some specific areas, significantly adverse impacts on the protected good "fauna, flora and biological diversity" were expected in Rostock (Table A5). The accompanying landscape conservation plan determined a compensation area of 83.44 ha and the required compensation measure (renaturation of parts of a polder). The Espoo report on the Swinoujscie/Szczecin waterway did not detail all impact analyses on the protected goods but focused more on potential cross-border impacts. The conclusion was that there are none, as the adverse impacts are locally restricted. Mitigation measures for noise impacts on marine mammals were mentioned, as well as compensation measures, including the extension of Natura 2000 areas and the renaturation of degraded habitats of white and grey dunes. In both cases, the potential impacts on Natura 2000 areas were assessed separately. The Polish report listed 39 Natura 2000 areas within a 20 km radius around the waterway deepening, but only carried out an "appropriate assessment" for seven areas, according to the methodological guidance for the assessment of plans and projects significantly affecting Natura 2000 sites [39]. The Polish report only concluded negligible, very weak and insignificant impacts on Natura 2000 sites [40]. In the case of Rostock, three Natura 2000 sites are in the vicinity of the development area. For each one, a screening report was prepared, which came to the conclusion that no further assessments were necessary as no significant impacts are expected [26].

2.3. Selection of Ecosystem Services and the Ecosystem Service Assessment Tool

The selected ecosystem services are based on the CICES 5.1 classification [41], which is officially used by the European Commission. This hierarchical classification follows the guidance of the United Nations Statistical Division and categorizes ecosystem services into "provisioning", "regulation and maintenance", and "cultural" services. The authors, in collaboration with other scientists, chose 27 ecosystem services relevant for the topic. Pre-tests demonstrated that this number allows for assessment by external stakeholders and experts within a manageable timeframe of approximately 30–45 min, followed by subsequent feedback discussions. The selected ecosystem services are illustrated in Figure 3a.

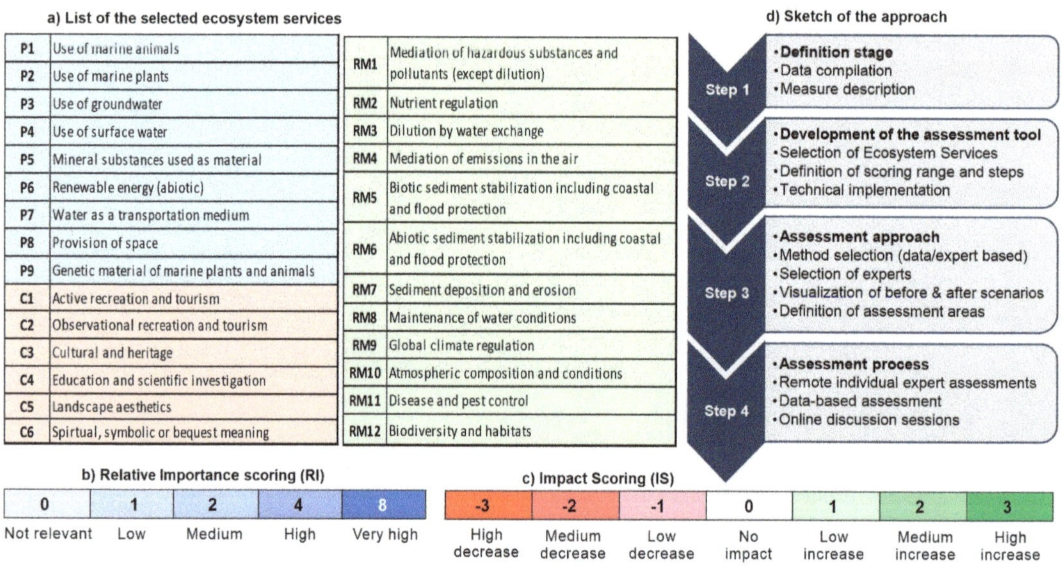

Figure 3. (**a**) Overview of the ecosystem services used in the assessment. The colors indicate provision (P), cultural C) and regulating and maintenance ecosystem services (RM). Explanations are provided in Appendix A. (**b**) The sketch provides an overview of the steps of the entire assessment. Scores and ranges used in the (**c**) assessment of changes between scenarios and (**d**) assessment of the relative importance of every ecosystem service.

The ecosystem services were compiled into tables and implemented in an EXCEL sheet to facilitate remote assessments by experts. This Coastal Ecosystem Service Assessment Tool (Co-ESAT) contains additional explanations about the ecosystem services, the scenarios, the task, and the scoring method and ranges. The tool is a further development of earlier approaches for other topics [20,42]. It allows the expert to enter personal information such as name, institution and field of expertise, as well as a self-assessment of different aspects of their expertise. The table allows one to enter scores for the relative importance (RI) of each service using the classes 0 (not relevant), 1, 2, 4 and 8 (very high relevance) (Figure 3d). It further asks the expert to score the scenario after compared to the situation before the waterway deepening for every ecosystem service. A scale was used ranging from −3 (very high decrease of a service) to +3 (very high increase) (Figure 3c). The tool further facilitates calculations, for example, the multiplication of the relative importance with the ecosystem service score for the change, which ends up in a weighted score, allows one to enter comments and questions and asks for the time spent for the assessment. The full tool is documented in Appendix B. The entire approach is sketched in Figure 3d.

2.4. The Scenarios, Assessment Areas and Selection of Experts

For both the Szczecin (German: Oder) Lagoon and the Warnow Estuary, two scenarios were developed. The scenarios describe the situation before and hypothetically after the waterway deepening. The scenarios were documented in Powerpoint presentations. On 18 slides, these presentations contained pictures, complementing information and compiled background data, e.g., on local setting, major utilizations, land-use and protection status, morphology and geology, chemical and biological status of the waters, and hydrodynamics, as well as detailed information on the deepening measure itself.

The area that is possibly affected by the waterway deepening was visualized, and the experts were asked to consider changes in this area only. In the case of the Oder Lagoon, this area covered about 80% of the Polish part of the lagoon, the Wielki Zalew. The assessment area in the Warnow case study covered 500 m left and right of the waterway, as well as the sediment dumping area. In both cases, the presentations served as information material for the experts and, besides the experts' own knowledge and experience, were a major source of information for the assessment. The areas are based on the EIA investigation area for human protection.

The choice of experts is crucial for the assessment and influences the later outcome. We defined three major fields of expertise, ecology and environmental sciences, ecosystem services, and planning, as well as coastal engineering, and searched for at least three experienced experts in each field. The minimum educational level was a completed master's study course and several years of work experience. Most experts had a doctorate. Another criterion was that the experts represent responsible federal state or national authorities, science, and NGOs. The third criterion was knowledge of the locality. Additionally, we invited coastal and environmental engineering as well as coastal management students to take part in the assessment. This fourth group represented educated citizens with a largely neutral view on the issue. In the Szczecin Lagoon case study, 15, and in the Warnow Estuary case study, 12 experts took place in the assessment. We contacted a much larger number of experts but failed to attract them, or they were not able to provide the assessment within the provided time frame. A shortcoming was that we were not able to attract Polish experts for the Szczecin Lagoon case study.

2.5. The Ecosystem Service Assessment Processes

We used two ecosystem service assessments approaches: a quantitative data and literature-based, subsequently called data-based, approach, and the expert-based approach. Both approaches exclusively focused on potential long-term changes. The data-based assessment was carried out in both case studies. Students with a suitable scientific background compiled data, literature, regional policy and planning documents, as well as monitoring data, and carried out the assessment based on this obtained knowledge and data, using the Coastal Ecosystem Service Assessment Tool. This took one to two months and was part of a master's and a bachelor's thesis [43,44].

In both case studies, the expert assessments were carried out remotely. After the experts indicated their readiness to take part in the assessment, they received the Powerpoint presentation, with the background information, as well as the Coastal Ecosystem Service Assessment Tool. The experts additionally received information on how to fill in the tool and explanations about its scientific purpose. They were asked to return their assessment within two weeks. The results were compiled and analyzed by the two diploma students responsible for the data-based assessments. Afterwards, the experts were contacted and invited to take part in several separate online meetings aiming to settle questions, address comments raised by the experts and discuss results where extreme differences between experts existed. Afterwards, the experts were asked to provide their view on the approach and its value. The online sessions had a duration of 1–1.5 h, were written down and took place in the weeks after the submission deadline of the individual assessments. Afterwards, detailed protocols were prepared. The sessions were prepared and moderated by the students that took care of the data-based assessment.

2.6. The 3D Ecosystem Model ERGOM—Szczecin/Oder Lagoon

The ecology of a largely enclosed brackish system like the Szczecin Lagoon is highly sensitive to changes in morphology and water exchange with the Baltic Sea. Additionally, as a Natura 2000 site, the effects of channel/waterway deepening must be examined in much greater detail [45]. Consequently, we applied the 3D ecosystem model ERGOM (Ecological Regional Ocean Model) to the Szczecin Lagoon, conducting a more in-depth analysis compared to the studies commonly carried out during the EIA process [46], as well as in case of the Oder Lagoon [40]. For the Warnow estuary case study, detailed hydrodynamic model simulations were conducted as part of the EIA. The results indicated only minimal changes in salinity and flow patterns and only in the vicinity of the channel. Consequently, significant impacts on the ecosystem, which could be captured by an ecosystem model, were deemed highly unlikely. As a result, the ecosystem model was not applied to the Warnow estuary.

We employed the coupled circulation and biogeochemical model detailed in Neumann et al. [47]. The circulation model used is the Modular Ocean Model (MOM5.1) [48] featuring a horizontal resolution of 150 m. Vertically, the model is divided into up to 28 layers, each with a thickness of 0.5 m, except for the surface layer, which had a thickness of 0.25 cm. The existing model bathymetry was modified according the EIA planning documents.

The biogeochemical model ERGOM is coupled with the circulation model via the tracer module, which is embedded within the Modular Ocean Model (MOM5.1) code. ERGOM simulates the cycles of nitrogen, phosphorus, carbon, oxygen, and partially sulfur. Primary production, driven by photosynthetically active radiation (PAR), is modeled using three functional phytoplankton groups (large cells, small cells, and cyanobacteria). Chlorophyll concentration in the optical model is derived from these phytoplankton groups [47]. Dead organic matter accumulates as detritus, while bulk zooplankton, representing the highest trophic level in the model, grazes on phytoplankton.

Phytoplankton and detritus can sink through the water column and accumulate in the sediment layer. Detritus undergoes mineralization into dissolved inorganic nitrogen and phosphorus, influenced by water temperature and oxygen concentration. Under oxic conditions, phosphate binds to iron oxide and is retained as sediment particles. These particles can be re-suspended by erosion events and transported by currents to deposition areas. Under anoxic conditions, the reduction of iron oxide liberates phosphate, making it available as dissolved phosphate. Oxygen is produced through primary production and consumed by various processes including metabolism and mineralization. The extracellular excretion of dissolved organic matter by phytoplankton can lead to non-Redfield carbon uptake. A comprehensive description and validation of the model can be found in Neumann et al. [47]. Our ecosystem model is not able to simulate effects on benthic flora, biodiversity and fauna or higher levels of organisms.

For atmospheric forcing, we utilized the High-RESolution Atmospheric Forcing Fields (HiResAFF) dataset provided by Schenk and Zorita [49]. Atmospheric and riverine loads were reconstructed following Gustafsson et al. [50], with both HiResAFF and load data analyzed. The model was forced by meteorological data from the coastDat-2 dataset [51].

3. Results

3.1. The Coastal Ecosystem Service Assessment Tool and Process

One objective was to develop and provide an ecosystem services assessment tool, specifically tailored for coastal engineering measures such as waterway deepening, dredging, and sediment dumping. The suitability of the tool can be assessed based on several factors: (1) the choice of ecosystem services; (2) the separation into an assessment of changes in ecosystem services, as well as an assessment of the relative importance of every ecosystem service with respect to waterway deepening; (3) the scoring steps and ranges; (4) the technical implementation in a spreadsheet (Excel); and (5) the time efficiency.

The involved experts were not explicitly asked about the suitability of the tool but had the possibility to comment on it during the online sessions and to enter notes in the assessment sheet. Neither the experts nor the two student moderators reported any issues or suggested improvements for the scoring steps, ranges, or technical implementation of the tool. The time required for remote scoring was noted on the assessment sheets. Among the experts, it ranged from 20 to 120 min with an average of 38 min for the Oder Lagoon and from 13 to 60 min with an average of 35 min for the Warnow. It can be assumed that the overall assessment, including reading the background information and completing the scoring, took about one hour on average. Therefore, selecting 27 ecosystem services appears to be a reasonable balance between detail and required assessment time.

The remote assessment provided flexibility for the experts and ensured comprehensive individual perspectives. The subsequent online discussion sessions lasted an additional 1 to 1.5 h. Meeting protocols indicate that these discussions were essential, particularly for clarifying the meaning of individual ecosystem services. Overall, the total duration of 2.5 h for the tool and the entire assessment process can be considered suitable and time-efficient. The experts' self-assessment of their expertise in different fields helped the moderator prepare for the online discussion and identify any misunderstandings.

The experts' assessment of the relative importance of the selected ecosystem services for evaluating waterway deepening measures provides insight into the suitability of this set of ecosystem services for such projects. Generally, none of the 27 ecosystem services were deemed irrelevant for the Oder Lagoon or the Warnow, although several were considered of low relevance. Perspectives varied significantly among experts from different fields, and even among those with similar backgrounds (Figure 4). Only the student groups and the two students who carried out an assessment based on data and literature considered several ecosystem services to be irrelevant. Specifically, the ecosystem services "Use of marine plants (P2)" and "Renewable energy" were generally considered to have low relevance.

Experts largely agreed on the high relevance of ecosystem services such as "Water as transport medium (P7)", "Sediment deposition and erosion (EM7)", "Biodiversity and habitats (RM12)", and "Active and observational recreation (C1 and C2)". The subsequent online discussions did not indicate that any ecosystem services were missing. Therefore, we can conclude that the chosen set of ecosystem services was suitable.

On average, cultural (C) services were considered the most important, followed by regulating and maintenance (RM) services, and then provisioning (P) services (Figure 4). This trend was observed for both the Warnow Estuary and, even more so, for the Oder Lagoon. The two data-based assessments also showed similar results. The larger number of RM services likely contributed to the lower relevance of each individual ecosystem service, but the findings also underscore the significance of cultural services.

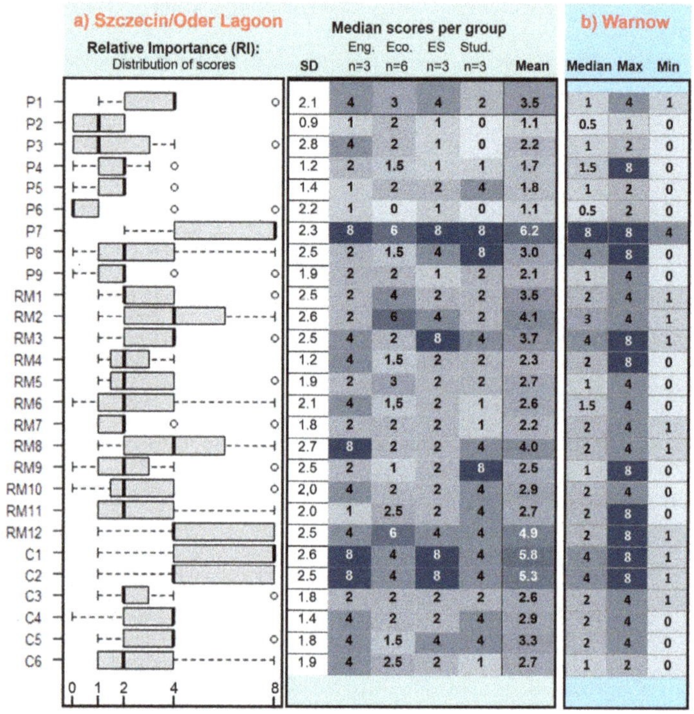

Figure 4. Relative importance (RI) of the selected ecosystem services for assessing waterway deepening measures based on expert judgements (scale: 0, not relevant, to 8, highly important. The expert groups are coastal engineers (Eng.), ecologists (Eco.), ecosystem service experts (ESs) and university students (Stud.). (**a**) Szczecin/Oder Lagoon case study; (**b**) Warnow Estuary/Rostock harbor case study.

3.2. *Warnow Estuary—Ecosystem Service Assessments*

Experts anticipate a significant increase only in the ecosystem service "Water as a transport medium" due to the deepening of the Warnow Estuary. Across all other ecosystem services, based on the median assessment of all 12 experts, minimal to no changes are expected (Figure 5). The inclusion of the student group in the calculation was justified by findings from the Oder case study, showing their scoring aligned closely with that of the experts.

Most ecosystem services show negative values, indicating that channel deepening slightly diminishes their provision. In summary, the projected changes in ecosystem services are modest, suggesting that channel deepening is perceived as having little impact on ecosystem service provision.

The scoring of changes in the data-based approach significantly differs from that conducted by experts. The higher positive and negative values cannot be fully explained, suggesting that the data-based assessment may be insufficient, although it does offer insights; for example, the higher score for the ecosystem service "dilution by water exchange" seems reasonable due to increased water exchange processes with deepening. Nevertheless, the data-based approach should be considered as supplementary rather than definitive.

No.	Ecosystem Services	Expert scores (n=12) Median	Max	Min	Data-based
P1	Use of marine animals	0	1	-1	0
P2	Use of marine plants	0	0	-2	0
P3	Use of groundwater	0	0	-1	-1
P4	Use of surface water	0	1	-2	0
P5	Mineral substances	0.5	3	0	2
P6	Renewable energy (abiotic)	0	0	0	0
P7	Water as a transportation medium	3	3	2	3
P8	Provision of space	0	2	-1	-1
P9	Genetic material	-0.5	0	-2	-2
RM1	Mediation of hazardous substance	-0.5	1	-2	-2
RM2	Nutrient regulation	-0.5	1	-3	-2
RM3	Dilution by water exchange	0	2	-2	2
RM4	Mediation of emissions in the air	-0.5	2	-2	0
RM5	Biotic sediment stabilization	-0.5	0	-2	-2
RM6	Abiotic sediment stabilization	-1	1	-2	-2
RM7	Sediment deposition and erosion	0	2	-2	0
RM8	Maintenance of water conditions	-1	2	-2	-1
RM9	Global climate regulation	0	0	-2	1
RM10	Atmospheric composition	0	0	-1	1
RM11	Disease and pest control	0	1	-2	-1
RM12	Biodiversity and habitats	-1	1	-2	0
C1	Active recreation and tourism	0	1	-1	-1
C2	Observational recreation	0	2	-1	-1
C3	Cultural and heritage	0	0	-1	1
C4	Education and sci. investigation	0	2	0	1
C5	Landscape aesthetics	0	0	-1	-1
C6	Spirtual or bequest meaning	0	0	-1	-1

Figure 5. Changes in ecosystem service provision between the two scenarios (before and after the waterway deepening measure) in the Warnow estuary (Rostock harbor) based on expert judgements (scale: 3 high increase; 3 high decrease after the measure) as well as results of the data-based assessment.

3.3. Szczecin/Oder Lagoon—Ecosystem Services Assessments

Apart from the ecosystem service "Water as a transport medium", experts generally perceived the impact of waterway deepening on the Oder Lagoon as modest (Figure 6). This consensus spans all four groups of experts, with consistent assessment results among them. The standard deviation of scores is typically close to one, indicating a high level of agreement among experts.

One notable discrepancy among the groups pertains to the ecosystem service "Biodiversity and habitats". Engineers and ecologists foresee insignificant changes, while experts in ecosystem services anticipate a moderate increase, and the student group predicts a moderate decrease. Variances also exist regarding "Landscape aesthetics", with ecologists expressing concerns about the presumed increase in ship traffic. Overall, the consensus suggests that waterway deepening is expected to marginally reduce the provision of ecosystem services.

The data-based assessment conducted by a student largely aligns with that of the experts, with a significant difference noted in their assessment of "Cultural and heritage" services, where the student assigns a high positive score.

In summary, the results are considered stable and reliable across different expert backgrounds, indicating that the number of experts involved was adequate for this study. The findings collectively suggest that the impact of waterway deepening on ecosystem service provision is generally perceived as modest, especially in the Warnow Estuary case study.

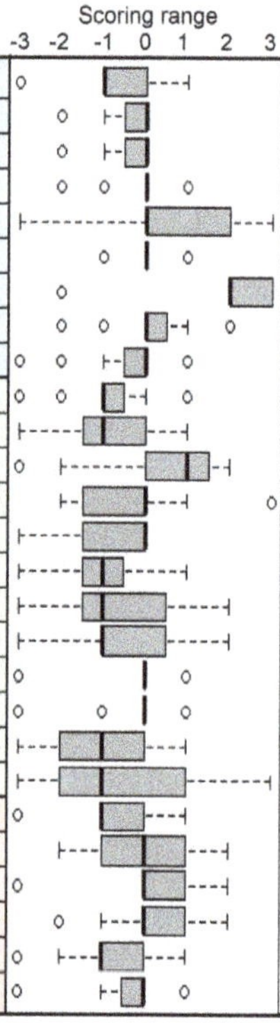

No.	Ecosystem Services	SD	Eng. n=3	Eco. n=6	ES n=3	Stud. n=3	Mean n=15	Data-based
			Median expert scores per group					
P1	Use of marine animals	1.0	-1	0	-1	-1	-0.5	-1
P2	Use of marine plants	0.6	-1	0	0	0	-0.3	0
P3	Use of groundwater	0.8	-1	0	0	0	-0.5	0
P4	Use of surface water	0.7	0	0	0	0	-0.1	0
P5	Mineral substances	1.5	0	0	1	2	0.7	2
P6	Renewable energy (abiotic)	0.5	0	0	0	0	0.1	0
P7	Water as a transportation medium	1.2	2	2	2	3	2.1	2
P8	Provision of space	1.2	0	0	0	2	0.2	-2
P9	Genetic material	1.1	0	0	0	1	-0.3	-1
RM1	Mediation of hazardous substance	1.0	-1	-1	-1	-1	-0.8	-1
RM2	Nutrient regulation	1.3	1	-1	-1	-1	-0.9	-1
RM3	Dilution by water exchange	1.5	1	0.5	1	1	0.5	1
RM4	Mediation of emissions in the air	1.4	0	-0.5	0	-1	-0.4	0
RM5	Biotic sediment stabilization	1.2	0	0	0	0	-0.7	0
RM6	Abiotic sediment stabilization	1.1	-1	-1	-1	-1	-1.0	-2
RM7	Sediment deposition and erosion	1.6	0	-0.5	1	-1	-0.4	1
RM8	Maintenance of water conditions	1.3	1	-0.5	-1	-1	-0.5	-2
RM9	Global climate regulation	0.8	0	0	0	0	-0.1	-1
RM10	Atmospheric composition	0.9	0	0	0	-1	-0.2	1
RM11	Disease and pest control	1.2	0	-1	-1	-2	-0.9	1
RM12	Biodiversity and habitats	1.7	-1	-0.5	2	-2	-0.5	1
C1	Active recreation and tourism	1.0	-1	-0.5	0	-1	-0.5	1
C2	Observational recreation	1.4	0	-0.5	1	-1	0.0	-1
C3	Cultural and heritage	1.1	0	0	1	0	0.2	3
C4	Education and sci. investigation	1.1	0	0	1	0	0.3	1
C5	Landscape aesthetics	1.3	0	-1.5	0	-1	-0.7	1
C6	Spirtual or bequest meaning	1.0	0	0	0	0	-0.2	0

Figure 6. Changes in ecosystem service provision between the two scenarios (before and after the waterway deepening measure) in the Oder Lagoon based on expert judgements (scale: 3 high increase; 3 high decrease after the measure). SD indicates the standard deviation. The expert groups are coastal engineers (Eng.), ecologists (Eco.), eosystem service experts (ESs) and university students (Stud.). Additionally, the results of the data-based assessment are shown.

3.4. Szczecin/Oder Lagoon—Ecosystem Model Results

Based on the model, the waterway deepening has an immediate effect on water exchange between lagoon and the Baltic Sea. More saline Baltic Sea water will enter the lagoon and increase the salinity especially in the Wielki Zalew by 11% and by 5.5% in the entire lagoon (Figure 7a). However, the absolute average salinity is only about 1 g/kg, and the changes are low. Water temperature and ice cover are practically not affected by the waterway deepening, and the effects on dissolved nitrogen and phosphorus concentrations are low. The model suggests a significant reduction in the chlorophyll-a and phytoplankton concentrations between 7% and 11%. The surface water temperatures and the surface oxygen concentrations and are practically not affected by the deepening. Altogether, the waterway deepening has only very limited effects on the lagoon. The absolute changes in the parameters do not change the overall ecological state of the system.

However, the model suggests strong changes in nitrogen, phosphorus and carbon processes (Figure 7b,c). Nitrogen fixation is reduced by 5% in the entire lagoon as a result of reduced cyanobacteria concentrations in summer. Denitrification is reduced between 6% and 8%. This can be explained by a slightly decreased water retention time. The model suggests important increases in the nutrient burial in the sediment. The waterway deepening increases the burial of N, P and C by 70% to 74%. In absolute numbers, it means that 807 t N/a, 112 t P/a and 4661 T C/a are stored additionally in the sediments. This takes place exclusively in the Wielki Zalew, while the shallower western bay (Kleines Haff) does not show any nutrient burial in the sediments.

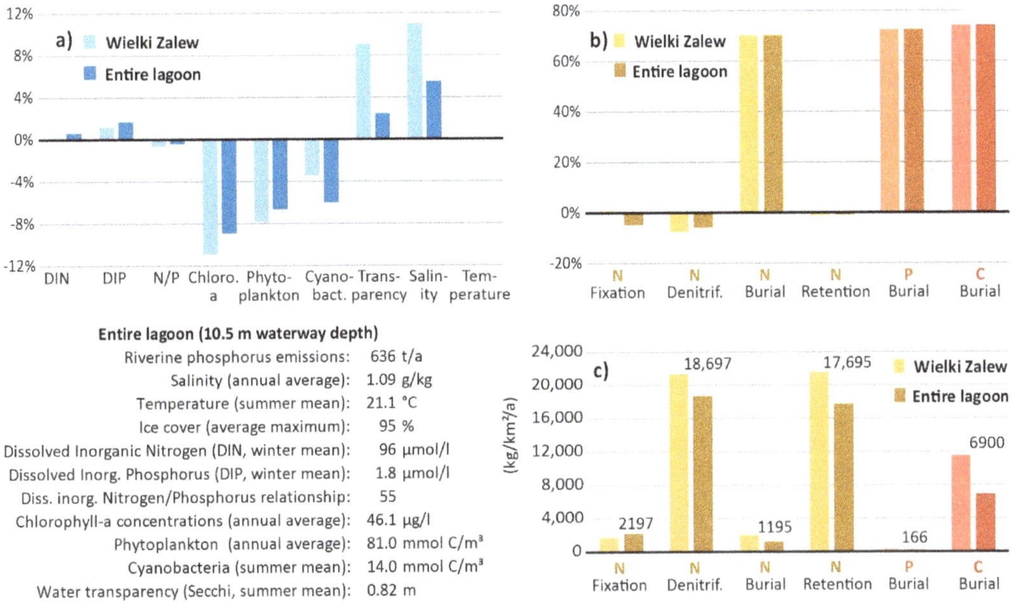

Figure 7. Results of the 3D ecosystem model simulations reflecting the situation before (10.5 m waterway depth/draught) and after the waterway deepening to 12.5 m draught, spatially averaged over the entire Oder Lagoon, as well as the eastern bay (Wielki Zalew). (**a**) Relative differences between selected parameters. Below are the absolute values for these parameters for the Oder Lagoon with 10.5 m waterway depth. (**b**) Relative differences for nitrogen-, phosphorus- and carbon-related processes before and after the waterway deepening and (**c**) the corresponding absolute values.

According to the 3D ecosystem model, the burial of organic matter (nutrients and carbon) in the sediments depends strongly on water depth. In a water depth down to about 8–10 m, storm-induced resuspension does not allow for a permanent burial. Re-suspended sediments leave the lagoon and are washed into the Baltic Sea or are trapped in deeper areas of the lagoon. The waterway is the deepest part of the lagoon, and the model suggests that the deepening to 12.5 m draught strongly increases this trap function. Consequently, more frequent sediment dredging is required to maintain the channel depth.

With respect to nitrogen, three processes, namely N-fixation, denitrification and burial in sediments, determine the total N retention in the lagoon. Quantitatively, denitrification is the most important process. Between 2010 and 2019, on average, 12,845 t N/a was removed via denitrification and 821 t N/a was buried in sediments, while N-fixation added 1510 t N/a to the water body. This means that the total N retention in the lagoon is 17,695 t N/a. The waterway deepening reduces this retention function by about 1% (144 t N/a). However, these numbers indicate that the lagoon is an important sink for nitrogen.

4. Discussion

4.1. The Ecosystem Service Assessment Tool—Lessons Learned

The 27 ecosystem services selected appeared to have a reasonable balance between detail and the required assessment time and were considered as suitable for this topic. In general, more ecosystem services were considered relevant for the Oder Lagoon compared to the Warnow Estuary. Additionally, the average Relative importance was higher, at 2.9

for the Oder Lagoon versus 2.1 for the Warnow Estuary. This clearly indicates that the two case studies represent different environmental and socio-economic settings, namely an urban environment and a largely natural protected system. These differences suggest that the tool and the included ecosystem services are generally applicable and transferable to other thematically comparable case studies.

The imbalance in the number of ecosystem services across the three ecosystem service categories—cultural, regulating/maintenance, and provisioning—combined with the different scores for their relative importance, complicates comparisons between categories and individual ecosystems. To address this, the tool principally allows for weighting, where the relative importance is multiplied by the scores of changes in ecosystem services. However, this option has not been used since the absolute results were not in focus. In other settings, for example, a stakeholder discussion process, a weighting might make sense to better visualize which ecosystem services have the highest importance. This would allow a prioritization between the single ecosystem services.

Examining the minimum and maximum expert scores for each ecosystem service reveals a wide range, highlighting diverse expert perceptions. This underscores the necessity of involving a group of experts to balance these differing views. The involvement of nine experts (and three students) proved necessary but also sufficient to achieve stable results. This number of experts appears generally suitable for waterway deepening projects.

Overall, the assessment approach of ecosystem service changes resulting from waterway deepening proved effective in both case studies, suggesting potential applicability to comparable projects worldwide. While these assessments reflect expert perceptions well, the results are associated with high uncertainties and hardly provide reliable data.

4.2. Environmental Impact Assessments—State and Deficits

In the European Union, major construction and development projects must undergo environmental impact assessments (EIAs) according to a defined legal framework, specifically the Environmental Impact Assessment Directive (2014/52/EU). In EU member states, such as Germany, additional specific guidelines exist for particular projects, including those involving navigation waterways [38]. An evaluation of EIA implementation in Germany concludes that EIAs generally meet quality criteria, are comprehensive and complete, and are relevant and effective for environmental protection, providing societal benefits [52].

However, the report also highlights several shortcomings [52]. For instance, the EIA process is not always well suited to the specific type of project, and the consideration of protected assets and spatial coverage can sometimes be inadequate or poorly justified. The early involvement of specialists and approval authorities is crucial for success, but it often occurs too late, and the authorities involved may lack the necessary expertise. Similarly, public involvement and information dissemination, which are essential, sometimes occur too late in the process. The report emphasizes the importance of a well-presented EIA summary for the acceptance of a project, noting that a transparent evaluation process enhances public support. Therefore, greater attention should be paid to these aspects.

Looking at the international literature, several challenges and deficiencies in EIA processes are identified. These issues mostly relate to procedural aspects, the quality of environmental information, and the integration of this information into decision-making. A European Union report has called for improvements in screening, scoping, the consideration of alternatives, monitoring, public participation, and EIA quality control [53]. Similarly, researchers advocate for improved procedures based on high-quality data, impact analyses, mitigation measures, and holistic, integrated monitoring plans. They also note weaknesses, such as the subjective evaluation of impacts and inconsistencies [54]. A low quality of the EIA reports may accrue due to a lack of training and capacity building for those responsible [53], and even if the presented environmental information is sound, it does not necessitate integration into decision-making [55]. Jay et al. [55] found that, in many cases, the results of EIA are considered as useful additional information but hardly influence the development decisions, which are rather based on political considerations.

A common concern regarding European environmental policy is the issue of over-regulation [56]. Concerns with respect to EIAs are particularly due to the costs and delays they impose on project implementation. Although the direct cost of an EIA typically accounts for around or below 1% of the total project cost [57], the necessary studies can significantly increase overall investment expenses. In Germany, the median duration of the EIA approval process, from application to decision, is nearly two years, with some cases extending beyond three years [58]. For complex projects, such as the deepening of the Weser River, the timeline from planning to implementation spanned over two decades, with the EIA being an important contributing factor to this prolonged timeline. Attempts to address this issue in Germany, such as the "Maßnahmengesetzvorbereitungsgesetz", a legislative act, have not been successful [59]. Therefore, enhancing the time and cost-efficiency of EIAs is crucial. To overcome some of these deficiencies, we explore two topics: the use of ecosystem models and/or ecosystem service assessments.

4.3. Benefits of Ecosystem Model Applications in EIAs

The EIA focuses on both the short- and long-term consequences of a project on the environment, addressing a hypothetical future state. The local data and knowledge used are based on the current state, and the quality of the EIA depends largely on the knowledge of the experts involved and the methods they employ. These methods can vary between different projects, as they are not predefined in legal framework documents. However, with respect to waterways in Germany, detailed guidelines [38] and a wide range of generally accepted procedural rules and publications exist. They serve as a framework and ensure a procedure according to the current state of the art.

The experts and companies involved in environmental assessments typically possess extensive experience, which gives them a competitive advantage. However, this experience is often kept internal, limiting knowledge sharing and transparency. The quality of expertise and decision-making processes could greatly benefit from well-documented, comparable case studies. To achieve this, systematic retrospective analyses are needed to extract lessons learned for future projects, along with thorough documentation of the results.

An alternative transparent approach for gaining insights into a future environmental state is the application of models. In navigation channel deepening and dredging projects, the use of 3D hydrodynamic models is particularly important and is often employed, as in the case of the Warnow estuary deepening. For coastal water deepening projects of this kind, a hydrodynamic model may be sufficient, since the ecological effects are generally considered low. However, in closed systems like the Oder Lagoon, which has a high environmental protection status and sensitive ecology, the use of a 3D ecosystem model may be more appropriate, providing new insights and additional data.

A comparison between the EIA and 3D ecosystem model results for the Oder Lagoon project generally shows that the EIA was sufficiently comprehensive. Both approaches suggest that the environmental impact of the waterway deepening on the lagoon is limited, even in such an ecologically sensitive area. However, the model provides new insights, such as the quantification of increased organic sediment burial resulting from the two-meter channel deepening. Between 1950 and 2000, approximately 1.5 million cubic meters of wet organic and mineral sediment were dredged and dumped annually from the channel across the Oder Lagoon [60]. An estimated 70% increase in inorganic sediments, as suggested by the model, would result in additional dredging and dumping costs. The model further indicates that the channel's sediment trap function prevents accumulation in other parts of the lagoon, increasing the risk of coastline and bottom erosion. However, these aspects could be covered by hydrodynamic models with an integrated sediment transport module, as well, which are often employed in large projects.

Our ecosystem model further indicates the ecological implications of nitrogen, phosphorus and carbon burial in the sediments for both the lagoon ecosystem and the downstream Baltic Sea. The model results emphasize the lagoon's role as a retention basin for nutrients, particularly nitrogen. The total nitrogen retention in the lagoon is 17,695 tons per year, and the waterway deepening would reduce this retention by about 1% (144 tons per year). Considering the total flow-normalized nitrogen loads to the Oder Lagoon between 2010 and 2014 of 61,950 tons per year [61], this means that 28.6% of all external nitrogen loads are retained in the lagoon. The effects of the waterway deepening on nutrient retention and budgets, as well as the implications for Baltic Sea management, were not considered in the EIA for the Oder Lagoon.

Another benefit of using models in EIAs is the ability to assess the effects of alternative measures or modified project implementation strategies, which, when based solely on expert knowledge, remain speculative. For this kind of study, a high depth of investigation is not necessarily required. This means that a high model performance is not mandatory, as long as the model effectively reflects the differences between alternative environmental states.

Our application illustrates that ecosystem models are valuable tools for environmental assessments. Moreover, using these models can significantly strengthen the data foundation for evaluating ecosystem services. A variety of ecosystem services can be directly or indirectly calculated using data generated by ecosystem models, offering new insights into the benefits of different environmental states—before and after a given measure—for both human use and society [62].

The use of 3D ecosystem models is currently limited by the availability of reliable models and computational capacity. However, in the Baltic Sea region, these limitations are expected to be overcome within the next decade, provided that reliable input data and expertise are available. One argument against the application of 3D ecosystem models in EIAs is the relatively low ecological impact of channel deepening on coastal systems. Additionally, there is a risk that the qualitative improvements offered by these models could lead to increased time and costs for EIAs. This would contradict ongoing efforts to streamline and reduce the costs and duration of the EIA process.

4.4. Ecosystem Service Assessments in EIAs

Despite a large number of studies on ecosystem service assessments, there are still complaints that the concepts remain theoretical and rarely support decision-making processes and policy implementation [63]. However, impact assessment processes are generally considered highly suitable for being linked to and supported by ecosystem service assessments [64]. Several conceptual frameworks exist for embedding ecosystem services into existing impact assessment practices [10,13,65].

In 2021, the International Association of Dredging Companies (IADC) released a report that presented a conceptual framework, with a particular focus on dredging and marine construction projects [66]. The report suggests that considering ecosystem services might lead to more sustainable and adaptive solutions and support the project decision-making process. The provided framework incorporates ecosystem service assessments at various stages of the project cycle, including baseline, prospective, retrospective, and adaptive assessments for different case studies. However, for applications in the European Union, where specific requirements and guidelines for EIAs are in place, this global framework is too general. In the European Union, the potential role of ecosystem service assessments in EIAs requires a more detailed investigation.

Ecosystem service assessments and EIAs share several similarities. Both are broad and comprehensive, cover a wide range of topics, are spatially flexible, and can be applied to comparable spatial units. EIAs address, besides environmental aspects, other factors such as human health, cultural values, and material assets. For example, in the Warnow case study, factors like noise emission, bathing water quality, and air quality were considered. However, in practice, EIAs typically focus on the consequences of a measure on nature and the environment. In contrast, ecosystem service assessments adopt an anthropocentric perspective. While EIAs emphasize potential damages and risks, ecosystem service assessments focus more on human benefits. Ecosystem services are usually divided into three categories: provisioning and regulating services, which focus on the natural system and its functioning, and cultural services, which directly describe the value to humans and society. Recent studies highlight that measures and environmental changes in a region particularly affect cultural ecosystem services [20,22]. Complementing especially information on cultural services might be one of the benefits of ecosystem service assessments in EIAs. This implies that ecosystem service assessments are carried out during the EIA process. However, if ecosystem service assessments are conducted in addition to an EIA, there is a risk of redundant analysis. Challenges also remain in properly integrating ecosystem services, quantifying them, developing impact indicators for prediction and assessment, and identifying beneficiaries at appropriate scales [67].

Many studies have shown that public participation is a crucial factor in improving EIA procedures [53–55] and that focusing on ecosystem services can enhance public engagement [68] and improve stakeholder engagement [67]. The benefits include identifying and prioritizing ecosystem services and beneficiaries who might experience negative impacts if service provision declines, thereby supporting the determination of the EIA's scope and the delineation of the study area [67,69].

4.5. Benefits of Expert-Based Ecosystem Service Assessments in EIAs

What are the benefits, strengths, and weaknesses of our expert-based ecosystem service assessments, and what role can they play in EIAs?

Our approach, which involves the structured engagement of an expert group, the comparative assessment of ecosystem services between the present state and alternative scenarios, and the selection of relevant ecosystem services, is generally suitable for similar coastal and hydraulic engineering projects. Moreover, this approach and the chosen ecosystem services are transferable to other regions. The experts involved considered all the included ecosystem services to be relevant, and the approach is fundamentally applicable within an EIA.

In general, the expert opinions on the consequences of waterway deepening align well with the findings of environmental impact reports and 3D ecosystem model results. In general, we can summarize that models and experts consider the effects of waterway deepening to be very limited. The experts' views, on average, seem reasonable, and utilizing their knowledge is particularly valuable when data are lacking.

However, our approach has its weaknesses. The quality of the expert-based ecosystem assessment depends on the background, knowledge, and perception of the involved experts, which introduces subjectivity and limits the reliability of the results. We involved 12 to 15 experts, which seems sufficient to stabilize the results, reduce subjectivity, and capture a full spectrum of opinions. Nevertheless, the selection and motivation of experts are crucial aspects that require careful consideration and expert mapping before the assessment. The problem of subjectivity is underscored by other expert-based assessments as well, namely that the results reflect opinions rather than facts [20] and possibly a bias [70]. Moreover, when applied to hypothetical future scenarios with high levels of uncertainty, these assessments risk being not only misleading but potentially counterproductive [22]. Other case studies highlight limitations arising from the unbalanced representation of stakeholders and interest groups, as well as an insufficient number of participants with adequate expertise [71]. Even cultural background and nationality can affect expert-based assessment results [72]. On the other hand, expert-based assessments can broaden the view of involved persons and especially emphasize the role of cultural changes [70], serve as a tool visualize trade-offs, analyze patterns and processes at regional scales and hence facilitate decision-making [73].

If the ecosystem service assessment is conducted alongside an EIA, it could benefit from the environmental studies already prepared. If 3D ecosystem model results are available, the database serving as background information for the experts would be even more robust. Additionally, ecosystem services could be partially converted into monetary terms, offering financial insights into the benefits of an investment and quantifying the connections between ecology and the economy [62]. All these data help reduce subjectivity in expert-based assessments and are expected to reduce the observed high variability in expert scores, leading to more stable results. A sound data basis provides a foundation for a data-driven ecosystem service assessment as well.

In our retrospective ecosystem service assessment, the experts involved identified cultural services as highly important. This suggests that cultural aspects are particularly relevant when evaluating measures, and this dimension of our assessment could effectively complement an EIA. It is possible that cultural services play a more significant role in urban systems compared to natural ones, as the benefits of ecosystem services to humans may be more pronounced in urban environments. Consequently, ecosystem service assessments might hold greater importance in these settings. The added value of incorporating an ecosystem service assessment into the EIA process would likely be similar to that observed in our retrospective analysis.

The alternative approach is to conduct an expert-based ecosystem service assessment early in the EIA process, specifically during the scoping phase. Our method is designed for flexible adaptation to the unique needs of coastal engineering and hydraulic projects and is both easy and quick to apply. The entire process does not require specialized knowledge and can be carried out by educated laypersons with moderation skills. When implemented during the scoping phase of an EIA, only general documents and background information need to be compiled and prepared; the EIA results themselves are not necessary at this stage. Assessing the importance of each ecosystem service allows experts to tailor the system to better reflect their understanding and perspective on the project.

If formally applied at an early stage of an EIA, our approach would need to be adapted from an expert-based to a stakeholder-based model. In this scenario, stakeholders identified and defined by the responsible licensing authority and the planning commissioner would participate in the ecosystem service assessment, replacing the experts. These stakeholder-based assessments could be valuable for capturing stakeholder views and concerns early on, helping to define investigation topics, protective assets, and spatial coverage. This approach could also foster public participation by highlighting the project's impact on the human usability of an ecosystem, supporting dialogue forums, and gathering public opinions beyond just the stakeholders if provided in an easily accessible online format. Additionally, it could serve as a complementary structural framework for summary presentations by incorporating a human-use perspective. Different to data-based ecosystem service assessments, the assessments carried out with and by stakeholders will remain qualitative. The results hardly enable a translation of changes into monetary values.

Stakeholder-based ecosystem service assessments could be used to evaluate both the long-term and short-term effects of a project. In contrast, data-based assessments are limited to assessing long-term effects. These assessments focus on comparing tangible, measurable changes that have occurred or are expected to occur as a result of the project, impacting the human use of the system.

Figure 8 summarizes our experiences in a Strength–Weaknesses–Opportunities–Threats (SWOT) diagram, incorporating insights from additional literature [74–76]. We believe that especially stakeholder-based ecosystem service assessments could be formally integrated into EIAs, even within the highly structured European Union EIA framework, particularly during the early scoping stage. This integration could enhance cooperation and communication with stakeholders, help to reduce conflicts, and potentially save time. Additionally, it could improve the compilation and addressing of stakeholder concerns, potentially reducing costs associated with unnecessary studies.

Strengths	**Opportunities**
• Complementing holistic, interdisciplinary approach with anthropocentric view on the nature • Identification of beneficiaries, social consequences and impacts on communities • Flexible, conceptually simple and enables fast applications • Suitable for different spatial scales • Supports public participation and engagement of stakeholders	• Increase awareness of the importance of the biophysical environment • Increased acceptance of measures • Reduced conflicts between project developers and local population • Detection of additional impacts • Joint holistic understanding among planning levels and involved actors • Framework for communication and meetings • Improved quality when complemented by data and models
Weaknesses	**Threats**
• Depends on perception of the measure and provided background information • Depends on knowledge/expertise of experts • Outcome depends on considered spatial scale • Results with limited reliability	• Redundant and duplicate analyses • Lack of experience and unclear responsibility for execution • No legal obligation to include ESA in EIA • Heterogeneous approach when based on experts, data and models

Figure 8. SWOT analysis of the use of expert/stakeholder-based ecosystem service assessments (ESA) within Environmental Impact Assessment (EIA) studies.

5. Conclusions

In the European Union, there is a well-established framework for Environmental Impact Assessments, which is considered successful but faces demands for reduced time and costs. We investigate the potential of ecosystem models and ecosystem service assessments to enhance EIA processes. Our findings suggest that ecosystem models are valuable tools for environmental assessments and offer new insights. However, some argue against the use of 3D ecosystem models in EIAs due to the relatively low ecological impact of channel deepening on coastal water ecosystems and the potential increase in time and costs, which could be counterproductive.

Comparing model results with expert-based ecosystem service assessments reveals a good agreement, and experts' views generally appear reasonable. The use of ecosystem models strengthens the data foundation for evaluating ecosystem services, reduces subjectivity in expert assessments, and is expected to lessen the observed variability in expert scores. Data and model results also provide a robust basis for a data-driven ecosystem service assessment.

In our retrospective assessment of case studies, experts highlighted the importance of cultural services, suggesting that information on these services can complement EIAs. Ecosystem service assessments can be formally integrated into the European Union EIA framework and carried out within the EIA process. Our method is adaptable to the needs of coastal engineering and hydraulic projects and is both easy and quick to apply. The process does not require specialized knowledge and can be executed by educated laypersons with moderation skills. However, our approach has weaknesses and suffers from subjectivity.

The primary benefit of applying our expert-based ecosystem service assessments during the early scoping stage of an EIA is the potential to enhance cooperation and communication with stakeholders, reduce conflicts, and save time. Additionally, it could improve the compilation and addressing of stakeholder concerns, potentially reducing costs associated with unnecessary studies. For this purpose, our approach would need to be adapted into a stakeholder-based model.

The case studies in the literature focus on retrospective assessments or hypothetical future management options. Future research should investigate whether the proposed benefits of expert-based ecosystem service assessments in stakeholder involvement [20,71,73] align with expectations in 'real-world' early-stage applications. In the context of our study, this implies that our approach needs to be tested during the actual scoping phase of an Environmental Impact Assessment for a coastal water engineering project.

Author Contributions: Conceptualization, G.S., S.S. and M.J.; methodology, G.S., S.S., M.J. and T.N.; software, T.N.; validation, G.S. and T.N.; formal analysis, G.S., S.S., M.J., M.v.T., T.N. and F.K.; investigation, G.S., S.S., M.J., M.v.T., T.N. and F.K.; writing—original draft preparation, G.S.; writing—review and editing, S.S., M.J., M.v.T., T.N. and F.K.; visualization, G.S., S.S., M.J. and T.N.; supervision, G.S.; funding acquisition, G.S. and M.v.T. All authors have read and agreed to the published version of the manuscript.

Funding: This work was financially supported by the German Federal Ministry of Education and Research, projects "Coastal Futures II" (grant number 03F0980B) and "Prime Prevention" (grant number 03F0911B).

Data Availability Statement: No new data were created or analyzed in this study. Data sharing is not applicable to this article.

Acknowledgments: We thank all experts for taking part in the assessment. We thank Martin Heindl for comments on the manuscript.

Conflicts of Interest: Author Swantje Steffen was employed by the company Heinrich Hirdes GmbH. The company employment started only after the research was finished. The remaining authors declare that the research was conducted in the absence of any commercial or financial relationships that could be construed as a potential conflict of interest.

Appendix A. Complementary Data on EIA's in Both Case Studies

Table A1. Protected goods that were subject to the EIA in Rostock and Swinoujscie/Szczecin.

Rostock	Swinoujscie/Szczecin
Humans, human health	Humans, human health
Fauna, flora and biological diversity	Fauna, flora and biological diversity
Soil/sediment	Soil
Water	Water
Air	Air/noise
Climate	Landscape
Landscape	Cultural heritage and other tangible assets
Cultural assets and other material assets	

Table A2. Spatial extent, or study area, for each protected good investigated for impacts in the Rostock waterway deepening.

Protected Good		Spatial Extent of Investigated Impacts
Humans		Sound emissions of the 35 dB(A) night isophone
Fauna, flora and biological diversity	Marine biotopes	200 m around the waterway and the deposit area
	Riverside biotopes	Breitling and Unterwarnow, up to the paved riverbank or up to 1 m contour line for unpaved banks
	Macrozoobenthos and macrophytes	200 m around the waterway and the deposit area
	Fish and cyclostomi	500 m around the waterway and the deposit area, plus Unterwarnow area
	Marine mammals	3000 m around the waterway and the deposit area
	Resting birds	Estuarine area + 500 m around the waterway and the deposit area
	Breeding birds	Pagenwerder, parts of the Breitling riverside system with comparatively near-natural habitat characteristics or unpaved bank boundaries as well as anthropogenically influenced parts on both sides of the development route
Soil		Under-water soils of the estuary and 200 m around the waterway and the deposit area
Water		The estuary and 200 m around the waterway and the deposit area
Climate and air		100 m around the waterway and the deposit area
Landscape		500 m around the waterway and the deposit area
Cultural assets and other material assets		200 m around the waterway and the deposit area

Table A3. Assessment of the potential impairments or improvements of changes accruing from the waterway deepening in Rostock.

Impairment					Improvement			
Very High (−16)	**High (−8 to −12)**	**Medium (−4 to −6)**	**Small (−1 to −3)**	**None (0)**	**Small (1–3)**	**Medium (4–6)**	**High (8–12)**	**Very High (16)**
Significantly adverse		Insignificantly adverse		Neither adverse nor beneficial	Insignificantly beneficial		Significantly beneficial	

Table A4. The process from application to final approval of the Rostock waterway deepening.

14 April 2009	Application of the state of Mecklenburg–Western Pomerania to deepen the seaward approach to the port of Rostock
December 2010	Forecast of expansion-related changes in ship-generated loads Statement on the environmental risk assessment
31 January 2011	Feasibility study (impacts on hydrodynamic)
March 2011	Environmental risk assessment with an FFH impact assessment
7 May 2013	The Federal Ministry of Transport, Building and Urban Development approves the main investigation for two variants of the channel deepening (after preliminary investigations of the expansion measure)
13 March 2014	Scoping meeting
10 October 2014	Determination of the scope of investigation
2016–2018	Inventories (ecological, geo-technical, pollutant load report, noise & air pollution, 3D-HN modeling, modeling of ship-generated wave loads
18 July 2017	Information event
2018–2019	Preparation of the application documents for the planning approval procedure (environmental impact study, landscape conservation plan, FFH preliminary assessment, expert contribution on species protection, technical contributions to the WFD and MSFD, technical planning, dredged material disposal concept
26 March 2019	Information event
30 August 2019	Application for planning approval
23 September–22 October 2019	Publication of the documents
23 September–22 November 2019	Opportunity for the public and public authorities to submit objections and comments
25 February 2020	Discussion meeting/public hearing
19 May 2021	Planning approval decision

Table A5. Conclusions on the expected impacts from the EIA report in Rostock for each protected good.

Protected Good	Conclusion on Expected Impacts
Humans and human health	No significant impacts are expected. However, the noise required a dredging prohibition during night.
Fauna, flora and biological diversity	Impacts are predominantly insignificantly detrimental or neither detrimental nor beneficial. Some impacts are insignificantly beneficial. In some areas, significantly adverse impacts are expected.
Soil	Insignificant adverse impacts are expected
Water	Mostly no adverse nor beneficial effects; some insignificant adverse as well insignificant beneficial impacts are expected. However, several avoidance measures were taken.
Air	No significantly adverse impacts are expected
Climate	No significantly adverse impacts; one insignificantly beneficial impact
Landscape	No significantly adverse impacts
Cultural assets and other material assets	No significantly adverse impacts

Appendix B. The EXCEL—Ecosystem Service Assessment Tool for Coastal Engineering Applications

Personal Information:

Expert´s name	
Institution	
Background / Field of Expertise	

Expert´s self-assessment:	*1 = low, 2 = medium, 3 = high*
Knowledge in ecosystem service assessments	
Knowledge in coastal engineering	
Knowledge in Envir. Impact Assessments	
Local expertize	

oring for Relative Importance (RI)

0	1	2	4	8
Not levant	Low	Moderate	High	Very high

Impact Scoring (=IS)

-3	-2	-1	0	1	2	3
High decrease	Medium decrease	Low decrease	No impact	Low increase	Medium increase	High increase

Ecosystem Services (ES)	Description/ Examples	Relative Importance (RI) of each ES (0,1,2,4 or 8)	Weighting Factor	Scenario after compared to the situation before the waterway deepening Impact Score (IS) (-3 to +3)	Weighted Score (RI*IS)	Comments or questions
1 Use of marine animals	Use of wild and reared (from aquaculture) animals, e.g. fish, mussels as source for human nutrition, for energy conversion or material purposes		0.00	0	0.00	
2 Use of marine plants	Use of wild and cultivated plants as human food source or supplements, as material e.g. as fertilizer in agriculture or reed for thatched roofs or as biomass for energy conversion		0.00	0	0.00	
3 Use of groundwater	Using groundwater for drinking and process water			0		
4 Use of surface water	Using surface water for example as cooling water in the industry or river water for drinking purposes		0.00	0	0.00	
5 Mineral substances used as material	Material that can be used for construction, coastal protection or beach nourishment, e.g. the dredged material		0.00	0	0.00	
6 Renewable energy (abiotic)	Using abiotic characteristics of the ecosystem for Wind energy, solar power or flow and wave energy		0.00	0	0.00	
7 Water as a transportation medium	Use of water e.g. by cargo ships or ferries as a transport route		0.00	0	0.00	
8 Provision of space	Provision of space that can be used, e.g. for building or for the construction of fairways.		0.00	0	0.00	
9 Genetic material of marine plants and animals	Seeds, spores, plants or marine animals (e.g. fish or mussels) that can be used to maintain or establish a new population		0.00	0	0.00	
1 Mediation of hazardous substances and pollutants (except dilution)	Mediation processes like filtration, decomposition, accumulation, sequestration and storage by microorganisms, algae, plants, animals or sediment e.g. filtration by mussels or decomposition of hazardous substances (e.g. oil) by bacteria		0.00	0	0.00	
2 Nutrient regulation	Decomposition processes like denitrification or fixing processes and their effect on sediment quality, e.g. sequestration and storage of phosphor in sediment enhancing remineralization processes (e.g. formation of vivianite)		0.00	0	0.00	
3 Dilution by water exchange	Dilution of pollutants and nutrients by water exchange between the Lagoon and the Baltic Sea		0.00	0	0.00	
4 Mediation of emmisions in the air	Mediation of Emmisions in the air like: polluants e.g.Sulfur dioxide or particulate matter, noise e.g. from cargo ships or smell by plants and trees or wind		0.00	0	0.00	
5 Biotic sediment stabilization including coastal and flood protection	Controlling or preventing costal erosion by decreasing flow and wave energy by biotic factors like reed belts or seagrass meadows		0.00	0	0.00	
6 Abiotic sediment stabilization including coastal and flood protection	Regulation of wave energy and flood protection by hydromorphology e.g. by sandbanks, slowly rising foreshore or natural dunes		0.00	0	0.00	
7 Sediment deposition and erosion	Transport and storage of sediment by currents		0.00	0	0.00	
8 Maintenance of water conditions	Maintenance of chemical composition of seawater column and sediment like oxygen level, temperature and salinity to ensure favorable living conditions for biota or processes like remineralization of phosphorus and denitrification		0.00	0	0.00	
9 Global climate regulation	Ecosystem contribution to carbon sequestration and regulation of greenhouse gas		0.00	0	0.00	
10 Atmospheric composition and conditions	Regulation of air, temperature and humidity, including evapotranspiration		0.00	0	0.00	
11 Disease and pest control	Control of the development and occurrence of pathogens such as blue-green algae or vibrions and spread and dominance of invasive species		0.00	0	0.00	
12 Biodiversity and habitat	Maintaining nursery populations and habitats (incl. Breeding and feeding grounds) for wild plants or animals		0.00	0	0.00	
1 Active recreation and tourism	Use of the ecosystem for recreational and sports activities such as swimming, sailing, fishing, boating or wind- and kitesurfing		0.00	0	0.00	
2 Observational recreation and tourism	Using nature to relax, e.g. with opportunities for landscape viewing such as lookout towers or excursion boating		0.00	0	0.00	
3 Cultural and heritage	Using the ecosystem to understand the local history or culture, such as a historic village, a tradtional fishing ground, or finding historic items that contribute to cultural heritage		0.00	0	0.00	
4 Education and scientific investigation	Using nature for scientific studies or research and for educational purposes like excursions for students or tourists		0.00	0	0.00	
5 Landscape aesthetics	The inherent beauty of nature		0.00	0	0.00	
6 Spiritual, symbolic or bequest meaning	Elements of living system that have symbolic, bequest or spiritual meaning, like red list or iconic species or special landscapes etc.		0.00	0	0.00	
			0.00		**0.00**	

Time spent for the assessment (in minutes)	

References

1. Eurostat. Maritime Transport Statistics—Short Sea Shipping of Goods. Available online: https://ec.europa.eu/eurostat/statistics-explained/index.php?title=Maritime_transport_statistics_-_short_sea_shipping_of_goods (accessed on 19 June 2024).
2. Bodewing, K. European Maritime Space. 2022. Available online: https://transport.ec.europa.eu/system/files/2022-09/2022-mos-dip.pdf (accessed on 19 June 2024).
3. Kerbiriou, R. Modernisation of Container Ship Fleets: State of Play and Consequences for the Baltic Sea. *TransNav* **2024**, *18*, 211–217. [CrossRef]
4. HELCOM. The Baltic Marine Environment 1999–2002. Baltic Sea Environment Proceedings 87, 2003. Available online: https://helcom.fi/wp-content/uploads/2019/10/BSEP87.pdf (accessed on 19 June 2024).
5. HELCOM. Economic and Social Analyses in the Baltic Sea Region. 2017. Available online: https://stateofthebalticsea.helcom.fi/wp-content/uploads/2017/07/HELCOM_Economic_and_social_analyses_Supplementary_report_first_version_2017.pdf (accessed on 19 June 2024).
6. European Commission. D6 Sea-Floor Integrity. Available online: https://mcc.jrc.ec.europa.eu/main/dev.py?N=24&O=135&titre_chap=D6%20Sea-floor%20integrity (accessed on 19 June 2024).
7. European Commission. Environmental Impact Assessment. Available online: https://environment.ec.europa.eu/law-and-governance/environmental-assessments_en (accessed on 6 October 2024).
8. World Resources Institute. *Millennium Ecosystem Assessment. Ecosystems and Human Well-being: Synthesis*; Island Press: Washington, DC, USA, 2005; Available online: https://www.millenniumassessment.org/documents/document.356.aspx.pdf (accessed on 18. November 2023).
9. Bouwma, I.; Schleyer, C.; Primmer, E.; Winkler, K.J.; Berry, P.; Young, J.; Preda, E. Adoption of the ecosystem services concept in EU policies. *Ecosyst. Serv.* **2018**, *29*, 213–222. [CrossRef]
10. Karjalainen, T.P.; Marttunen, M.; Sarkki, S.; Rytkönen, A.-M. Integrating ecosystem services into environmental impact assessment: An analytic–deliberative approach. *Environ. Impact Assess. Rev.* **2013**, *40*, 54–64. [CrossRef]
11. Baker, J.; Sheate, W.R.; Phillips, P.; Eales, R. Ecosystem services in environmental assessment—Help or hindrance? *Environ. Impact Assess. Rev.* **2013**, *40*, 3–13. [CrossRef]
12. Koellner, T.; de Baan, L.; Beck, T.; Brandão, M.; Civit, B.; Margni, M.; i Canals, L.M.; Saad, R.; de Souza, D.M.; Müller-Wenk, R. UNEP-SETAC guideline on global land use impact assessment on biodiversity and ecosystem services in LCA. *Int. J. Life Cycle Assess* **2013**, *18*, 1188–1202. [CrossRef]
13. Partidario, M.R.; Gomes, R.C. Ecosystem services inclusive strategic environmental assessment. *Environ. Impact Assess. Rev.* **2013**, *40*, 36–46. [CrossRef]
14. Sousa, P.; Gomes, D.; Formigo, N. Ecosystem services in environmental impact assessment. *Energy Rep.* **2020**, *6*, 466–471. [CrossRef]
15. Hasan, S.S.; Zhen, L.; Miah, M.G.; Ahamed, T.; Samie, A. Impact of land use change on ecosystem services: A review. *Environ. Dev.* **2020**, *34*, 100527. [CrossRef]
16. Liquete, C.; Piroddi, C.; Drakou, E.G.; Gurney, L.; Katsanevakis, S.; Charef, A.; Egoh, B. Current Status and Future Prospects for the Assessment of Marine and Coastal Ecosystem Services: A Systematic Review. *PLoS ONE* **2013**, *8*, e67737. [CrossRef]
17. Müller, F.; Bicking, S.; Ahrendt, K.; Bac, D.K.; Blindow, I.; Fürst, C.; Haase, P.; Kruse, M.; Kruse, T.; Ma, L.; et al. Assessing ecosystem service potentials to evaluate terrestrial, coastal and marine ecosystem types in Northern Germany—An expert-based matrix approach. *Ecol. Indic.* **2020**, *112*, 106116. [CrossRef]
18. Marcos, C.; Díaz, D.; Fietz, K.; Forcada, A.; Ford, A.; García-Charton, J.A.; Goñi, R.; Lenfant, P.; Mallol, S.; Mouillot, D.; et al. Reviewing the Ecosystem Services, Societal Goods, and Benefits of Marine Protected Areas. *Front. Mar. Sci.* **2021**, *8*, 613819. [CrossRef]
19. Inácio, M.; Schernewski, G.; Pliatsika, D.A.; Benz, J.; Friedland, R. Assessing changes in ecosystem services provision in coastal waters. *Sustainability* **2019**, *11*, 2632. [CrossRef]
20. Schernewski, G.; Paysen, P.; Robbe, E.; Inacio, M.; Schumacher, J. Ecosystem services and the EU Water Framework Directive: Comparative assessments in urban and rural estuarine water bodies. *Front. Mar. Sci.* **2019**, *6*, 183. [CrossRef]
21. Schernewski, G.; Voeckler, L.N.; Lambrecht, L.; Robbe, E.; Schumacher, J. Building with Nature—Ecosystem Service Assessment of Coastal-Protection Scenarios. *Sustainability* **2022**, *14*, 15737. [CrossRef]
22. Schernewski, G.; Konrad, A.; Roskothen, J.; von Thenen, M. Coastal Adaptation to Climate Change and Sea Level Rise: Ecosystem Service Assessments in Spatial and Sectoral Planning. *Appl. Sci.* **2023**, *13*, 2623. [CrossRef]
23. European Commission. Environmental Impact Assessment of Projects, 2017. Available online: https://infrastructure.planninginspectorate.gov.uk/wp-content/ipc/uploads/projects/TR010044/TR010044-001667-TAN%20-%20EIA_guidance_EIA_report_final.pdf (accessed on 6 October 2024).
24. Schernewski, G.; Robbe, E. Ecosystem Service Assessment in European Coastal and Marine Policies. In *Southern Baltic Coastal Systems Analysis*; Schubert, H., Müller, F., Eds.; Ecological Studies; Springer: Cham, Switzerland, 2023; Volume 246, pp. 347–366. [CrossRef]
25. TÜV Nord. Anpassung der Seewärtigen Zufahrt Zum Seehafen Rostock; Umweltverträglichkeitsuntersuchung. 2019. Available online: https://www.uvp-portal.de/sites/default/files/2021-11/Planfeststellungsbeschluss%20vom%2019.%20Mai%202021.pdf (accessed on 6 October 2024).
26. WSV. Anpassung der Seewärtigen Zufahrt Zum Seehafen Rostock. Available online: https://www.kuestendaten.de/OSK/DE/Projekte/Ausbau_Seekanal_Rostock/Ausbau_Seekanal_Rostock_node.html (accessed on 19 June 2024).

27. Hansestadt Rostock. Seehafenzufahrt Rostock: Bauarbeiten Zur Anpassung Gestartet. Available online: https://rathaus.rostock.de/de/rathaus/aktuelles_medien/seehafenzufahrt_rostock_bauarbeiten_zur_anpassung_gestartet/331922 (accessed on 19 June 2024).
28. Port Monitor. Polish Ports in 2022—Summary and Future Outlook. Available online: https://actiaforum.pl/wp-content/uploads/2023/05/Polish-ports-in-2022.pdf (accessed on 1 October 2024).
29. CMW LEGAL. Larger Ships to Berth in Szczecin-Świnoujście Port. The Deepening of the Waterway Has Finished. Available online: https://cmwlegal.pl/en/larger-ships-to-berth-in-szczecin-swinoujscie-port-the-deepening-of-the-waterway-has-finished/ (accessed on 19 June 2024).
30. SWS Kancelaria Prawna. The Modernization of Świnoujście-Szczecin Waterway Completed—Legal Services of SWS Law Firm. Available online: https://sws.com.pl/en/the-modernisation-of-swinoujscie-szczecin-waterway-completed-legal-services-of-sws-law-firm/ (accessed on 19 June 2024).
31. European Commission. Increased Capacity for the Świnoujście-Szczecin Waterway to Boost Growth in Western Pomerania. Available online: https://ec.europa.eu/regional_policy/projects/projects-database/increased-capacity-for-the-swinoujscie-szczecin-waterway-to-boost-growth-in-western-pomerania_en (accessed on 19 June 2024).
32. European Commission, Directorate-General for Environment. Guidelines on the Implementation of the Birds and Habitats Directives in Estuaries and Coastal Zones—With Particular Attention to Port Development and Dredging. Publications Office, 2012. Available online: https://data.europa.eu/doi/10.2779/44024 (accessed on 6 October 2024).
33. European Commission. EU Strategy for the Baltic Sea Region (EUSBSR). Available online: https://ec.europa.eu/regional_policy/sources/policy/cooperation/macro-regional-strategies/baltic/factsheet_eusbr_en.pdf (accessed on 6 October 2024).
34. European Commission. Environmental Impact Assessment. Available online: https://environment.ec.europa.eu/law-and-governance/environmental-assessments/environmental-impact-assessment_en#overview (accessed on 19 August 2024).
35. Umweltprüfungsportal des Bundes. Wie läuft Eine Umweltverträglichkeitsprüfung ab? Available online: https://www.uvp-portal.de/de/node/250 (accessed on 19 August 2024).
36. EUR-Lex. Convention on Environmental Impact Assessment in a Transboundary Context (Espoo Convention). Available online: https://eur-lex.europa.eu/EN/legal-content/summary/convention-on-environmental-impact-assessment-in-a-transboundary-context-espoo-convention-and-the-protocol-on-strategic-environmental-assessment-sea-protocol.html (accessed on 19 August 2024).
37. MV. UVP-Dokumentation Für den Geplanten Bau Eines Containerterminals im Außenhafen in Swinemünde. Available online: https://www.regierung-mv.de/Landesregierung/wm/Infrastruktur/Aktuelle-Planfeststellungsverfahren/Bau-eines-Containerterminals-im-Aussenhafen-Swinemuende/?racr=a (accessed on 19 August 2024).
38. BMVBS. Leitfaden Zur Umweltverträglichkeitsprüfung An Bundeswasserstraßen. 2022. Available online: https://izw.baw.de/publikationen/umwelt-handbuch/0/UVP-Leitfaden%202022.pdf (accessed on 19 August 2024).
39. European Commission. Assessment of Plans and Projects Significantly Affecting Natura 2000 Sites. Available online: https://op.europa.eu/en/publication-detail/-/publication/2b6c4b16-e867-42da-b604-f67ee6fe60c3 (accessed on 19 August 2024).
40. GDA SK S—SZCZECIN. Modernization Swinoujscie—Szczecin to a Depth of 12.5 m. Volume II 2015. Available online: https://altadiusdutchstatebusiness.nl (accessed on 19 August 2024).
41. Haines-Young, R.; Potschin, M. Common International Classification of Ecosystem Services (CICES) V5.1 and Guidance on the Application of the Revised Structure, 2018. Available online: www.cices.eu (accessed on 25 May 2024).
42. Inácio, M.; Schernewski, G.; Nazemtseva, Y.; Baltranaite, E.; Friedland, R.; Benz, J. Ecosystem services provision today and in the past: A comparative study in two Baltic Lagoons. *Ecol. Res.* **2018**, *33*, 1255–1274. [CrossRef]
43. Steffen, S. Ecosystem Service Analyses for the Deepening of the Navigation Channel in the Oder Lagoon. Masterarbeit Thesis, Studiengang Umweltingenieurwissenschaften an der Universität Rostock, Rostock, Germany, 23 May 2023, *unpublished*.
44. Jekat, M.-S. Ecosystem Service Analysis of Waterway Channel Deepening: The Example of the Warnow. Bachelorarbeit Thesis, Studiengang Umweltingenieurwissenschaften an der Universität Rostock, Rostock, Germany, 16 June 2023, *unpublished*.
45. European Commission. *Guidance on Inland waterway transport and Natura 2000*; Publications Office of the European Union: Luxembourg, 2018. [CrossRef]
46. Bundesanstalt Für Gewässerkunde. Fachliche Bewertung vorhabenbedingter Auswirkungen bei Umweltverträglichkeitsprüfungen an Bundeswasserstraßen, BfG-Bericht 2072, Koblenz, 2022; 140p. Available online: https://izw.baw.de/publikationen/umwelt-handbuch/0/Anl4_Bewertungsverfahren.pdf (accessed on 19 August 2024).
47. Neumann, T.; Radtke, H.; Cahill, B.; Schmidt, M.; Rehder, G. Non-Redfieldian carbon model for the Baltic Sea (ERGOM version 1.2)—Implementation and budget estimates. *Geosci. Model Dev.* **2022**, *15*, 8473–8540. [CrossRef]
48. Griffies, S.M. *Fundamentals of Ocean Climate Models*; Princeton University Press: Princeton, NJ, USA, 2004; ISBN 9780691118925.
49. Schenk, F.; Zorita, E. Reconstruction of high resolution atmospheric fields for Northern Europe using analog-upscaling. *Clim. Past Discuss.* **2012**, *8*, 819–868. [CrossRef]
50. Gustafsson, B.G.; Schenk, F.; Blenckner, T.; Eilola, K.; Meier, H.E.M.; Müller-Karulis, B.; Neumann, T.; Ruoho-Airola, T.; Savchuk, O.P.; Zorita, E. Reconstructing the Development of Baltic Sea Eutrophication 1850–2006. *AMBIO* **2012**, *41*, 534–548. [CrossRef]
51. Geyer, B.; Rockel, B. coastDat-2 COSMOCLM Atmospheric Reconstruction [Data Set], 2013. Available online: https://www.wdc-climate.de/ui/entry?acronym=coastDat-2_COSMO-CLM (accessed on 9 May 2024).

52. Sonderforschungsgruppe Insitutionenanalyse. Evaluation des UVPG des Bundes. 2008. Available online: https://www.bmuv.de/fileadmin/Daten_BMU/Download_PDF/Umweltpruefungen/uvp_evaluation_abschlussbericht.pdf (accessed on 19 August 2024).
53. Morgan, R.K. Environmental impact assessment: The state of the art. Impact Assess. *Proj. Apprais.* **2012**, *30*, 5–14. [CrossRef]
54. Nita, A.; Fineran, S.; Rozylowicz, L. Researchers' perspective on the main strengths and weaknesses of Environmental Impact Assessment (EIA) procedures. *Environ. Impact Assess. Rev.* **2022**, *92*, 106690. [CrossRef]
55. Jay, S.; Jones, C.; Slinn, P.; Wood, C. Environmental impact assessment: Retrospect and prospect. Environ. *Impact Assess. Rev.* **2007**, *27*, 287–300. [CrossRef]
56. Sachverständigenrats Für Umweltfragen" Die Zukunft der Europäischen Umweltpolitik. 2020. Available online: https://www.umweltrat.de/SharedDocs/Downloads/DE/01_Umweltgutachten/2016_2020/2020_Umweltgutachten_Kap_08_Zukunft_Umweltpolitik.html (accessed on 25 August 2024).
57. Oosterhuis, F. Costs and Benefits of the EIA Directive. 2007. Available online: https://cuadernosdederechoparaingenieros.com/wp-content/uploads/06_Infcoste-beneficioEIA.pdf (accessed on 25 August 2024).
58. Umweltbundesamt. Genehmigungsverfahren 2014–2023. Available online: https://www.umweltbundesamt.at/uvpsup/verfahrensmonitoring/vm-dauer/gv-dauer (accessed on 25 August 2024).
59. Wasserstraßen- und Schifffahrtsverwaltung des Bundes. Available online: https://www.gdws.wsv.bund.de/DE/wasserstrassen/planfeststellung/Verfahren_MgvG.html (accessed on 25 August 2024).
60. Minning, M. Der Schifffahrtskanal im Oderhaff. Eine Sediment-, NÃ¤hr- und Schadstofffalle?.Diplomarbeit. Christian-Albrechts-Universität, Kiel, 2003. Available online: https://eucc-d-inline.databases.eucc-d.de/files/documents/00000695_Diplomarbeit_Minning2.pdf (accessed on 19 August 2024).
61. Friedland, R.; Schernewski, G.; Gräwe, U.; Greipsland, I.; Palazzo, D.; Pastuszak, M. Managing Eutrophication in the Szczecin (Oder) Lagoon-Development, Present State and Future Perspectives. *Front. Mar. Sci.* **2019**. [CrossRef]
62. Schernewski, G.; Neumann, T.; Bučas, M.; von Thenen, M. Ecosystem services of the Baltic Sea—State and changes during the last 150 years. *Environments* **2024**, *11*, 200. [CrossRef]
63. Ruckelshaus, M.; McKenzie, E.; Tallis, H.; Guerry, A.; Daily, G.; Kareiva, P.; Polasky, S.; Ricketts, T.; Bhagabati, N.; Wood, S.A.; et al. Notes from the field: Lessons learned from using ecosystem service approaches to inform real-world decisions. *Ecol. Econ.* **2013**, *115*, 11–21. [CrossRef]
64. Geneletti, D. Reasons and options for integrating ecosystem services in strategic environmental assessment of spatial planning. *Int. J. Biodivers. Sci.* **2011**, *7*, 143–149. [CrossRef]
65. Geneletti, D. Ecosystem services in environmental impact assessment and strategic environmental assessment. *Environ. Impact Assess. Rev.* **2013**, *40*, 1–2. [CrossRef]
66. IADC. Available online: https://www.iadc-dredging.com/wp-content/uploads/2021/09/terra-et-aqua-164-4.pdf (accessed on 25 August 2024).
67. Rosa, J.C.S.; Sánchez, L.E. Is the ecosystem service concept improving impact assessment? Evidence from recent international practice. *Environ. Impact Assess. Rev.* **2015**, *50*, 134–142. [CrossRef]
68. Landsberg, F.; Treweek, J.; Stickler, M.M.; Henninger, N.; Venn, O. Weaving Ecosystem Services into Impact Assessment—A Step-By-Step Method 46. 2013. Available online: http://pdf.wri.org/weaving_ecosystem_services_into_impact_assessment.pdf (accessed on 25 August 2024).
69. Rosa, J.C.S.; Sánchez, L.E. Advances and challenges of incorporating ecosystem services into impact assessment. *J. Environ. Manag.* **2016**, *180*, 485–492. [CrossRef]
70. von Thenen, M.; Effelsberg, N.; Weber, L.; Schernewski, G. Perspectives and Scenarios for Coastal Fisheries in a Social-Ecological Context: An Ecosystem Service Assessment Approach in the German Baltic Sea. *Sustainability* **2023**, *15*, 15732. [CrossRef]
71. Robbe, E.; Abdallah, L.B.; El Fels, L.; Chaher, N.E.H.; Haseler, M.; Mhiri, F.; Schernewski, G. Towards Solving the Beach Litter Problem: Ecosystem Service Assessments at North African Coasts. *Sustainability* **2024**, *16*, 5911. [CrossRef]
72. Ritzenhofen, L.; Schumacher, J.; Karstens, S.; Schernewski, G. Ecosystem Service Assessments within the EU Water Framework Directive: Marine Mussel Cultivation as a Controversial Measure. *Appl. Sci.* **2022**, *12*, 1871. [CrossRef]
73. Karstens, S.; Inácio, M.; Schernewski, G. Expert-Based Evaluation of Ecosystem Service Provision in Coastal Reed Wetlands Under Different Management Regimes. *Front. Environ. Sci.* **2019**, *7*, 69. [CrossRef]
74. Gallardo, A.L.C.F.; Rosa, J.C.S.; Sánchez, L.E. Addressing ecosystem services from plan to project to further tiering in impact assessment: Lessons from highway planning in São Paulo, Brazil. Environ. *Impact Assess. Rev.* **2022**, *92*, 106694. [CrossRef]
75. Hugé, J.; Rochette, A.-J.; Janssens de Bisthoven, L.; Dahdouh-Guebas, F.; Koedam, N.; Vanhove, M.P.M. Utilitarian framings of biodiversity shape environmental impact assessment in development cooperation. *Environ. Sci. Policy* **2017**, *75*, 91–102. [CrossRef]
76. Bull, J.; Jobstvogt, N.; Böhnke-Henrichs, A.; Mascarenhas, A.; Sitas, N.; Baulcomb, C.; Lambini, C.; Rawlins, M.; Baral, H.; Zähringer, J.; et al. Strengths, weaknesses, opportunities and threats: A SWOT analysis of the ecosystem services framework. *Ecosyst. Serv.* **2016**, *17*, 99–111. [CrossRef]

land

MDPI

Article

Spatiotemporal Evolution and Influencing Factors of Coupling and Coordination between the Ecosystem Service Value and Economy in the Pearl River Delta Urban Agglomeration of China

Zeduo Zou, Xiaodie Yuan, Zhuo Zhang, Xingyan Li and Chunshan Zhou *

School of Geography and Planning, Sun Yat-Sen University, Guangzhou 510006, China;
zouzd@mail2.sysu.edu.cn (Z.Z.); yuanxd8@mail2.sysu.edu.cn (X.Y.); zhangzh567@mail2.sysu.edu.cn (Z.Z.);
lixingy6@mail2.sysu.edu.cn (X.L.)
* Correspondence: zhoucs@mail.sysu.edu.cn

Abstract: In the context of pursuing high-quality development, the coupling and coordination of the ecosystem and economy has become the fundamental goal and inevitable choice for achieving the sustainable development of urban agglomerations. Based on remote sensing and statistical data for the Pearl River Delta (PRD) region from 2005 to 2020, in this paper, we construct an index system of the ecological and economic levels to assess the ecosystem service value (ESV). We use the equivalent factor method, entropy method, coupling coordination model, and relative development model to systematically grasp the spatial pattern of the levels of the two variables, analyse and evaluate their spatial and temporal coupling and coordination characteristics, and test the factors influencing their coupling and coordination using the geographical and temporal weighted regression (GTWR) model. The results show that ① the ESV in the PRD exhibited a fluctuating decreasing trend, while the level of the economy exhibited a fluctuating increasing trend; ② the coordination degree of the ESV and economy in the PRD exhibited a fluctuating increasing trend, and the region began to enter the basic coordination period in 2007; ③ in terms of the spatial distribution of the coordination degree, there was generally a circular pattern, with the Pearl River Estuary cities as the core and a decrease in the value towards the periphery; ④ the coordinated development model is divided into balanced development, economic guidance, and ESV guidance, among which balanced development is the major type; ⑤ the results of the GTWR reveal that the influencing factors exhibited significant spatial–temporal heterogeneity. Government intervention and openness were the dominant factors affecting the coordination, and the normalised difference vegetation index was the main negative influencing factor.

Keywords: ESV; sustainable development; coupling coordination; GTWR; PRD

1. Introduction

Sustainable development is the fundamental path through which China could achieve high-quality development. Since the reform and opening-up, China has made outstanding achievements in urbanisation and the economy. Its urbanisation rate has increased from 17.9% in 1978 to 63.89 in 2020 and its gross domestic product (GDP) has increased from 367.870 billion RMB in 1978 to 101.36 trillion RMB in 2020 [1]. However, the rapid increases in these two values have been accompanied by a series of ecological and environmental problems [2], such as haze, water pollution, and a shortage of land resources. One of the major manifestations of the urbanisation process is the occupation of a massive number of natural landscapes, especially large areas of high-quality land, resulting in the degradation of ecosystems, which in turn impairs the supply of ecosystem services [3]. Because the United Nations actively advocates sustainable development [4], it is necessary to explore the characteristics of the coordination between the regional economy and the ecosystem and to clarify their coordination characteristics so as to promote the realisation of an ecological civilisation and achieve high-quality development in China.

Scholars have shown great interest in the coordination of ecology and socio-economic development. Ecosystem services provide material information about, and intangible well-being for, human life, reflecting the benefits that people obtain from the natural environment [5,6]. They are life-supporting products and services that act directly or indirectly through the ecosystem's structures, operation processes, and functions, and they link the economic activities of human society and the ecological processes of natural resources [7]. However, the growing desire for economic profit depletes ecosystem services, leading to the deterioration of the ecological environment and endangering the sustainability of development [8,9]. Moreover, considering that urban agglomerations are significant engines for the rapid growth of the regional socio-economy, whether they can achieve sustainable development and promote the high-quality development of a country has also attracted the attention of the academic community. Currently, there is no recognised way to evaluate the sustainable or high-quality development of urban agglomerations. In the early stages of such research, it was pointed out that sustainable development of urban agglomerations was a system that included the economy, society, and ecology [10]. For this reason, some studies have discussed the interaction and coupling between urbanisation and the ecological environment, and the essentialness of the socio-economy and environment in sustainable development has also been emphasised in studies of megalopolises [11,12]. Some scholars have proposed and utilised relevant indicators to evaluate the development quality of urban agglomerations [13]. Previous research has mostly focused on the coupling and coordination of the economy and ecology in urban agglomerations, as well as the implementation of a large number of empirical analyses [14–17].

Coupling and coordination has become a popular topic in global change research. As a key way to reflect the level of coordination between the regional ecological environment and social and economic development, it quantifies the coordination and its trends and can be used to analyse the impact of human activities on an ecosystem. Thus, policies conducive to regional sustainable development can be formulated accordingly [18,19]. Related global studies have mainly explained the theory and policy of environmental Kuznets from different perspectives [20] or have analysed the relationship between the regional ecological environment and the social–economy from the pressure–state–response perspective (PSR) using the PSR model [21]. Some scholars have used an input–output model [22] and a CGE model [23] to explore the relationship between environmental protection and socio-economic development, providing a theoretical foundation for studying the relationship between the ecological environment and socio-economic changes. With the breakthroughs in quantitative research on the ecosystem service value (ESV) made by Costanza et al. [24] and other researchers, studies on the ESV's association with the social–economy have begun to be favoured. China's relevant achievements have mainly relied on index systems of the two dimensions. As research on the human–environment relationship is deepening [25], research on the coupling relationship between the social–economy and ecosystem has gradually matured. Theories of a human–environment coupling circular system [26] and a social–economic–natural complex ecosystem [27] have been proposed, which are of great importance when studying the coupling relationship between the ESV and the social–economy. Since Xie et al. [28,29] promoted quantitative studies of the ESV, the interactions between socio-economic development and the ESV have become a popular topic in academia in China. In terms of the commonly used techniques, the most important and widely recognised method is the coupling coordination model of physics. In terms of the content of such studies, many scholars only analysed the evolution process of the coupling coordination degree model in different time periods, while in-depth explorations of the influencing factors and driving mechanisms have rarely been reported. In terms of the research objects, the majority of existing studies were conducted at the province and prefecture scales, and little research has been conducted on combinations of urban agglomerations or comprehensive planning areas [30]. With the implementation of China's regional synergy strategy in the new era (starting in 2012), the scientific problem of strengthening the coupling and coordination mechanism between the ESV and social–economy in typical urban agglomerations has been promoted to a high position, providing a theoretical supplement and practical guidance [31].

Rapid urbanisation has caused dramatic changes in land use in the PRD region. The areas of cropland and water bodies have decreased significantly, while the area of construction land has rapidly increased. The stress on the ecological environment has increased correspondingly, making sustainable development a serious challenge. In view of this, in this study, we took the PRD urban agglomeration as the research object and used the coupling coordination degree model to measure its overall trend during 2005–2020, as well as the evolution characteristics of the different cities. The influencing factors and characteristics of the coupling coordination degree in response to the changes in and mechanism of the coupling coordination relationship between the ESV and social–economy were analysed using the spatiotemporal, geographically weighted regression model in order to provide a scientific basis for the coordinated and sustainable development of urban agglomeration in the PRD.

The purposes of this study were (a) to detect the change characteristics of the land use and ESV in the PRD region from 2005 to 2020; (b) to measure and analyse the spatiotemporal evolution characteristics of the economic level in the PRD region; (c) to explore the spatiotemporal coupling characteristics of the ESV and high-quality economic development level; and (d) to use the spatiotemporal, geographically weighted regression model to explore the factors influencing the coupling coordination degree of the ESV and economy. The main highlight of the paper is that it evaluates the influencing factors of CCD with the GTWR model. The advantage of this model is that it can directly quantify the spatial and temporal relationship between independent and dependent variables, strengthening the detailed and scientific results by observing the spatiotemporal non-stationarity of influencing factors. Additionally, the negative effects of construction land must be considered when evaluating the ecological environment. Therefore, referring to existing research, the introduced equivalent factor of construction land is negative, so that the ESV of the PRD region and its cities can be estimated more comprehensively and concretely. Last but not least, this study also constructs a comprehensive index system of economic development, differing from the traditional single-dimensional economic growth evaluation. The results not only provide policy management suggestions for the synergic development of the ecological environment and social–economy in the PRD region but also provide a reference case study for other regions for the promotion of sustainable development.

2. Literature Review

Scholars have shown great interest in the coordination of the ecology and socio-economic development. Ecosystem services provide material information about resources and intangibly improve human welfare, reflecting the benefits that people obtain from the natural environment [5,6]. They are life-supporting products and services that act directly or indirectly through the ecosystem's structures, operation processes, and functions, and they link the economic activities of human society and the ecological processes of natural resources [7]. However, the growing desire for economic profit will deplete ecosystem services, leading to the deterioration of the ecological environment and endangering the sustainability of development [8,9]. Moreover, considering that urban agglomerations are significant engines for the rapid growth of a region's socio-economy, whether they can achieve sustainable development and promote the high-quality development of a country has also attracted the attention of the academic community. Currently, there is no recognised way to evaluate the sustainable or high-quality development of urban agglomerations. In the early stages of such research, it was pointed out that the sustainable development of urban agglomerations was a system that included the economy, society, and ecology [10]. For this reason, some studies have discussed the interaction and coupling between urbanisation and the ecological environment, and the essentialness of the socio-economy and environment in sustainable development in studies of megalopolises has also been emphasised [11,12]. Some scholars proposed and utilised relevant indicators to evaluate the quality of the development of urban agglomerations [13]. Previous research has mostly focused on the coupling and coordination of the economy and ecology in urban agglomerations, as well as the implementation of a large number of empirical analyses [14–17].

Coupling and coordination has become a popular topic in global change research. As a key way to reflect the level of coordination between the regional ecological environment and social and economic development, it quantifies the trends in coordination and can be used to analyse the impact of human activities on an ecosystem. Thus, policies conducive to regional sustainable development can be formulated accordingly [18,19]. Related global studies have mainly explained the theory and policy of environmental Kuznets from different perspectives [20] or have analysed the relationship between the regional ecological environment and social–economy from the pressure–state–response perspective (PSR) using the PSR model [21]. Some scholars have used an input–output model [22] and a CGE model [23] to explore the relationship between environmental protection and socio-economic development, providing a theoretical foundation for studying the relationship between the ecological environment and socio-economic changes. With the breakthroughs in quantitative research on the ecosystem service value (ESV) made by Costanza et al. [24] and other researchers, studies on the ESV's association with the social–economy have begun to be favoured. China's relevant achievements have mainly relied on index systems of the two dimensions. As research on the human–environment relationship is deepening [25], research on the coupling relationship between the social–economy and ecosystem has gradually matured. Theories of a human–environment coupling circular system [26] and a social–economic–natural complex ecosystem [27] have been proposed, which are of great importance for studying the coupling relationship between the ESV and the social–economy. Since Xie et al. [28,29] promoted quantitative studies of the ESV, the interactions between socio-economic development and the ESV have become a popular topic in academia in China. In terms of the commonly used techniques, the most important and widely recognised method is the coupling coordination model of physics. In terms of the content, many scholars only analysed the evolution of the coupling coordination degree model in different time periods, while in-depth explorations of the influencing factors and driving mechanisms have rarely been reported. In terms of the research objects, the majority of existing studies were conducted at the province and prefecture scales, and little research has been conducted on combinations of urban agglomerations or comprehensive planning areas [30]. With the implementation of China's regional synergy strategy in the new era (starting in 2012), the scientific problem of strengthening the coupling and coordination mechanism between the ESV and the social–economy in typical urban agglomerations has attracted increasing interest, providing a theoretical supplement and practical guidance [31].

The main highlight of this paper is that it evaluates the influencing factors of CCD using the GTWR model. The advantage of this model is that it can directly quantify the spatial and temporal relationship between independent and dependent variables, strengthening the detailed and scientific results by observing the spatiotemporal non-stationarity of influencing factors. Additionally, the negative effects of construction land must be considered while evaluating the ecological environment. Therefore, referring to existing research, the introduced equivalent factor of construction land is negative, so that the ESV of the PRD region and its cities can be estimated more comprehensively and concretely. Last but not least, this study also constructs a comprehensive index system of economic development, differing from the traditional single-dimensional economic growth evaluation.

3. Materials and Methods

3.1. Study Area

The urban agglomeration in the PRD is located in the central–southern part of Guangdong Province, China (111°20′–115°25′ E, 21°27′–23°57′ N) (Figure 1a). The total land area is 5.47 million hectares, of which hills, mountains, and islands account for 30%. The annual average temperature is 21.4–22.4 °C, and the annual average precipitation is 1600–2300 mm. The ecosystem structure exhibits various characteristics, and forests and water bodies are the main components (Figure 1b). The PRD's economy has been advancing rapidly. From 2000 to 2020, the GDP increased from 84.13 billion to 895.24 billion RMB, the per capita GDP increased from 20,369 to 115,459 RMB, and the urbanisation ratio increased from 69.50% to 87.24%. It can be seen that the PRD has a solid spatial ecosystem pattern that supports rapid economic development. However, rapid urbanisation and economic development have brought great challenges to the ecology of PRD's urban agglomeration, and the synergic evolution of economy and ecology has become a significant issue in the region's future development. The coordinated development of the environment and the economy is an important issue in the development of contemporary society, and is also the focus of attention in the PRD region and even globally. The Pearl River Delta region, at the frontier of China's reform and opening-up and an important engine of economic development, is representative of a typical metropolitan area in terms of the evolution of land development intensity and ecological environment relationship. The interactive relationship between environmental conditions and economic activities has always been a hot research topic. While the economic is improving by leaps and bounds, the region is also facing increasingly severe environmental challenges. In the future, the PRD region needs to focus more on innovation-driven development, transformation and upgrading, and ecological environmental protection, while maintaining sustained and healthy economic growth, in order to achieve a coordinated development between the economy and the environment. Our research results provide a reference and support for the development of urban agglomerations across the entire country.

3.2. Data Sources

The land-use data for 2005–2020 used in this study were obtained from the annual China Land Cover Dataset (CLCD, https://zenodo.org/records/8176941 [accessed on 2 October 2023]), with an overall accuracy of 80%. The ENVI 5.1 software was used to preprocess the remote sensing images for each year. According to China's Land Use Status Classification standard (GB/T21010-2017), the visual interpretation was combined with actual field investigation and the utilisation of Google Earth. The maximum likelihood method was used to supervise and classify the research area, and the resampling was classified into six categories: cropland, forests, grassland, water bodies, construction land, and unused land. The digital elevation model (DEM) data were provided by the European Aviation Authority (EAA, https://panda.copernicus.eu/panda [accessed on 4 October 2023]), with a spatial resolution of 30 m × 30 m. The annual average temperature, average precipitation, and vegetation coverage data were obtained from the National Science and Technology Resource Sharing Service Platform-National Earth System Science Data Center (http://www.geodata.cn). The urban expansion was measured using the light index, and the data acquisition was conducted using a method described in a previous study [32]. The Defence Meteorological Program (DMSP) Operational Line-Scan System (DMSP-OLS) data for the required time series were obtained. The socio-economic and food production data for the urban agglomeration were obtained from the *China City Statistical Yearbook* [33], the *Guangdong Statistical Yearbook* [34], and statistical bulletins for the nine prefectures involved. Limited by the availability of the data, the missing data were replaced using the linear difference method. The data that could not be obtained directly were obtained via conversion of the relevant original data.

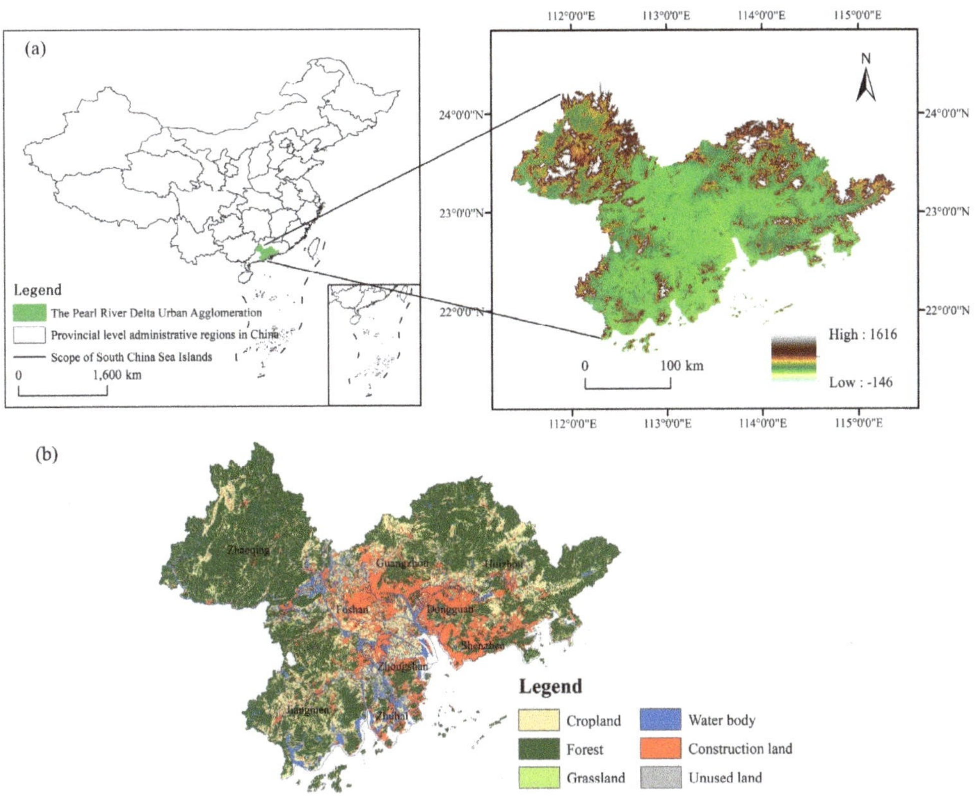

Figure 1. Location of the study region. (**a**) Location in China, and (**b**) the spatial distribution of different land cover types in PRD.

3.3. Research Methods

The socio-economic system and the natural ecosystem are closely related to the survival and development of human beings. There are nonlinear and open dynamic changes between them, causing them to interact and organically unify. The adjustment of the system's structure and optimisation of its function can be realised based on coupling coordination research. The coupling coordination relationship between the two systems is summarised in Figure 2. Ecosystem services include the supply services, regulation services, support services, and cultural services utilised by human beings and constitute the material basis for socio-economic development. Socio-economic development is the driver of the use of ecosystem services, and any increase in this development strengthens the degree of access to the ecosystem. If the demands of the social–economy exceed the limits of development, it will inevitably be restricted by the ecosystem. When socio-economic development reaches a certain level, human beings can promote regional synergic and high-quality development by formulating policies for the sustainable development of the local social–economy and ecosystem.

3.3.1. Method for Evaluation of the Comprehensive Economic Development Level

In this study, we constructed an evaluation index system for the economic development in the PRD's urban agglomeration (Table 1). The index system measures three aspects of the comprehensive level of economic development: development level, industry structure, and dynamics [35–37]. The specific calculation steps of each parameter were described in a previous study [35]. The detailed steps are as follows:

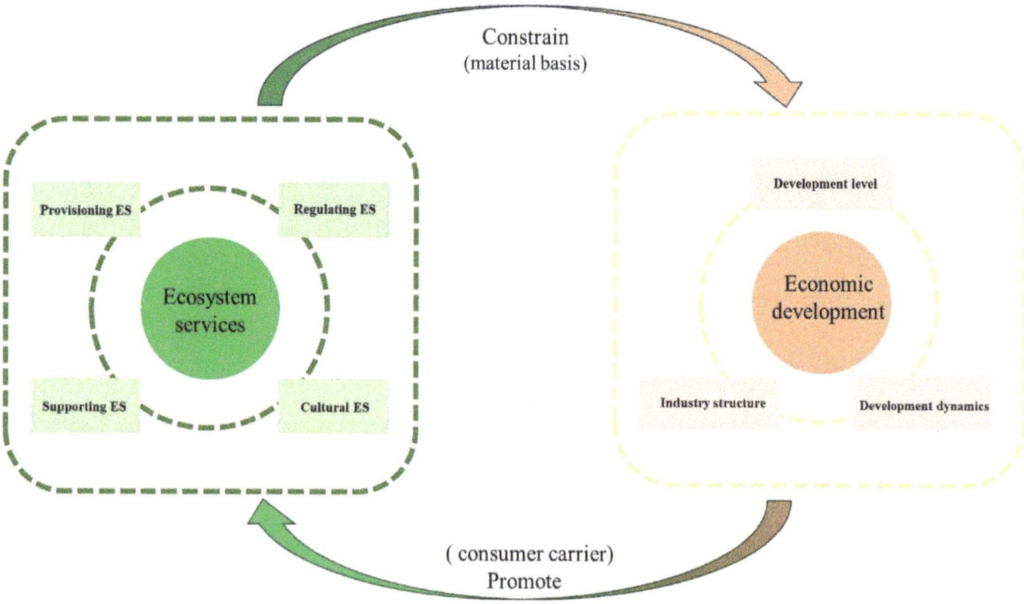

Figure 2. Coupling coordination relationship between ecosystem services and economic development.

Table 1. Economic development evaluation index system for the PRD's urban agglomeration.

Target Layer	Criterion Layer	Weight	Indicator Layer	Weight
Economic Development	Economic growth	0.2778	Per capita GDP	0.0952
			Per capita disposable income	0.0953
	Industrial structure	0.2636	Proportion of secondary industry	0.0491
			Proportion of tertiary industry	0.1316
	Dynamics	0.4586	Rate of increase in GDP	0.0727
			Per capita budgetary income	0.1594
			Per capita retail sales of social consumer goods	0.1049
			Per capita passenger volume	0.2919

① The index data shall be standardised using the following formula:

$$S'_{ij} = \frac{S_{ij} - min(S_{ij})}{max(S_{ij}) - min(S_{ij})} \text{ (positive indexes)} \quad (1)$$

$$S'_{ij} = \frac{max(S_{ij}) - S_{ij}}{max(S_{ij}) - min(S_{ij})} \text{ (negative indexes)} \quad (2)$$

where S_{ij} is the original data matrix of the index; S'_{ij} is the standardized data matrix; $max(S_{ij})$ and $\min(S_{ij})$ are the maximum and minimum values of the jth indicator, respectively.

② Calculation of the entropy σ_j of the indicator. The formula is as follows:

$$\sigma_j = -k\sum_{i=1}^{n} P_{ij}ln(P_{ij}) \quad (3)$$

where σ_j is the entropy value, $0 \leq \sigma_j \leq 1$; $k = \frac{1}{lnn}$, n is the number of evaluation subjects; $P_{ij} = \frac{S'_{ij}}{\sum_{i=1}^{n} S'_{ij}}$.

③ Calculation of the ω_j of the indicator. The formula is as follows:

$$\omega_{ij} = \frac{1 - \sigma_{ij}}{\sum_{i=1}^{n} 1 - \sigma_{ij}} \tag{4}$$

where ω_{ij} is the indicator weight; σ_{ij} is the indicator entropy value; and n is the number of indicators.

④ Calculation of the dimension layer scores. The formula is as follows:

$$X_i = \sum \omega_{ij} \times S'_{ij} \tag{5}$$

where X_i is the element level score and ω_{ij} is the indicator weight.

$$Y_i = \sum W_i \times X_i \tag{6}$$

where Y_i is the dimension layer score and W_i is the indicator weight.

⑤ Calculation of the target tier scores. The formula is as follows:

$$T_i = \sum Q_i \times Y_i \tag{7}$$

where T_i is the target layer score and Q_i is the indicator weight.

3.3.2. ESV Assessment Method

Costanza et al. [5,24] first proposed the theory and method of ESV evaluation in 1997, and it has been used ever since. However, this method has many defects and is not suitable for the conditions in China. There are significant differences in the ecosystem types and service functions in different research areas. Based on this, Xie et al. revised the equivalent table on the basis of Costanza's study and finally obtained the equivalent value of ecosystem services per unit area.

(i) Referring to the method of Xie et al. [28,29,38], in this study, we took 1/7 of the average economic value of grain production in Guangdong Province as the equivalent factor of the ESV. The formula is as follows:

$$E_a = \frac{1}{7} \times \frac{PQ}{S}, \tag{8}$$

where E_a is the economic value of grain per unit area of cropland (RMB/hm^2); P is the average grain purchase price (RMB); Q is the grain yield (kg); and S is the sown area of grain (hm^2). According to the weekly data on national grain variety purchase prices, the average value of P was calculated to be 3.05 RMB/kg and the economic value of the cropland per unit area was calculated to be 2418.58 RMB/hm^2; then, this was multiplied by the correction coefficient λ (λ = the average grain yield per unit area in each year; the average grain yield per unit area of the country = 0.968). Finally, the economic value of the cropland per unit area in the PRD was calculated to be 2341.52 RMB/hm^2.

(ii) Accordingly, the ecological service value coefficient (VC) can be calculated on the basis of the economic value of the cropland per unit area in the region according to the ESV equivalent table per unit area of the ecosystem. Previous studies have generally concluded that there is no ESV coefficient for construction land; however, it was recently found that construction land still has ecological functions, such as culture and entertainment. According to the literature [39–43], the service value per unit area of the PRD ecosystem is summarised in Table 2. The specific formula is as follows:

$$VC_{ij} = e_{ij} \times E_a \ (i, j = 1, 2 \cdots, n) \tag{9}$$

where e_{ij} is the ESV unit equivalent of class j for land-use type i.

Table 2. ESV per unit area in PRD.

Type #1	Type #2	Cropland	Forests	Grassland	Water Bodies	Construction Land	Wetlands	Unused Land
Provision services	Food products	3184.46	679.04	889.77	1873.21	0	1194.17	0
	Material products	210.73	1545.40	1311.25	538.54	0	1170.76	0
Regulating services	Water supply	−6158.19	796.11	725.87	19,411.20	−17,584.8	6064.53	0
	Gas regulation	2599.08	5081.09	4612.79	1802.97	−5666.48	4448.88	46.83
	Climate regulation	1334.66	15,219.88	12,199.32	5362.08	0	8429.47	0
	Environmental purification	398.05	4519.13	4027.41	12,995.44	−5760.14	8429.47	234.15
	Hydrological regulation	6368.93	11,098.80	8944.60	239,397.00	0	56,735.03	70.24
	Soil conservation	23.41	6205.02	5619.64	2177.61	46.83	5408.91	46.83
Supporting services	Maintenance of nutrient cycling	444.88	468.30	421.47	163.90	0	421.47	0
	Maintenance of biodiversity	491.71	5643.06	5104.51	5970.87	796.11	18,427.76	46.83
Cultural services	Aesthetic landscape	210.73	2482.01	2247.85	4425.47	23.41	11,075.39	23.41
Total		9108.45	53,737.84	46,104.48	294,118.29	−28,145.06	121,805.84	468.29

(iii) Based on the land-use data and the calculated unit area value table, each year's ESV accounting method for the PRD is as follows:

$$ESV = \sum_{i=1}^{n} VC_i \times A_i. \quad (10)$$

3.3.3. Coupling Coordination Degree Model

The coupling coordination degree model was used to study the co-evolution relationship between the ecological environment and economic development [44], and was used to quantitatively analyse the interaction between the PRD's ESV and economy. The specific calculation equations are as follows:

$$C = 2 \times \left[\sqrt[2]{\frac{f(x) \cdot g(x)}{(f(x) + g(x))^2}} \right], \quad (11)$$

$$T = \alpha f(x) + \beta g(x), \quad (12)$$

$$D = \sqrt{C \times T}, \quad (13)$$

where C is the coupling degree; T is the comprehensive coordination index; D is the coupling coordination degree; and $f(x)$ and $g(x)$ are the scores of the ecological environment and economic development, respectively. The larger the value, the higher the corresponding ecological environment and economic development degree. α and β are undetermined weights, reflecting the influence coefficient of the two dimensions. Based on the existing research results [43,45], we believe that the ecological environment and economic development were equally important in the study area, so we set $\alpha = \beta = 0.5$. There is no unified standard for classifying the coupling coordination degree at present. In this study, we referred to the existing literature on classifying the criterion of coupling and the coupling coordination degree [46,47] and divided them into six levels (Tables 3 and 4).

Table 3. Classification of ESV and economic coupling degree.

C Value Interval	Coupling Phase	Characteristic
[0, 0.3)	Low coupling	The relationship between the ESV and economy begins to emerge, but when $C = 0$, the two dimensions are in a state of unrelated interaction and are varied and disorderly.
[0.3, 0.5)	Rivalry	The interaction between the ESV and the economy gradually increases, implying the phenomenon of the stronger the strong, the weaker the weak.
[0.5, 0.8)	Matching	The variations in the ESV and economy gradually cooperate with each other, and the positive impact is significantly improved.
[0.8, 1]	High coupling	The orderly development of the ESV and economy occurs, and when $C = 0$, the two dimensions achieve coordination and tend toward a new orderly structure.

Table 4. Classification of ESV and economic coupling coordination degree.

Serial Number	Coordination Degree Interval	Classification
1	[0, 0.2)	Severe imbalance
2	[0.2, 0.4)	Moderate imbalance
3	[0.4, 0.6)	Basic coordination
4	[0.6, 0.8)	Moderate coordination
5	[0.8, 1]	High coordination

3.3.4. Geographical and Temporal Weighted Regression Model

After introducing the temporal dimension, the limitation of the traditional geographical weighted regression model that only considers non-spatial stationarity is overcome. The GTWR model helps to more accurately identify the spatial distribution and temporal dynamics of the influencing factors. The mathematical expression of the model is as follows [48]:

$$Y_i = \beta_0(u_i, v_i, t_i) + \sum_{j=1}^{n} \beta_j(u_i, v_i, t_i) X_{ij} + \varepsilon_i, \quad (14)$$

where Y_i is the coupling coordination degree (CCD) between the ESV of city *i* and economic development. $u_i, v_i,$ and t_i are the spatiotemporal coordinates of point *i*. $\beta_0(u_i, v_i, t_i)$ is the regression constant of point *i*, that is, the intercept term. $\beta_j(u_i, v_i, t_i)$ is the parameter estimation of the *j*th explanatory variable at the *i*th point in year *t*. X_{ij} is the value of the explanatory variable X_j at point *i*. ε_i is the random error term. *n* is the number of explanatory variables.

In order to analyse the influencing factors regarding the ESV and economy in the PRD's urban agglomeration, we referred to studies on the actual situation in the PRD [37,49,50] and selected ten social–economic and natural indicators as driving factors (Table 5). According to the research perspectives provided in literatures, the significance of factors selected include government intervention [51,52], opening-up [50,51,53], urbanisation [54], population [10,55], urban expansion [42,56] and ecological environment factors [51,57].

Table 5. Classification of ESV and economic coupling coordination degree.

Driving Force Type	Driving Factor	Measurable Indicator	Numbering	Unit
Socio-economic factors	Economy scale	GDP	X_1	10^4 RMB
	Population scale	Number of permanent residents	X_2	10^4
	Industry scale	Output value of industries above a designated size	X_3	10^4 RMB
	Urban expansion	Light data index	X_4	-
	Industrial structure	Industrial structure advanced index	X_5	-

Table 5. *Cont.*

Driving Force Type	Driving Factor	Measurable Indicator	Numbering	Unit
	Urbanisation level	Urbanisation ratio	X_6	%
	Environmental protection	Proportion of environmental protection use expenditure of total public expenditure	X_7	%
Natural factors	Temperature	Average annual temperature	X_8	°C
	Precipitation	Average annual precipitation	X_9	mm
	Vegetation coverage	NDVI	X_{10}	-

4. Analyses of Results

4.1. Comprehensive Analysis of ESV and Economic Development Level

4.1.1. Continuous Decreasing Trend of the Total ESV in the PRD

Ecosystem services exhibit different characteristics in different urbanisation stages [58]. Compared with the existing research on the stages of urbanisation [57,59], the urbanisation process in the PRD can be divided into a stage III mature period (70–80%; 2005–2007) and a stage IV post-mature period (>80%; after 2008). During the research period, the urbanisation process in the PRD was in stages III and IV (Figures 3 and 4). From the perspective of the temporal evolution, the total value of the ecosystem services in the PRD exhibited a decreasing trend. Among them, the total ESV increased in stage III and decreased significantly in stage IV. The ESV of the different land types exhibited different evolution characteristics; that is, in stage III, the ESV of the cropland continued to decrease, the absolute ESV of the construction land services continued to increase, and the ESV of the forests, grassland, water bodies, and unused land continued to increase. The forests and water bodies had a significant positive effect on the total ESV, and the construction land had a significant negative effect on the total ESV. In stage IV, the ESV of the construction land and unused land increased, and the ESV of the cropland, forests, grassland, and water bodies decreased. During the entire study period, the cropland, forests, water bodies, and construction land greatly influenced the total ESV in the PRD, indicating that the occupation of cropland, forests, and water bodies by urban construction during the urbanisation process was the core factor leading to the decrease in the total ESV.

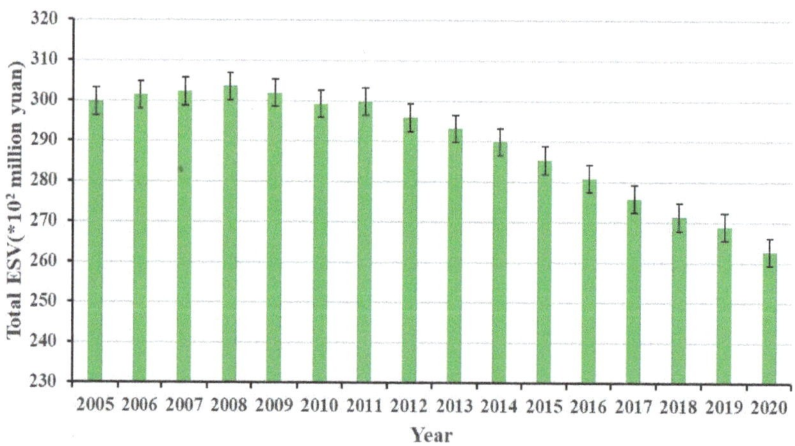

Figure 3. Trend of the total ESV in the PRD from 2005 to 2020.

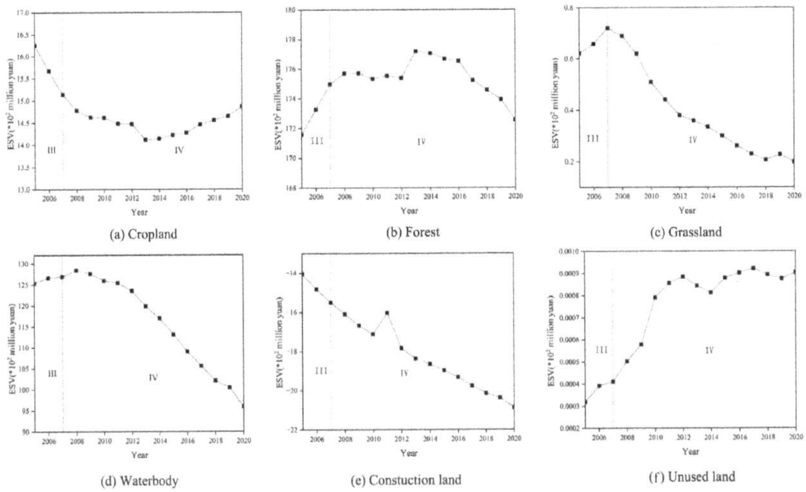

Figure 4. Trends of the ESV of the various land-use types in the PRD from 2005 to 2020.

From the perspective of the evolution of the ESV from 2005 to 2020, the spatial pattern of the cities in the PRD urban agglomeration remained relatively stable (Figure 5). In 2005, Zhaoqing's total ESV was greater than that of the other cities because of the high ESV of the forests and water bodies. Huizhou and Jiangmen ranked high in terms of the total ESV because of the relatively high ESV of their cropland, forests, and water bodies. The ESV of the water bodies in Guangzhou, Foshan, and Zhuhai were relatively high, so their total ESV was at a medium level. Due to the high absolute values of the ESV of the water bodies and construction land, the total ESV in Dongguan and Zhongshan was at a medium–low level. Shenzhen's total ESV was relatively very low due to the low ESV of its forests and water bodies, even though the absolute value of the ESV of its construction land was high. In 2020, the spatial pattern of the ESV in the PRD exhibited a trend of stable fine-tuning. Foshan's ESV was mainly due to the decrease in the ESV of its water bodies and the increase in the absolute value of the ESV of its construction land, and its total ESV was reduced to the medium–low level. Dongguan's ESV was mainly due to the decrease in the ESV of its water bodies and cropland and the increase in the absolute value of the ESV of its construction land, and its total ESV was degraded to a very low level.

4.1.2. Fluctuating Increasing Trend of the Economic Development Level in the PRD

A comprehensive index of economic development was obtained using the weighted sum of the entropy method. The results show that the vitality of the development had a great influence on the comprehensive level of the economic development. Combined with Chenery's economic stage division standard [57], the economic development in the PRD was divided into five stages: stage I—primary product production (before 1993); stage II—early industrialisation (1994–2001); stage III—medium industrialisation (2002–2005); stage IV—late industrialisation (2006–2011); and stage V—early developed economy (after 2012). Throughout the study period, the economic development in the PRD occurred in the later part of stage IV (industrialisation; 2005–2011) and stage V (developed economy; 2012–2020). From the perspective of the time series, the comprehensive level of economic development in the PRD significantly improved, especially in stages IV and V, after 2008. The main reasons for the slowdown in stage V include the low demand for foreign trade, a decline in real estate growth, the weakening of the domestic trade driving force, the erosion of corporate profits by structural inflation, and a rise in commodity prices (Figure 6).

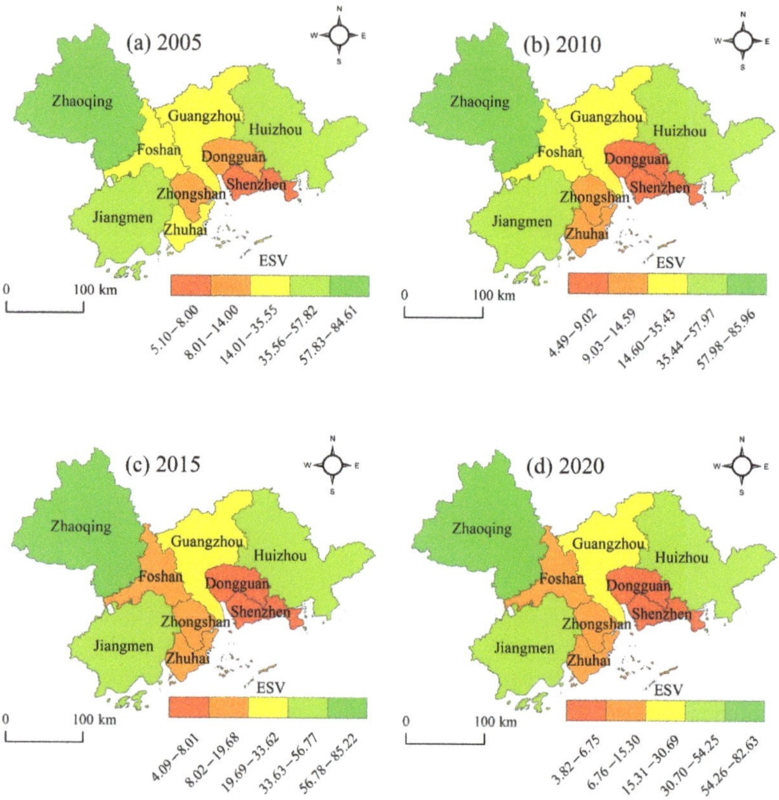

Figure 5. Spatial distribution of ESV in the PRD urban agglomeration from 2005 to 2020.

Regarding the evolution of the comprehensive level of economic development from 2005 to 2020 (Figure 7), the cities in the PRD's urban agglomeration were quite different and the spatial pattern remained relatively stable. In 2005, due to the high level of the industrial structure and development vitality, the comprehensive levels of economic development in Guangzhou and Shenzhen were ahead of the levels in the other cities. The comprehensive levels of economic development in Dongguan and Zhuhai were higher due to their higher levels of development vitality. Due to their relatively high levels of development vitality, the overall economic development in Foshan and Zhongshan was at the middle level. The comprehensive levels of economic development in Huizhou and Jiangmen were relatively low due to the low levels of their industrial structures and development vitality. Zhaoqing's comprehensive economic level was very low due to the low values of the three dimensions. In 2020, the levels of these three dimensions were higher in Guangzhou, Shenzhen, and Zhuhai, and their comprehensive economic levels were ahead of those in the other cities. Foshan, Dongguan, Zhuhai, Huizhou, and Jiangmen were in the middle comprehensive level of economic development due to the balance between these three dimensions. The comprehensive level of economic development in Zhaoqing lagged behind that of the other cities. Its development level, industrial structure, and development vitality were the lowest in the PRD's urban agglomeration.

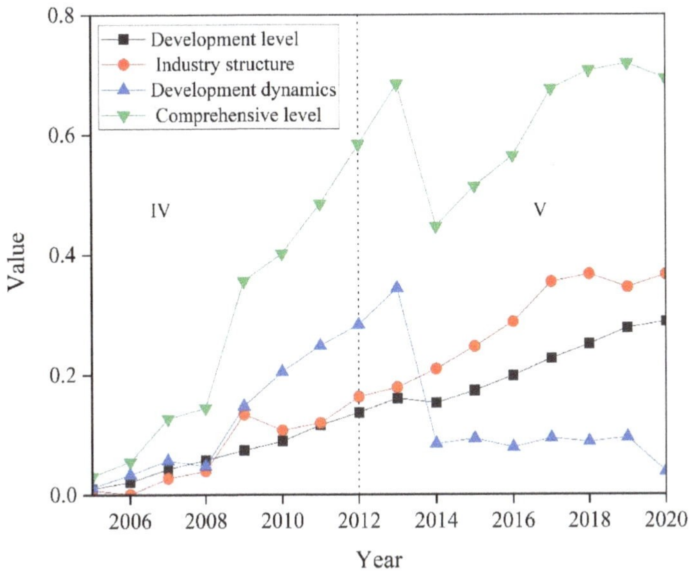

Figure 6. Comprehensive level of economic development in the PRD from 2005 to 2020.

4.2. Spatiotemporal Evolution Characteristics and Models of the Coordination Degree

4.2.1. Fluctuating Increasing Trend of the Coordination Degree in the PRD

According to the results, the coupling relationship between the ESV and the economy in the PRD from 2005 to 2020 was always in a low coupling state (Table 6). The coordination degree of the ESV and the economy increased in a fluctuating manner (Figure 8). These two phenomena indicate that the coordination of the two dimensions was reinforced. The turning point regarding the coordination degree occurred in 2007. At the earlier stage, the mutual promotion between the ESV and the economy was weak and the coordination degree was low. At the later stage, the economy developed rapidly and its promotion effect on the ESV was enhanced. The year 2007 was not only an important change point regarding urbanisation in the PRD from the mature stage to the post-mature stage, but it was also a critical period for the transition of economic development from the middle stage of industrialisation to the late stage of industrialisation.

4.2.2. Spatial Distribution of the Coordination Degree, Taking Estuary Cities as the Core and Decreasing in an Outwards Direction

From 2005 to 2020, the two dimensions of the coordination degree in the PRD exhibited a spatial distribution, with the cities along the Pearl River Estuary serving as the core, showing gradually decrease in this degree towards the periphery of the circle. The regional differences gradually decreased, and the coordination level evolved from moderate imbalance to basic coordination (Figure 9). In 2005, the coordination degree in the PRD urban agglomeration ranged from 0.31 to 0.46, indicating a state of mild imbalance to basic coordination, and the overall level was poor. Among the cities, Shenzhen, Zhuhai, and Dongguan exhibited basic coordination, and the others exhibited moderate imbalance. By 2020, the coordination degree in all of the cities in the PRD's urban agglomeration had improved, ranging from 0.35 to 0.49, and the overall change was from moderate imbalance to basic coordination. Among them, Guangzhou, Foshan, and Zhongshan rapidly attained basic coordination, while Zhaoqing, Jiangmen, and Huizhou remained in a state of moderate imbalance.

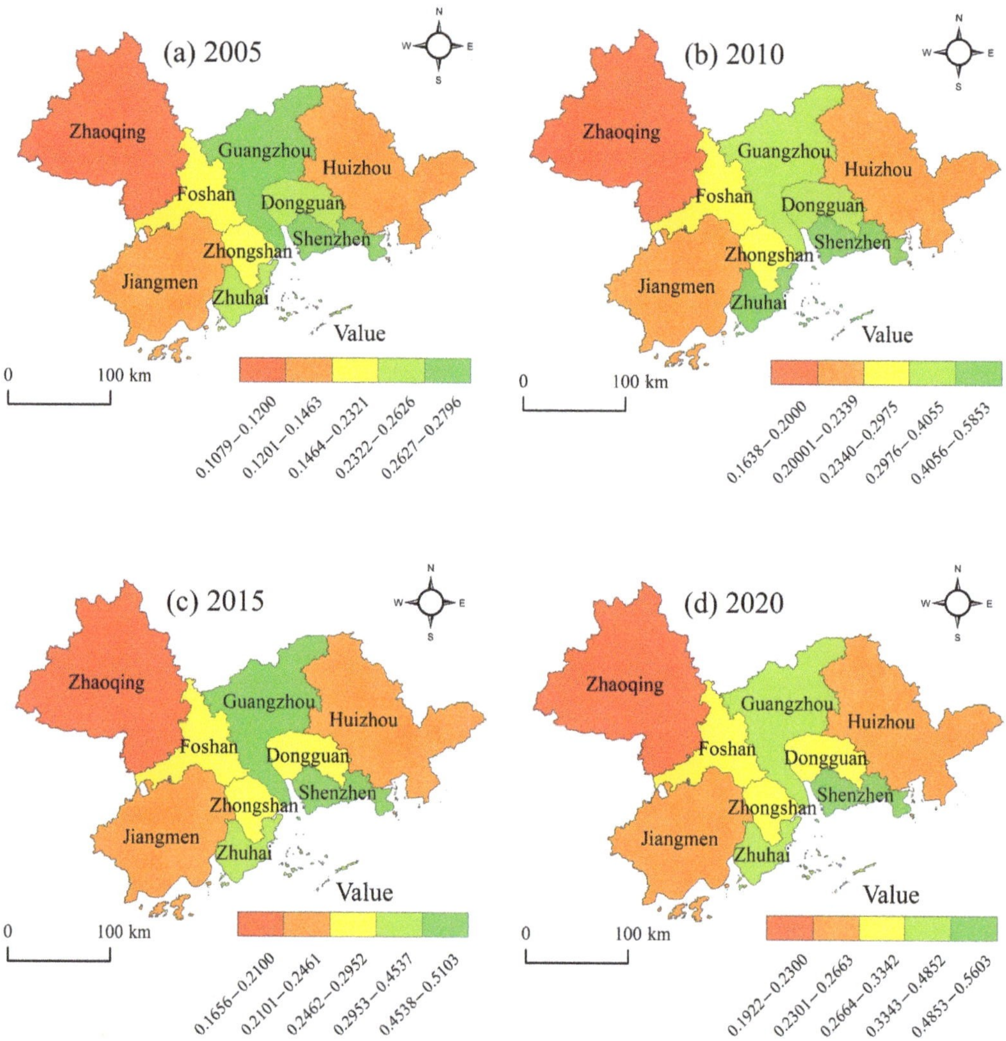

Figure 7. Spatial distribution of economic development in the PRD from 2005 to 2020.

4.2.3. Types and Evolution Models of the Coordination Relationship between the ESV and Economic Development

To explore the evolution of the coordination of the ESV and economic development, referring to a relevant study [45], the order of the ESV and the economic development level during the study period were compared to divide the evolution into types. ESV guidance was defined as the case when the ESV of the economic development was more than two. Economic guidance was defined as when the ESV of the economic development was more than two. Balanced development was defined as the case when the difference between the two positions was less than or equal to 2 (Figure 10).

Table 6. Coupling stage and coordination classification of ESV and economic level in the PRD.

Year	ESV	Economic Level	Coupling Stage	Coordination Classification	Year	ESV	Economic Level	Coupling Stage	Coordination Classification
2005	299.7452	0.0080	Low coupling	Moderate imbalance	2013	293.1343	0.6841	Low coupling	Basic coordination
2006	301.3780	0.0544	Low coupling	Moderate imbalance	2014	289.8818	0.4471	Low coupling	Basic coordination
2007	302.2502	0.1270	Low coupling	Moderate imbalance	2015	285.3290	0.5141	Low coupling	Basic coordination
2008	303.5062	0.14474	Low coupling	Basic coordination	2016	280.7839	0.5650	Low coupling	Basic coordination
2009	301.9126	0.3567	Low coupling	Basic coordination	2017	275.7713	0.6757	Low coupling	Basic coordination
2010	299.2613	0.4021	Low coupling	Basic coordination	2018	271.3035	0.7065	Low coupling	Basic coordination
2011	299.8094	0.4847	Low coupling	Basic coordination	2019	268.8801	0.7181	Low coupling	Basic coordination
2012	295.8922	0.5840	Low coupling	Moderate imbalance	2020	262.6957	0.6931	Low coupling	Basic coordination

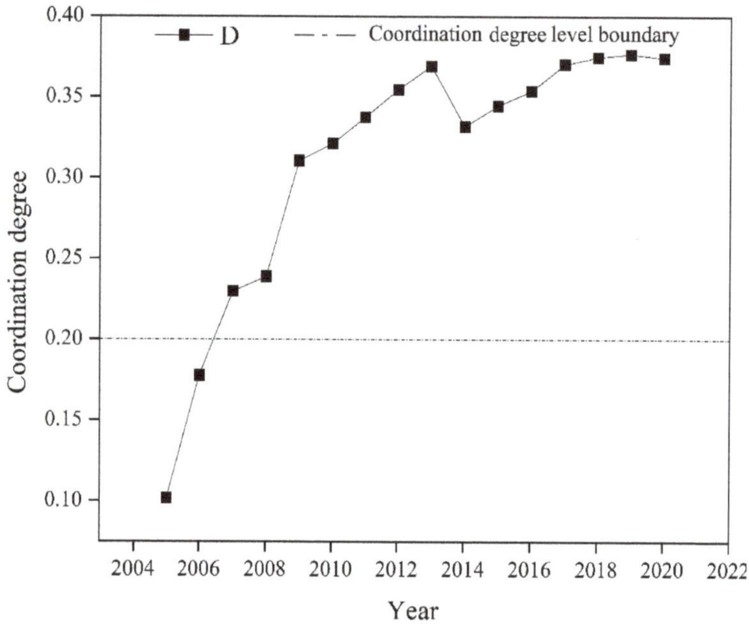

Figure 8. Trend of the coordination degree of ESV and the economic level in the PRD.

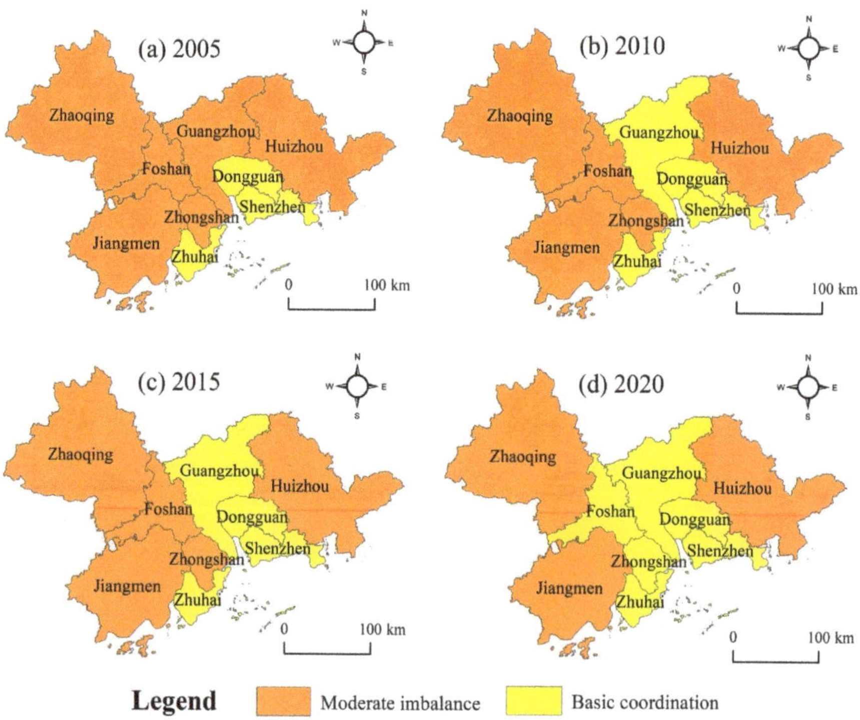

Figure 9. Spatial distribution of coordination degree in the PRD.

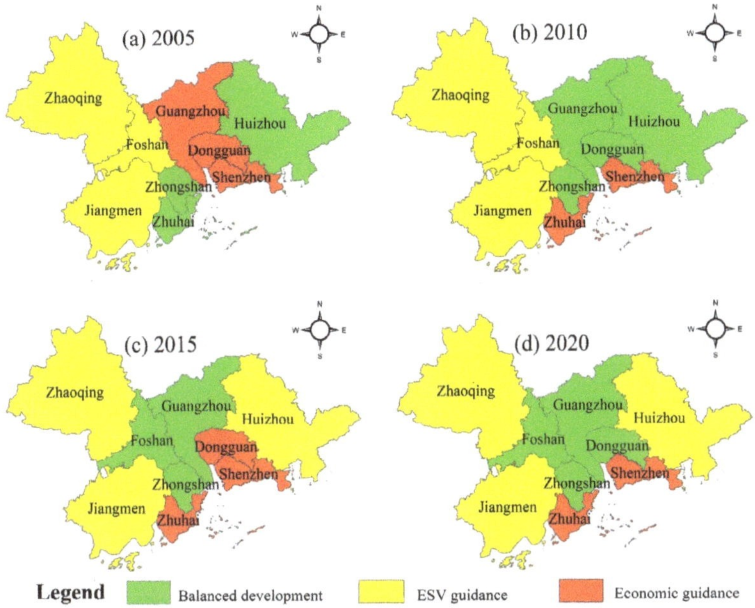

Figure 10. Types of coordination evolution for cities in the PRD.

From 2005 to 2010, the number of balanced development cities increased slightly; the number of ESV guidance cities did not change; and the number of economic guidance cities decreased from three to two. From the perspective of the spatial distribution of the types of evolution, the balanced development cities were concentrated in the east, the ESV guidance cities were concentrated in the west, and the economic guidance cities were mainly distributed along the Pearl River Estuary. From 2010 to 2015, the number of balanced development cities decreased slightly. The number of ESV guidance cities did not change, but the spatial disparity was more significant. The number of economic guidance cities increased from 2 to 3. The characteristics of the spatial distribution of the types during this period showed that the area of balanced development cities shrank towards the west, the area of the economic guidance cities shrank towards the north, and the ESV guidance cities exhibited a dispersed pattern. From 2015 to 2020, only the balanced development and economic guidance cities changed. Among them, the area of the balanced development type spread southward, the economic guidance type exhibited a uniform distribution on the east and west sides of the Pearl River Estuary, and the ESV guidance type still exhibited a dispersed distribution.

4.3. Analysis of Factors Influencing the Coordination Degree

4.3.1. Fluctuating Increasing Trend of the Coordination Degree in the PRD

Using the GTWR model, the coupling coordination degree between the ESV and economic development was taken as the dependent variable, and the annual openness [60], population size [61], government intervention [62], urban expansion [63,64], urbanisation [63,64], precipitation [43,58], temperature [43,58], and normalised difference vegetation index (NDVI) [43,58] were taken as the explanatory variables. Before utilizing the GTWR model, all of the variables needed to be standardised. In order to avoid the phenomenon of pseudo-regression in the regression, the regression analysis method was used to test the multicollinearity of all of the variables after standardisation, and the variables with variance inflation factors (VIFs) of greater than 10 were eliminated. Finally, five indicators, namely, openness (OPEN), government intervention (GOV), urban expansion (UE), annual average temperature (TEM), and NDVI, were determined, serving as explanatory variables for the GTWR model. The regression coefficients were calculated by the spatiotemporal, geographically weighted regression analysis module of ArcGIS 10.8, with the bandwidth set to automatic optimisation, and the ratio of the space–time distance parameter set to 1. The results are shown in Table 7, where the residual square is the residual sum of the squares of the model, and σ is the square root of the normalised residual sum of the squares. The smaller both values are, the more explanatory they are. The residual squares and σ values in this study are 0.0019 and 0.0036, respectively, indicating that the model fits the observation data well. AICC is a measure of model performance, and models with lower AICC values usually fit the data better. Given that the AICC value in this study is −730.9710 and the adjusted R2 is 0.9942, the GTWR model is able to effectively measure the influence of the driving factors on the coupling coordination degree by combining the numerical results of each parameter.

Table 7. Relevant parameters of the GTWR model.

Parameter	Bandwidth	Residual Square	σ	AICC	R^2	Adjusted R^2
Value	0.1150	0.0019	0.0036	−730.9710	0.9945	0.9942

4.3.2. Temporal Heterogeneity of GTWR Fitting Coefficients

According to the GTWR model, a regression analysis of the driving factors of the coupling coordination degree between the ESV and the economy in the cities in the PRD during the study period was carried out, and box plots of its evolution were drawn based on the regression coefficients of these factors (Figure 11).

From the perspective of the socio-economic factors, first, the degree of opening to the outside world (OPEN) mainly exerted a positive driving effect, and its dispersion generally exhibited a convergence trend, indicating that a higher level of opening to the outside world attracted a large amount of foreign investment into the PRD. Foreign investment not only directly promotes economic growth but also brings advanced environmental protection technology and equipment, which helps to enhance the ESV. Foreign-funded enterprises often pay more attention to environmental protection and sustainable development in the production process, thus promoting the coordinated development of the ESV of the ecosystem and the economy [60]. Second, it was found that the regression coefficient of the opening (OPEN) had a significant impact on the coupling coordination in stage IV (later period of industrialisation), which confirms the division of the development stage in the PRD. It also indirectly shows that the degree of opening promoted economic–environmental coordination in the PRD mainly by affecting economic development. The degree of government intervention (GOV) mainly exerted an N-type positive driving effect, but its positive effect gradually decreased, and its dispersion exhibited an expanding trend. This implies that, in the processes of guiding the economic layout, supporting industry, and promoting ecological protection, the industrial structure in the PRD was optimised and upgraded, and growth in regional, green, low-carbon, and environmental protection industries was encouraged accordingly. The policies helped to reduce the damage to ecosystems, enhance the ESV, and help the economy transition to a sustainable economy [62]. Third, the change trend of the GOV coefficient was basically consistent with the economic development and urbanisation stage division for the PRD described in this paper, indicating that government intervention promoted the coordinated development of the economy and environment in the PRD by affecting the ESV and economy. Urban expansion (UE) mainly exerted a positive driving effect, and its overall positive effect gradually increased. Its dispersion exhibited an expanding trend. The process of urban expansion may be associated with a reduction in ecological land area; however, through reasonable planning, the win–win situation of improvements in both the ESV and the economy can be achieved. This coupling coordination model is not only conducive to the sustainable development of the city but also provides residents with a more liveable living environment [63,64]. Furthermore, it was found that the change trend of the urban expansion (UE) coefficient was at its largest in around 2011, indicating that the UE promoted economic–environmental coordination in the PRD by affecting economic development.

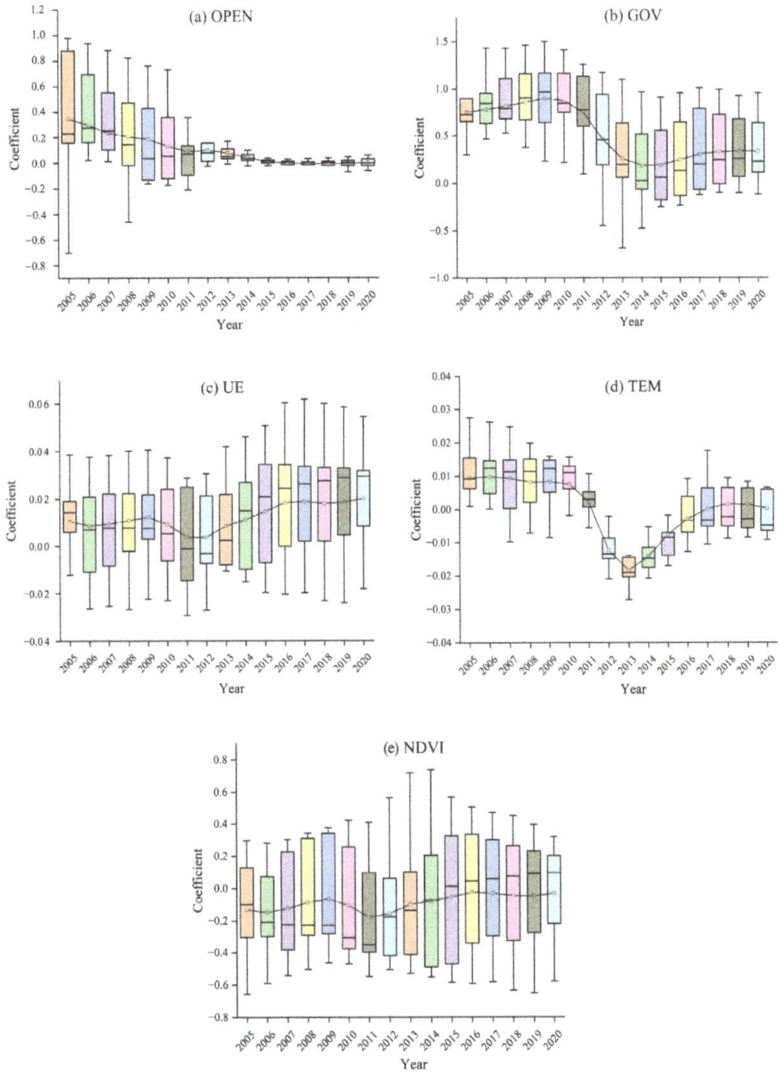

Figure 11. Trends of regression coefficients of driving factors.

From the perspective of natural factors, first, the temperature (TEM) mainly exerted a positive V-type driving effect and exhibited obvious stage characteristics. Before 2011, it had a promoting effect, while after 2011, this changed to a restraining effect. As an important factor of the natural environment, TEM can be affected by reasonable regulation, resource allocation optimisation, and ecological resilience enhancements [43,58], thus promoting sustainable development in the PRD. Second, the NDVI mainly exerted a negative driving effect during the study period, and its inhibitory effect generally exhibited a weakening trend. In general, the higher the vegetation coverage was, the stronger the stability and resistance of the ecosystem, and the impacts of natural disasters (such as soil erosion and desertification) weakened, so the ecosystem service functions were well reflected. In the wetlands, forests, and other ecosystems in the PRD, the high NDVI values reflected richer vegetation resources and ecological services, which can play roles in services such as water conservation, climate regulation, and air purification. The NDVI is usually positively correlated with the beauty of the natural environment and the ecological landscape, and ecotourism and green industry in the PRD can be used to obtain these development conditions accordingly [43,58]. The results show that the NDVI played an increasingly important role in the coupling and coordination of the ESV and economy in the PRD during the study period.

4.3.3. Spatial Heterogeneity of GTWR Fitting Coefficients

Using the spatial analysis tool, we applied the natural fracture method and set 0 as the breakpoint. Based on the results, the spatial heterogeneity of the local driving factors of the coupling coordination effect in the PRD was determined (Figure 12).

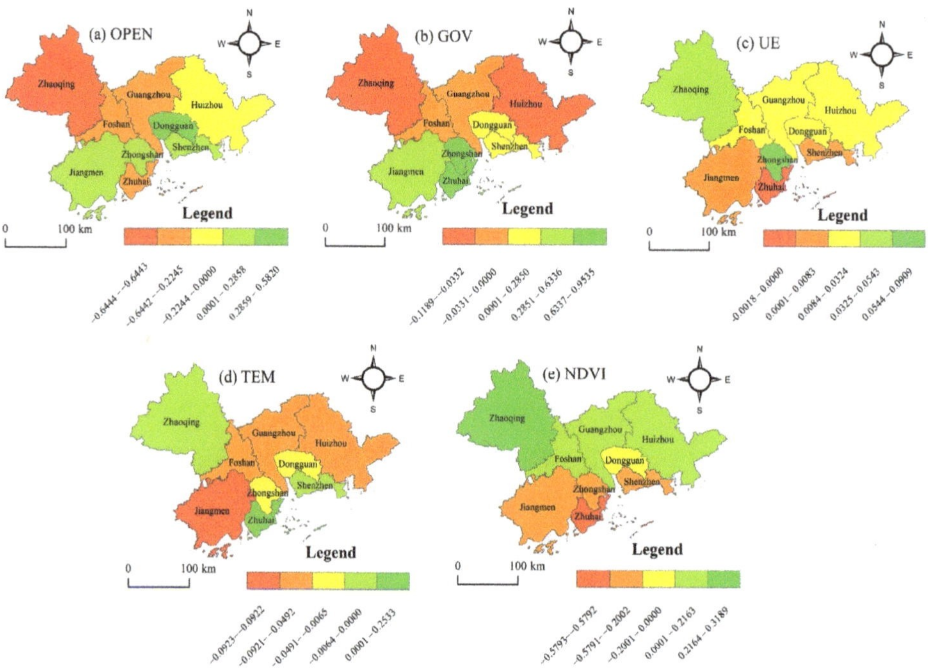

Figure 12. Spatial heterogeneity of regression coefficients of driving factors.

From the perspective of the socio-economic factors, the openness (OPEN) basically exerted a negative effect in the north and a positive effect in the south. The city with the lowest OPEN fitting coefficient was Huizhou. This indicates that following improvements in the level of openness, the coordination of the ESV and economy is more worthy of attention. The regression coefficient of government intervention (GOV) exhibited a negative double-winged spatial pattern. The high-value areas of its negative effects were located in Huizhou and Zhaoqing, indicating that these two cities need to improve the incentive compatibility mechanism of environmental governance among government departments, organically combine environmental regulation with economic transformation, and reduce the distortion impact of financial pressure and/or promotion incentives on environmental protection in the future, so as to reduce the loss of ecological efficiency due to imperfect institutional mechanisms and to ensure the coordination of the ecological environment and urban development. The spatial distribution of the UE regression coefficients revealed that the expansion of most of the cities in the PRD was positively correlated with the coordination of the ecology and economy, with the exception of Zhuhai, in which ecological deficits and environmental damage occurred during the urbanisation process, presenting obstacles to coordination and sustainable development.

From the perspective of the natural factors, the coefficients of the TEM were mostly negative, with the exception of those in Shenzhen, Zhuhai, and Zhaoqing. An increase in temperature often has an inhibitory effect on grasslands and forests [65], which leads to the heat island effect and restricts coupling and coordinated development. The coefficients of the NDVI were positive in the north and negative in the south, exhibiting significant agglomeration characteristics. The promoting effect extended from Zhaoqing to the east, and the inhibiting effect extended from Zhuhai to the surrounding cities.

5. Discussion

The PRD has long played a leading role in China's economic development in the past, but it has also sacrificed its ecological environment. At present, under the requirements of high-quality development, the PRD is not only committed to maintaining a strong economic position but also to achieving leading levels of innovation, coordination, greenness, openness, and sharing (known as the new development philosophy). After 2005, development in the Pearl River Basin entered a golden age, and Guangdong Province's economy made a great leap forward. During this stage, urbanisation began to occupy a large amount of land-use types with a high ESV [66]. The traditional economic model of the excessive consumption of natural resources had a serious negative impact, resulting in a decrease in the ESV [38–41]. In order to restore the damaged ecosystem, Guangdong Province and the PRD have issued a number of ecological protection plans since 2005, aiming to provide guidance for and a guarantee of regional environmental improvements and sustainable development. The implementation of this planning has reasonably controlled the intensity of regional land space development and the scale of pollutant emissions, increased the area of green space coverage, achieved cleaner production, and promoted eco-economic synergy in the PRD. From 2005 to 2020, the number of cities characterised by basic coordination and above increased significantly. It should be noted that by 2020, the basic coordination type was dominant in the PRD region, and the overall proportion of moderate imbalance was small, indicating that the coupling and coordination level of the ESV and economic development in each city had greatly improved.

The coupling coordination degree of the ESV and economic development in each city exhibited the characteristics of aggregation, and the higher-level cities were mainly concentrated in the Pearl River Estuary. This is consistent with previous research results [67]. From 2005 to 2020, the area of the high-value agglomeration areas continued to expand. These areas included two national central cities, Guangzhou and Shenzhen, the GDPs of which accounted for 58.86% of the GDP of the PRD. This had a dual-core driving effect on the regional pattern and led to the formation of an economic layout that was closely related to the neighbouring areas. The spillover of production factors and technology could promote the coordinated development of the ecological environment and economic growth in the surrounding areas. The moderate imbalance area was mainly concentrated in Jiangmen, Zhaoqing, and Huizhou, which are located on the periphery of the PRD, far from the growth pole. Their low levels of economic development restricted the coordination in this area. The moderate imbalance area had a significant negative spatial transmission effect and also affected the coordinated development of the region [68].

As one of the most economically dynamic and influential regions in China, the PRD is facing both ecological and economic challenges, but there are also opportunities for improvement. The key to achieving prosperity and harmony is to integrate the two dimensions into a comprehensive framework. Our research shows that the coordinated development of the PRD's urban agglomeration is affected by the comprehensive influence of socio-economic and natural environmental factors. Based on this, the following strategies and recommendations are proposed:

(1) We should deepen international cooperation and broaden the degree of opening up. According to the results for the openness, the PRD's urban agglomeration should strengthen economic and trade cooperation with neighbouring countries and regions along the Belt and Road, promote the diversified development of foreign trade, and enhance the competitiveness of the international market. We should take advantage of the policy advantages of the Guangdong–Hong Kong–Macao Greater Bay Area, deepen cooperation with Hong Kong and Macao, jointly create a world-class business environment, and attract foreign investment and advanced technology. We should actively participate in the reformation and construction of the global economic governance system, promote the liberalisation and facilitation of trade and investment, and create a good external environment for economic development.

(2) We should strengthen policy guidance and promote economic transformation and upgrading. According to the results regarding government intervention, scientific economic development plans should be formulated to clarify the industrial positioning and development direction in each city and avoid industrial isomorphism and vicious competition. More policies and measures need to be introduced to promote economic transformation and upgrading, encourage enterprises to increase R&D investment, and promote the development of industry in the direction of high-end, intelligent, and green production. The government's investment in and allocation of innovation resources should also be strengthened to support the construction of high-level scientific and technological innovation platforms and research and development institutions.

(3) We should accelerate the formation of green productivity and promote the development of regional economic cooperation. According to the results for urban expansion, temperature, and NDVI, urban expansion should strictly follow the principle of ecological protection, avoid damage to the ecological environment, and pay attention to greening and ecological restoration to improve the quality of the urban ecological environment. Green buildings and low-carbon travel modes are worth promoting to help reduce the negative impacts of urban expansion. The cities in the PRD should establish a regional coordinated development mechanism to jointly promote integrated development, especially the joint prevention and control of ecological protection. The best pathway is to rely on resource sharing and complementary advantages to enhance the overall competitiveness of the region. The PRD should also adhere to the construction of an intensive and efficient development pattern, give full play to the exemplary role of advanced cities in management, innovation, industry, and green development, and cooperate with Hong Kong and Macao to build a world-class urban agglomeration.

Against the background of the increasing proportion of urban areas, the negative impact of construction land on ecology must be considered. Based on relevant research [29,69,70] and the actual situation in the PRD, we introduced the negative value of the equivalent factor to make a more comprehensive and specific estimation of the ESV of each city in the PRD. This paper analyses the coordinated development of ecology and the economy in the PRD from a geographic perspective, and our empirical results broaden the understanding of the evolution of the coupling and coordination of the watershed environment and economy at the urban agglomeration scale. Our research also provides a new perspective when studying the response of the ecology to social and economic development. Based on our analysis of the spatial and temporal evolution characteristics and driving factors of the coupling coordination degree, we suggest important measures that can be taken to achieve environmental protection and economic development. Our approach has broad applicability in determining the impacts of land-use changes, ecological changes, and socio-economic changes in the study area, and it can be used to determine the impacts of relevant policies and management practices. The results of this study provide new evidence for the formulation and implementation of ecological restoration and environmental protection policies in the PRD urban agglomeration, as well as a theoretical and practical basis for realising the sustainable development of the ecological society and economy in the Pearl River Basin.

In this study, we constructed an ecological–economic index system suitable for the PRD and used it to optimise the values of the traditional single economic evaluation index into categories. We estimated the ESV in the PRD as a quantitative indicator of the ecological environment and analysed the impacts of external factors on the sustainable and high-quality development of the PRD's urban agglomeration.

However, some limitations of this study need to be improved in future studies. First, due to the availability of the data and previous research, the individual indicators of some of the cities could not be fully included in the evaluation system. Therefore, establishing a more comprehensive evaluation method will be a focus of future research. Second, the classification of land-use types in most studies is relatively extensive. For example, forest land was not specifically divided into broad-leaved forests or shrub forests in this study. However, in practice, the ESV of specific types of forests are different. The subdivision of land-use types could better reveal the supply of various ecological functions. In addition, due to the lack of evaluation criteria for construction land and the relative lack of background data, the accuracy of the measurement results cannot be fully verified. Relevant research methods need to be improved to improve the accuracy of the evaluation.

6. Conclusions

Focusing on urban agglomeration, this study measures the coordinated development of the economy and ecology in the PRD region. The results can not only expand the understanding of the high-quality development of urban agglomerations, but provide a new research perspective regarding the response of ecology to socio-economic development. By analysing the spatiotemporal evolution of CCD in the PRD's urban agglomeration and its driving factors, this study proposes important measures for environmental protection and sustainable development. Our approach has broad applicability in determining the impact of changes in regional land use, ecology, and the socio-economy, as well as changes in relevant policies and management practices. Additionally, the results provide new evidence for the formulation and implementation of ecological restoration and environmental protection policies within the PRD, helping to introduce a theoretical and practical basis for achieving sustainability in other regions of the Pearl River Basin. The main conclusions are as follows:

(1) From 2005 to 2020, the areas of cropland, water bodies, and grassland in the PRD generally decreased, and the area of construction land continued to increase, resulting in a decrease in the total ESV.

(2) In terms of the spatiotemporal evolution characteristics of the ESV and economic development, the coordination degree of the two exhibited trends of increasing volatility, and the study area began to enter the basic adjustment stage after 2007. The coordination degree between the two exhibited a pattern of gradually decreasing from the coastal cities of the Pearl River Estuary towards the periphery of the study area. The evolution mode of the coordination relationship was mainly balanced development, among which the balanced development areas were concentrated in Guangzhou, Foshan, Dongguan, and Zhongshan. The ESV guidance areas were mostly scattered within Zhaoqing, Huizhou, and Jiangmen. The economic guidance areas were concentrated in Shenzhen and Zhuhai along the Pearl River Estuary.

(3) The influence direction and degree of each factor were different in distinct spatiotemporal positions. In terms of the time scale, the degree of external development, government intervention, and urban expansion all exerted positive impacts. The order of their influence was as follows: government intervention > openness > urban expansion. The NDVI always had a negative impact during the study period. The regression coefficient of the annual average temperature exhibited obvious stage characteristics, and exerted a positive–negative–positive effect trend during the study period. From the perspective of the spatial scale, openness generally exerted a negative effect in the northern part of the region and a positive effect in the southern part of the region. The regression coefficient of GOV exhibited a negative double-winged spatial pattern. The UE had a positive impact in

most of the study area, and it only had a negative impact in Zhuhai. The TEM exhibited a southeast–southwest negative effect pattern in the study area. The regression coefficient of the NDVI was positive in the north and negative in the south.

Author Contributions: Conceptualization, C.Z.; Methodology, Z.Z. (Zeduo Zou) and X.L.; Software, X.Y.; Formal analysis, Z.Z. (Zeduo Zou) and Z.Z. (Zhuo Zhang); Investigation, X.Y.; Data curation, Z.Z. (Zhuo Zhang); Writing—original draft, Z.Z. (Zeduo Zou); Visualization, X.L.; Supervision, C.Z.; Funding acquisition, C.Z. All authors have read and agreed to the published version of the manuscript.

Funding: This research received no external funding.

Data Availability Statement: The raw data supporting the conclusions of this article will be made available by the authors on request.

Conflicts of Interest: The authors declare no conflict of interest.

References

1. Zou, Z.; You, M.; Zhao, W.; Fu, C.; Zhang, W.; He, Z. Changes in the "Production-Living-Ecological Space" Pattern in the Interlocking Mountain and River Zones of the Yellow River Basin—Taking Xinxiang City as an Example. *J. Resour. Ecol.* **2023**, *14*, 479–492.
2. Cheng, Y.; Kang, Q.; Liu, K.; Cui, P.; Zhao, K.; Li, J.; Ma, X.; Ni, Q. Impact of Urbanization on Ecosystem Service Value from the Perspective of Spatio-Temporal Heterogeneity: A Case Study from the Yellow River Basin. *Land* **2023**, *12*, 1301. [CrossRef]
3. Hong, X.; Peng, Q.; Zheng, R.; Lin, W.; Fan, S.; Su, K. Evaluating the Spatial Evolution of the Eco-Economy Harmony in Anxi County, China, Based on Ecosystem Services Value. *Sustainability* **2024**, *16*, 1491. [CrossRef]
4. Li, S.; Ma, S. A Quantitative Analysis on the Coordination of Regional Ecological and Economic Development Based on the Ecosystem Service Evaluation. *Land* **2024**, *13*, 207. [CrossRef]
5. Costanza, R.; de Groot, R.; Braat, L.; Kubiszewski, I.; Fioramonti, L.; Sutton, P.; Farber, S.; Grasso, M. Twenty years of ecosystem services: How far have we come and how far do we still need to go? *Ecosyst. Serv.* **2017**, *28*, 1–16. [CrossRef]
6. Daily, G.C.; Soderqvist, T.; Aniyar, S.; Arrow, K.; Dasgupta, P.; Ehrlich, P.R.; Folke, C.; Jansson, A.; Jansson, B.-O.; Kautsky, N.; et al. Ecology. The value of nature and the nature of value. *Science* **2000**, *289*, 395–396. [CrossRef]
7. Wu, M.; Kang, Q.; Cheng, Y.; Mi, W. Analysis of the Evolution of Ecosystem Service Value and the Coordinated Development of Eco-Economy in Xi'an Metropolitan Area. *Yellow River* **2023**, *45*, 97–102.
8. Bateman, I.J.; Harwood, A.R.; Mace, G.M.; Watson, R.T.; Abson, D.J.; Andrews, B.; Binner, A.; Crowe, A.; Day, B.H.; Dugdale, S.; et al. Bringing ecosystem services into economic decision-making: Land use in the United Kingdom. *Science* **2013**, *341*, 45–50. [CrossRef]
9. Kaletova, T.; Loures, L.; Castanho, R.A.; Aydin, E.; da Gama, J.T.; Loures, A.; Truchy, A. Relevance of Intermittent Rivers and Streams in Agricultural Landscape andTheir Impact on Provided Ecosystem Services-A Mediterranean Case Study. *Int. J. Environ. Res. Public Health* **2019**, *16*, 2693. [CrossRef]
10. Yao, S.; Li, Q.; Wu, Q.; Roger, C.; Chen, K.; Zhang, L. Study on the development trend and direction of Chinese urban agglomerations. *Geogr. Res.* **2010**, *29*, 1345–1354.
11. Fang, C.; Zhou, C.; Wang, Z. Sustainable development strategy and priorities of spatially differentiated development of urban agglomerations along the Yangtze River Economic Belt. *Prog. Geogr.* **2015**, *34*, 1398–1408.
12. Fang, C.; Zhou, C.; Gu, C.; Chen, L.; Li, S. Theoretical analysis of interactive coupled effects between urbanization and eco-environment in mega-urban agglomerations. *Acta Geogr. Sin.* **2016**, *71*, 531–550. [CrossRef]
13. Wang, D. Comprehensive Measurement and Evolution Law of Urbanization Development Quality of Urban Agglomeration in China. *Chin. J. Popul. Sci.* **2018**, 46–59, 127.
14. Liu, C.; Duan, D.; Yu, R.; Wang, T. Spatio-temporal Structure of Coupling of the Economy-Resources-Environment System in Wuhan Metropolitan Area. *China Popul. Resour. Environ.* **2014**, *24*, 145–152.
15. Ma, J.; Li, X.; Zhang, H. The Coordination of Urban Development Quality System in Urban Agglomeration in the Middle Reaches of the Yangtze River. *Econ. Geogr.* **2016**, *36*, 53–61.
16. Sun, H.; Huang, Z.; Xu, D.; Shi, X.; Liu, H.; Tan, L.; Ge, J. The Spatial Characteristics and Drive Mechanism of Coupling Relationship between Urbanization and Eco-Environment in the Pan Yangtze River Delta. *Econ. Geogr.* **2017**, *37*, 163–170, 186.
17. Li, F.; Zhang, M. Coupling and Coordination Differences of the Economy-Resources-Environment System in the Yangtze River Delta Megalopolis. *East China Econ. Manag.* **2018**, *32*, 59–65.
18. Shi, T.; Yang, S.; Zhang, W.; Zhou, Q. Coupling coordination degree measurement and spatiotemporal heterogeneity between economic development and ecological environment-Empirical evidence from tropical and subtropical regions of China. *Cleaner Prod.* **2020**, *244*, 118739. [CrossRef]
19. Tomal, M. Evaluation of coupling coordination degree and convergence behaviour of local development: A spatiotemporal analysis of all Polish municipalities over the period 2003–2019. *Sustain. Cities Soc.* **2021**, *71*, 102992. [CrossRef]

20. Clark, W.C. Sustainability science: A room of its own. *Proc. Natl. Acad. Sci. USA* **2007**, *104*, 1737–1738. [CrossRef]
21. Levrel, H.; Kerbiriou, C.; Couvet, D.; Weber, J. OECD pressure-state-response indicators for managing biodiversity: A realistic perspective for a French biosphere reserve. *Biodivers. Conserv.* **2009**, *18*, 1719–1732. [CrossRef]
22. Allan, G.; Hanley, N.; McGregor, P.; Swales, K.; Turner, K. The impact of increased efficiency in the industrial use of energy: A computable general Equilibrium analysis for the United Kingdom. *Energy Econ.* **2007**, *29*, 779–798. [CrossRef]
23. Galinis, A.; Van Leeuwen, M.J. A CGE model for Lithuania: The future of nuclear energy. *J. Policy Model.* **2000**, *22*, 691–718. [CrossRef]
24. Costanza, R.; d'Arge, R.; de Groot, R.; Farber, S.; Grasso, M.; Hannon, B.; Limburg, K.; Naeem, S.; O'Neill, R.V.; Paruelo, J.; et al. The value of the world's ecosystem services and natural capital. *Ecol. Econ.* **1998**, *25*, 3–15. [CrossRef]
25. Qiu, J.; Liu, Y.; Chen, C.; Huang, Q. Spatial structure and driving pathways of the coupling between ecosystem services and human well-beings: Acase study of Guangzhou. *J. Nat. Resour.* **2023**, *38*, 760–778.
26. Ma, F.; Guo, Y.; Yuen, K.F.; Woo, S.; Shi, W. Association between new urbanization and sustainable transportation: A symmetrical coupling perspective. *Symmetry* **2019**, *11*, 192. [CrossRef]
27. Fang, C. Interaction theory for regional eco-environmental circle and development circle. *Stud. Dialect. Nat.* **1999**, *15*, 32–34.
28. Xie, G.; Lu, C.X.; Leng, Y.; Zheng, D.; Li, S. Ecological assets valuation of the Tibetan Plateau. *J. Nat. Resour.* **2003**, *18*, 189–196.
29. Xie, G.; Zhang, C.; Zhang, L.; Chen, W.; Li, S. Improvement of the evaluation method for ecosystem services value based on per unit area. *J. Nat. Rescour.* **2015**, *30*, 1243–1254.
30. Kong, F.B.; Yang, W.C.; Xu, C.Y. Coordinated relationship and influencing factors of ecological environment and socio-economic coupling of urban agglomeration around Hangzhou Bay in China. *Acta Ecol. Sin.* **2023**, *43*, 2287–2297.
31. Duan, D.; Chen, Y.; Du, D. Regional Integration Process of China's Three Major Urban Agglomerations from the Perspective of Techmolog. *Trnsfer. Sci. Geogr. Simica* **2019**, *39*, 1581–1591.
32. Wu, Y.; Shi, K.; Chen, Z.; Liu, S.; Chang, Z. Developing Improved Time-Series DMSP-OLS-Like Data (1992–2019) in China by Integrating DMSP-OLS and SNPP-VIIRS. *IEEE Trans. Geosci. Remote Sens.* **2022**, *60*, 4407714. [CrossRef]
33. National Bureau of Statistics. *China Urban Statistical Yearbook (2006~2022)*; China Statistical Publishing House: Beijing, China, 2021.
34. Guangdong Provincial Bureau of Statistics. *Guangdong Statistical Yearbook (2005~2021)*; Guangdong Provincial Bureau of Statistics: Guangzhou, China, 2021.
35. Tang, X.; Qiu, F.; Zhu, C.; Zhang, C.; Yao, X. Temporal-spatial change of coordinated development between economy and environment in Jiangsu province. *World Reg. Stud.* **2015**, *24*, 68–77.
36. Liu, K.; Qiao, Y.; Shi, T.; Zhou, Q. Study on coupling coordination and spatiotemporal heterogeneity between economic development and ecological environment of cities along the Yellow River Basin. *Environ. Sci. Pollut. Res. Int.* **2020**, *28*, 6898–6912. [CrossRef] [PubMed]
37. Zhang, Q.; Shen, J.; Sun, F. Spatiotemporal differentiation of coupling coordination degree between economic development and water environment and its influencing factors using GWR in China's province. *Ecol. Model.* **2021**, *462*, 109794. [CrossRef]
38. Jiang, X.; Yang, Q.; Geng, Q.; Wang, X.; Liu, J. Spatial-temporal Differentiation and Driving Mechanism of Coordinated Development of Ecological-Economic-Society Systems in the Yangtze River Economic Belt. *Resour. Environ. Yangtze Basin* **2019**, *28*, 493–504.
39. Zhang, G. *Impact Assessment of Land Use Change on Ecosystem Service Value under Multi-Scenario*; Guangdong University of Technology: Guangzhou, China, 2022.
40. Xuege, W.; Fengqin, Y.; Fenzhen, S. Impacts of Urbanization on the Ecosystem Services in the Guangdong-Hong Kong-Macao Greater Bay Area, China. *Remote Sens.* **2020**, *12*, 3269. [CrossRef]
41. Ye, Y.; Zhang, J.; Wang, T.; Bai, H.; Wang, X.; Zhao, W. Changes in Land-Use and Ecosystem Service Value in Guangdong Province, Southern China, from 1990 to 2018. *Land* **2021**, *10*, 426. [CrossRef]
42. Wang, X.; Yan, F.; Zeng, Y.; Chen, M.; Su, F.; Cui, Y. Changes in Ecosystems and Ecosystem Services in the Guangdong-Hong Kong-Macao Greater Bay Area since the Reform and Opening Up in China. *Remote Sens.* **2021**, *13*, 1611. [CrossRef]
43. Zhang, H.; Wang, Y.; Wang, C.; Yang, J.; Yang, S. Coupling analysis of environment and economy based on the changes of ecosystem service value. *Ecol. Indic.* **2022**, *144*, 109524. [CrossRef]
44. Dong, W.; Wang, C.; Zhang, Z. The space-time coupling between rural revitalization strategy and rural tourism development in Shandong Province. *Sci. Geogr. Sin.* **2020**, *40*, 628–636.
45. Zhang, G.; Wang, J.; Wu, K.; Xu, Z. Spatial-temporal characteristics and influencing factors of coordination between economic and environmental development of three major urban agglomerations in China. *Geogr. Res.* **2020**, *39*, 272–288.
46. Yin, J.; Shi, P.; Huang, W.; Shi, Z.; Li, Y. Spatiotemporal differentiation characteristics and influencing factors of coupling coordinated development of rural revitalization and new urbanization at the county level in Gansu province. *J. Nat. Resour.* **2023**, *38*, 2148–2168. [CrossRef]
47. Zhao, J.; Liu, Y.; Zhu, Y.; Qin, S.L.; Wang, Y.H.; Miao, C. Spatiotemporal differentiation and influencing factors of the coupling and coordinated development of new urbanization and ecological environment in the Yellow River Basin. *Resour. Sci.* **2020**, *42*, 159–171. [CrossRef]
48. Huang, B.; Wu, B.; Barry, M. Geographically and temporally weighted regression for modeling spatio-temporal variation in house prices. *Int. J. Geogr. Inf. Sci.* **2010**, *24*, 383–401. [CrossRef]

49. Zhang, S.; Huang, C.; Li, X.; Song, M. The spatial-temporal evolution and influencing factors of the coupling coordination of new-type urbanization and ecosystem services value in the Yellow River Basin. *Ecol. Indic.* **2024**, *166*, 112300. [CrossRef]
50. Huang, Z.; Chen, Y.; Zheng, Z.; Wu, Z. Spatiotemporal coupling analysis between human footprint and ecosystem service value in the highly urbanized Pearl River Delta urban Agglomeration, China. *Ecol. Indic.* **2023**, *148*, 110033. [CrossRef]
51. Liu, W.; Zhan, J.; Zhao, F.; Yan, H.; Zhang, F.; Wei, X. Impacts of urbanization-induced land-use changes on ecosystem services: A case study of the Pearl River Delta Metropolitan Region, China. *Ecol. Indic.* **2019**, *98*, 228–238. [CrossRef]
52. Chengjing, C.; Yihua, L. Spatiotemporal changes of ecosystem services value by incorporating planning policies: A case of the Pearl River Delta, China. *Ecol. Model.* **2021**, *461*, 109777.
53. Feng, P.; Growe, A.; Shen, Y. The Middle-aged and Knowledge Workers: Demographic and Economic Changes in the Pearl River Delta, China. *Chin. Geogr. Sci.* **2022**, *32*, 1–17. [CrossRef]
54. Song, M.; Zhu, M.; Wang, H.; Ji, Y.; Li, T. Comparing the farmland preservation goals of the government and the public from insights of ecosystem services trade-offs: Evidence from Hubei, China. *Habitat Int.* **2024**, *145*, 103014. [CrossRef]
55. Wang, K.; Liew, S.K.V. The Impact of Vertical Fiscal Imbalances and Local Government Tax Efforts on the Quality of Economic Development—A Study Based on Threshold Regression and Simultaneous Equation Models. *Sustainability* **2024**, *16*, 3173. [CrossRef]
56. Zhou, C.; Huang, W.; Liu, Y. Spatio-temporal Evolution and Influencing Factors of City Economic Openness in China. *Areal Res. Dev.* **2018**, *37*, 1–8.
57. Zhang, G.; Deng, H. Spatial-temporal Evolution of the Coordination Relationship between Service Industry and Urbanization in the Pearl River Delta Region. *Sci. Geogr. Sin.* **2018**, *38*, 1118–1128.
58. Yang, M.; Gao, X.; Siddique, K.H.; Wu, P.; Zhao, X. Spatiotemporal exploration of ecosystem service, urbanization, and their interactive coercing relationship in the Yellow River Basin over the past 40 years. *Sci. Total Environ.* **2022**, *858*, 159757. [CrossRef] [PubMed]
59. Yuan, X.; Chen, B.; He, X.; Zhang, G.; Zhou, C. Spatial Differentiation and Influencing Factors of Tertiary Industry in the Pearl River Delta Urban Agglomeration. *Land* **2024**, *13*, 172. [CrossRef]
60. Cai, B.; Zhao, W.; Li, Y.; Li, Z. Spatial Pattern and Influencing Factors of Coupling Coordination Degree Between Regional Innovation and Regional Economy in China. *Sci. Technol. Manag. Res.* **2019**, *39*, 96–105.
61. Zhang, G.; Deng, M.; Yao, Y.; Li, X. Comprehensive level of the green development of industry in Guangdong province and spatial econometric analysis of the influencing factors. *J. Nat. Resour.* **2019**, *34*, 1593–1605. [CrossRef]
62. Huang, H.; Huang, H.; Xiao, Y.; Xiang, X. Industrial structure upgrading, government's attention to ecological environment and the efficiency of green innovation: Evidence from 115 resource-based cities in China. *J. Nat. Resour.* **2024**, *39*, 104–124. [CrossRef]
63. Li, X.; Wang, L.; Zhang, Y. Spatial-temporal Evolution Pattern and Influencing Factors of Development of New Productivity Forces Along the Yangtze River Economic Belt. *Resour. Environ. Yangtze Basin* **2024**, *33*, 909–923.
64. Ma, H.; Lu, S.; Zhang, W. Coupling process and mechanism of urbanization and innovation in Beijing-Tianjin-Hebei Urban Agglomeration. *Geogr. Res.* **2020**, *39*, 303–318.
65. Liu, Z.; Chang, Y.; Pan, S.; Zhang, P.; Tian, L.; Chen, Z. Unfolding the spatial spillover effect of urbanization on composite ecosystem services: A case study in cities of Yellow River Basin. *Ecol. Ind.* **2024**, *158*, 111521. [CrossRef]
66. Liu, W.; Zhan, J.; Zhai, Y.; Zhao, F.; Asiedu Kumi, M.; Wang, C.; Bai, C.; Wang, H. Linking Ecosystem Service Supply and Demand to Evaluate the Ecological Security in the Pearl River Delta Based on the Pressure-State-Response Model. *Int. J. Environ. Res. Public Health* **2023**, *20*, 4062. [CrossRef] [PubMed]
67. Liang, G.; Fang, F.; Li, J. Spatial and temporal evolution and drivers of eco-efficiency in the Pearl River Delta urban agglomeration. *Chin. J. Ecol.* **2023**, *42*, 1973–1981.
68. Tian, Y.; Zhou, D.; Jiang, G. Conflict or Coordination? Multiscale assessment of the spatio-temporal coupling relationship between urbanization and ecosystem services: The case of the Jingjinji Region, China. *Ecol. Indicat.* **2020**, *117*, 106543. [CrossRef]
69. Ye, Y.; Bryan, B.A.; Zhang, J.; Connor, J.; Chen, D.; Qin, Z.; He, M. Changes inland-use and ecosystem services in the Guangzhou-Foshan Metropolitan Area, China from 1990 to 2010: Implications for sustainability under rapid urbanization. *Ecol. Ind.* **2018**, *93*, 930–941. [CrossRef]
70. Ye, C.; Dong, Y. Effects of Land Use Change on Ecosystem Service Value of the Pearl River Delta. *Trop. Geogr.* **2010**, *30*, 603–608, 621.

land

Article

Monitoring and Evaluation of Coastal Ecological Carrying Capacity in the Context of Sustainable Development: A Case Study of Shandong Province

Hao Li [1], Yuxin Zhang [2,3,*], Xiaoli Wang [4], Peng Guo [1,*] and Kai Liu [4,5]

1 School of Information Science and Engineering, Shandong Agricultural University, Taian 271018, China; 2022110549@sdau.edu.cn
2 School of Management, Shandong Technology and Business University, Yantai 264005, China
3 International Research Centre for Big Data for Sustainable Development Goals, Beijing 100094, China
4 Yantai Institute of Coastal Zone Research, Chinese Academy of Sciences, Yantai 264003, China; xlwang@yic.ac.cn (X.W.); kailiu@yic.ac.cn (K.L.)
5 University of Chinese Academy of Sciences, Beijing 100049, China
* Correspondence: 202414173@sdtbu.edu.cn (Y.Z.); guopeng@sdau.edu.cn (P.G.)

Abstract: The research on coastal ecological carrying capacity holds great significance for the sustainable development of coastal areas and is a focal point of the United Nations Sustainable Development Goals (SDGs). This study coupled multi-source data and ecological analysis models to construct a multi-level evaluation system and analysis method for the coastal ecological carrying capacity of Shandong Province so as to realize the dynamic monitoring and evaluation of the coastal ecological carrying capacity of Shandong Province from 2010 to 2020. The results indicated: (1) The ecological carrying capacity of the coastal zone in Shandong Province showed a "U"-shaped development trend, with 2016 being a turning point. (2) The economic development–social support system gradually became the main force driving the overall improvement of coastal ecological carrying capacity. (3) The system coupling coordination degree of ecological carrying capacity in the coastal areas of Shandong Province showed a trend of first decreasing and then increasing, with a high level of internal coupling coordination of carrying capacity. (4) Per capita GDP, environmental protection investment, per capita water resources, and other indicators were the main factors driving the changes in the ecological carrying capacity of the coastal zone. This study aims to provide methodological reference and data support for coastal ecosystem monitoring, assessment, and climate change response.

Keywords: coastal zone; ecological carrying capacity; monitoring and evaluation; contribution degree; coupling coordination degree; sustainable development

1. Introduction

The coastal zone, as a transitional zone between land and sea, possesses both marine and terrestrial environmental characteristics, serving as a frontier for achieving integrated sustainable development of land and sea [1–4]. The coastal zone is also considered to be an important focus for the study of sustainable development [5–8]. The coastal zone of Shandong Province plays an important role in China's economic development and ecological protection. However, the dual effects of human activities and climate change make the natural resources and ecological environment of the Shandong coastal zone face great challenges [9,10]. In terms of human activities, the acceleration of industrialization and urbanization has led to the deterioration of marine water quality, the fragmentation of the coastal landscape, and the loss of coastal biodiversity [11]. At the same time, sea level rise, marine disasters, and climate warming lead to frequent phenomena such as coastal erosion, soil salinization, and groundwater pollution, thus damaging the carrying capacity of coastal ecosystems and seriously hindering sustainable development [12,13].

Therefore, ensuring the stability of the coastal zone's ecological environment while supporting sustainable socio-economic development has become a pressing issue for scholars, government agencies, and other stakeholders [14,15].

Human understanding of "carrying capacity" has evolved over several centuries, with the widely accepted definition of carrying capacity being the maximum biomass or number of individuals that an ecosystem can support under specific environmental conditions [16,17]. In the mid-20th century and beyond, the subjects of carrying capacity have become increasingly diverse, leading to concepts such as population carrying capacity, land comprehensive carrying capacity, resource–environment carrying capacity, and urban comprehensive carrying capacity. For instance, population carrying capacity is defined as the maximum population that a natural system in a certain region can sustain while maintaining natural, social, and economic stability [18,19]. Diansheng [20] defined resource–environment carrying capacity as the capacity of the resource–environment to support a certain population size and corresponding social–economic total amount within different scale regions under the premise of ensuring the benign circulation of natural resources such as atmospheric resources, water resources, and marine biological resources. Cui et al. [21] defined land comprehensive carrying capacity as the threshold of the maximum human activities that an ecological–economic–social coupled system, with land as the carrier, can withstand under certain conditions. Ecological carrying capacity is a derivative of this concept. For example, Wu and Hu [22] defined comprehensive ecological carrying capacity as the ability of ecosystems to provide resources and environments for social and economic development. Wu et al. [23] defined ecological carrying capacity as the ability of the regional ecological environment to support sustainable social development while ensuring the coordinated development of the ecological environment and the social–economic system. In summary, coastal zone ecological carrying capacity (CECC) manifests as the maximum capacity of the ecological environment to support or provide for sustainable social development, serving as a crucial link between regional development and environmental protection.

However, due to the complex environmental characteristics of coastal zones, dynamic monitoring and evaluation of CECC coupled with multi-source data and ecological analysis models has become a major difficulty [24]. In terms of CECC monitoring, quantitative methods for assessing ecological carrying capacity include the Ecological Footprint method [25,26], the "Pressure–State–Response (PSR)" framework model [27], the State Space model [28], and multidimensional evaluation indicator systems [29]. Among them, the combination of a multidimensional evaluation index system and state–space model can make full use of multi-source data to achieve efficient monitoring of CECC [30], which is also the quantitative method of CECC in this study. In terms of CECC evaluation, CECC evaluation research focuses on judging the load state or maximum carrying capacity of coastal ecosystems [22,31]. The exploration of the coupling relationship between the coastal pressure system, the coastal resource system, and the socio-economic system and the exploration of the driving mechanism of CECC change can effectively solve the contradiction between man and nature and optimize the allocation and utilization of coastal resources [32,33]. Unfortunately, there is a lack of research on the above content. Therefore, this study combined multi-source data and an ecological analysis model to explore the dynamic monitoring and evaluation of CECC in Shandong Province.

Taking the coastal area of Shandong Province as an example, this study aims to solve a scientific problem: how can multi-source data and an ecological analysis model be combined to realize dynamic monitoring and evaluation of CECC? Therefore, the main research contents of this paper are as follows: (1) Based on multi-source data, a multi-level CECC evaluation system for Shandong Province was constructed, and the coastal ecological carrying capacity of Shandong Province was calculated by a state–space model. (2) The contribution degree model was used to analyze the contribution degree of CECC subsystems to CECC. (3) Using the system coupling degree model to analyze the coupling coordination degree between the carrying capacity subsystems. (4) Exploring the driving factors of the change in CECC with the help of geographical detectors.

2. Study Method

2.1. Study Region

The study area covers the coastal regions of Shandong Province, including Binzhou City, Dongying City, Weifang City, Yantai City, Weihai City, Qingdao City, and Rizhao City (Figure 1). The longitude range is from 117° E to 121° E, and the latitude range is from 35° N to 39° N. The coastal areas of Shandong Province are situated between the Yellow Sea and the Bohai Sea, facing the Liaodong Peninsula across the Bohai Strait. The study area is surrounded by sea on three sides and has a temperate monsoon climate with four distinct seasons. The terrain is flat, dominated by plains, hills, and mountains. The main industries include manufacturing, marine industry, and port transportation. The coastal areas of the study area are rich in natural resources, such as tidal flats, minerals, and islands. In addition, the study area has 51 excellent harbor sites suitable for constructing deep-water berths, providing favorable conditions for the vigorous development of the marine economy in Shandong Province.

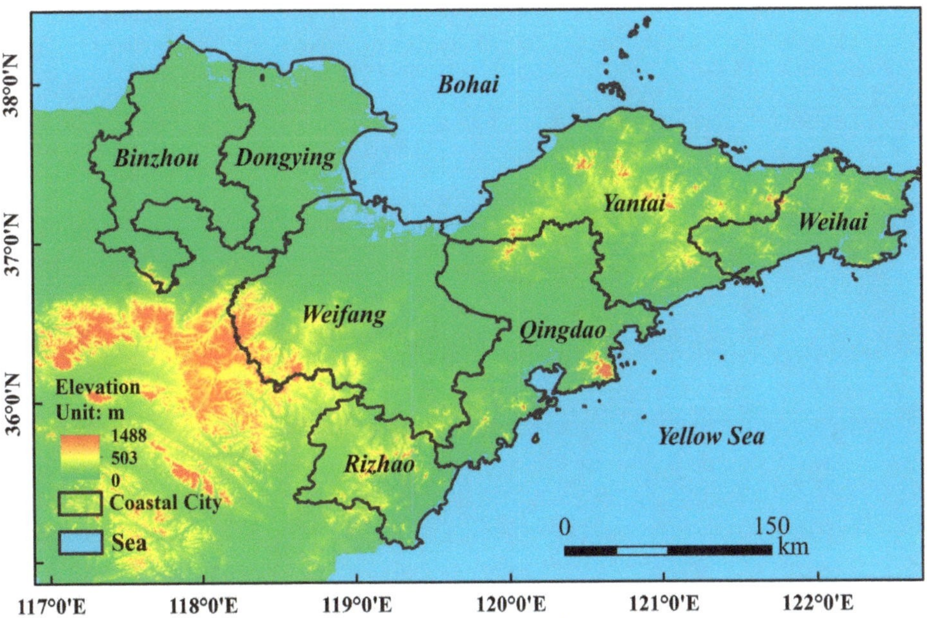

Figure 1. The study area.

In the past decade, Shandong Province has actively responded to China's strategy of becoming a maritime power. By 2023, its marine gross domestic product (GDP) exceeded 1.7 trillion yuan, ranking second in the country. However, the ecological vulnerability of the coastal areas in Shandong Province has been declining [34], and the ecological environment is facing significant pressure. Issues such as high fragmentation of coastal ecological landscapes, water scarcity, and intensified shoreline development are becoming more prominent. Therefore, the assessment and analysis of CECC in Shandong Province are crucial.

2.2. Data Source

This study constructed a CECC assessment system for the coastal areas of Shandong Province, including three criterion layers (pressure layer, state layer, and response layer), four primary indicators (environmental pressure, population pressure, coastal resources, and economic development–social support), and 20 secondary indicators (Table 1). The data sources for these indicators include the statistical yearbooks of Shandong Province and various cities, a municipal statistical bulletin of national economic and social development, a bulletin on the state of China's marine ecological environment, and remote sensing image data (Table 2).

Table 1. Evaluation index system of CECC.

Criterion Layers	Factor Layers	No.	Indicators	Attributes	Units
Pressure	Environmental pressure	X1	Industrial wastewater discharge	Negative	10^4 ton
		X2	Land use intensity	Negative	/
		X3	Coastline utilization degree	Negative	/
		X4	Million GPD energy consumption	Negative	tce/10^4 yuan
	Population Pressure	X5	Population density	Negative	capita/km^2
		X6	Passenger traffic	Negative	10^4 capita
State	Coastal resource	X7	Per capita wetland area	positive	m^2/capita
		X8	Net primary productivity of vegetation	positive	gC/m^2
		X9	Fishery resources	positive	10^4 ton
		X10	Per capita water resources	positive	m^3/capita
		X11	Vegetation coverage	positive	%
		X12	Nearshore water quality	positive	/
Response	Economic development–Social support	X13	Per capita GDP	Negative	10^4 yuan/capita
		X14	Port throughput	Negative	10^4 ton
		X15	Urbanization rate	Negative	%
		X16	Marine economy	positive	10^8 yuan
		X17	Social security and employment input	positive	10^4 yuan
		X18	Educational input	positive	10^4 yuan
		X19	Scientific and technological input	positive	10^4 yuan
		X20	Environmental protection input	positive	10^4 yuan

Table 2. Data type and resource.

Indicators	Data Type	Data Period	Data Source
X1, X4, X5, X6, X9, X10, X13, X15, X17, X18, X19 and X20	txt	2010–2020	«Shandong Statistical Yearbook»
X14 and X16	txt	2010–2020	«Municipal Statistical Bulletin of national economic and social development»
X12	txt	2010–2020	«Bulletin on the state of China's Marine ecological environment»
X2 and X7	Vector	2010–2020	https://www.resdc.cn/DOI/DOI.aspx?DOIID=54 (accessed on 2 July 2018)
X3	Grid	2010–2020	Extraction from Landsat8 OLI
X8	Grid	2005–2015	https://doi.org/10.3974/geodb.2019.03.02.V1
X11	Grid	2010–2020	https://doi.org/10.1016/j.rse.2019.111395.

2.3. Data Processing Method

Firstly, 20 indicators are standardized according to positive and negative attributes, then the weight of each indicator is determined according to the entropy method and analytic hierarchy process, and finally, the state space model is used to calculate the comprehensive ecological carrying capacity of the coastal zone. The specific research route is shown in Figure 2.

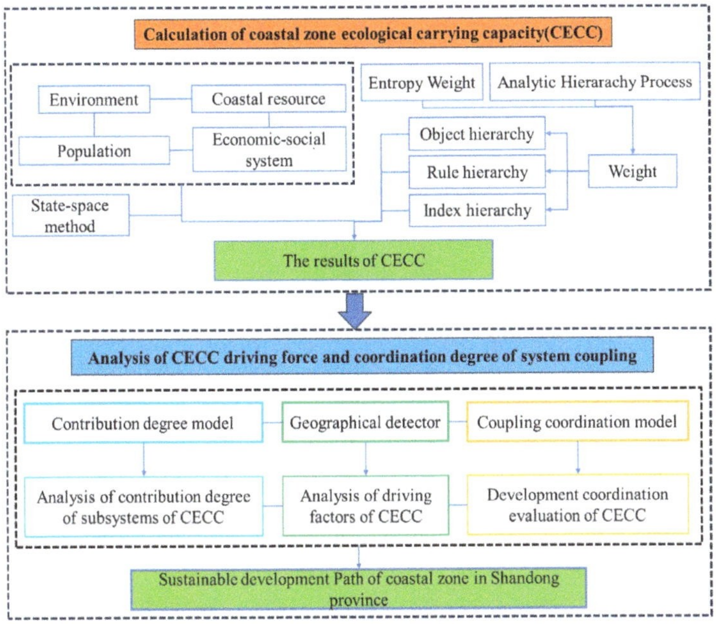

Figure 2. Research route.

2.3.1. Index Standardization

The adoption of range standardization for each indicator effectively eliminates the influence of indicator dimensions [35], ensuring that all indicators are normalized within the range of [0, 1]. Taking into account the positive/negative impact of indicators on the research content, standardization is carried out according to Formulas (1) and (2).

$$\text{For positive indicators}: \ X_{ij} = \frac{N_{ij} - N_{iMin}}{N_{iMax} - N_{iMin}} \tag{1}$$

$$\text{For positive indicators}: \ X_{ij} = \frac{N_{iMax} - N_{ij}}{N_{iMax} - N_{iMin}} \tag{2}$$

where i represents the number of years for indicators, j denotes the number of indicators, X_{ij} is the standardized result of the indicator, N_{ij} represents the actual value of each indicator in the corresponding year, and N_{iMin} and N_{iMax}, respectively, denote the minimum and maximum values of a certain indicator in the year i.

2.3.2. Weight Calculation

The determination of weights is a crucial step to ensure the quantification and objectivity of coastal ecological carrying capacity assessment. The entropy method, based on the concept of information entropy, calculates the weights of indicators by computing the entropy of each indicator, providing an objective weight calculation method [36,37]. The calculation principle is defined in Equations (3)–(5). Analytic Hierarchy Process (AHP) is a quantitative analysis method for multi-criteria decision-making with a certain degree of subjectivity. It involves constructing a hierarchy structure, establishing judgment matrices, calculating weights, and consistency testing to assist decision-makers in decision-making and evaluation [38–40]. Therefore, the combination of the entropy method and the analytic hierarchy process can obtain more scientific weights. The combination of them works as follows: Firstly, the weights W_{j1} and W_{j2} of each index are independently calculated by using the analytic hierarchy process and the entropy method, respectively. Then, these two groups of weights are averaged to obtain the final weights of each index. Considering the regional characteristics, we calculate the corresponding weights of all coastal cities in Shandong Province.

$$P_{ij} = \frac{X_{ij}}{\sum_{i=1}^{n} X_{ij}} \, (i = 1, 2, \cdots, n;\ j = 1, 2, \cdots m) \tag{3}$$

$$E_j = -\ln(n)^{-1} \sum_{i=1}^{n} P_{ij} \ln P_{ij} \ (P_{ij} = 0,\ \ln P_{ij} = 0) \tag{4}$$

$$W_{j2} = \frac{1 - E_j}{\sum_{j=1}^{m} (1 - E_j)} \tag{5}$$

In the equation, P_{ij} represents the proportion of the j-th indicator, X_{ij} is the standardized result of the indicator, E_j represents the information entropy of the j-th indicator, and W_{j2} reflects the weight of each indicator in the entropy method. The final weight of each indicator is defined as the average of the results from the AHP and the entropy method, i.e., $W_j = 0.5 \times (W_{j1} + W_{j2})$. The results are presented in Table 3.

Table 3. Weights of indicators in seven coastal cities.

Indicators	Weight						
	Qingda	Dongying	Yantai	Weifang	Weihai	Rizhao	BinZhou
X1	0.03	0.027	0.023	0.017	0.03	0.016	0.033
X2	0.075	0.065	0.066	0.055	0.063	0.068	0.054
X3	0.047	0.048	0.057	0.044	0.053	0.037	0.033
X4	0.024	0.027	0.024	0.022	0.025	0.023	0.025
X5	0.142	0.156	0.149	0.156	0.141	0.150	0.160
X6	0.036	0.043	0.040	0.037	0.038	0.036	0.043
X7	0.041	0.036	0.041	0.042	0.04	0.046	0.044
X8	0.031	0.038	0.039	0.041	0.046	0.033	0.051
X9	0.063	0.052	0.045	0.059	0.039	0.055	0.035
X10	0.077	0.074	0.081	0.084	0.074	0.089	0.076
X11	0.026	0.034	0.028	0.024	0.037	0.024	0.026
X12	0.035	0.027	0.041	0.094	0.053	0.028	0.027
X13	0.063	0.054	0.061	0.06	0.052	0.062	0.077
X14	0.03	0.041	0.037	0.026	0.027	0.027	0.031
X15	0.038	0.057	0.042	0.032	0.046	0.072	0.052
X16	0.079	0.075	0.079	0.072	0.074	0.076	0.076
X17	0.043	0.041	0.044	0.041	0.047	0.044	0.042
X18	0.032	0.034	0.032	0.03	0.033	0.033	0.031
X19	0.034	0.034	0.033	0.026	0.034	0.041	0.038
X20	0.054	0.038	0.038	0.037	0.05	0.039	0.047

2.3.3. Application of State–Space Method in CECC Calculation

The state–space method is a mathematical modeling approach that utilizes Euclidean geometric space to quantitatively describe and analyze the behavior of dynamic systems. It effectively provides a comprehensive and macroscopic perspective on the ecological carrying capacity of coastal zones [41,42]. This method's application principle in carrying capacity is represented by Equation (6).

$$CECC = \sqrt{\sum_{j=1}^{m} W_j X_{ij}^2} (i = 1, 2, \cdots, n) \tag{6}$$

In the equation, $CECC$ represents the coastal ecological carrying capacity of coastal zones, W_j reflects the weight of the $j-th$ indicator, and X_{ij} denotes the standardized results of each indicator.

2.3.4. Contribution Degree Model

The contribution degree model, which is used to evaluate and measure the contribution of individuals or factors to the overall outcome [43]. In this study, this model is employed to calculate the contribution of the four primary indicators of CECC, analyzing the reasons for changes in comprehensive ecological carrying capacity from a relatively macroscopic perspective. The calculation formulas are given by (7) and (8).

$$CECC_K = \sqrt{\sum_{j=1}^{o} W_j X_j^2}(K = 1,2,3,\cdots,m; j = 1,2,\cdots,n) \tag{7}$$

$$CD_K = \left(\frac{CECC_K}{CECC}\right)^2 \tag{8}$$

In the equation, $CECC_K$ represents the carrying capacity of the $K-th$ primary indicator, and CD_K denotes the contribution degree of the $K-th$ primary indicator to the overall carrying capacity.

2.3.5. Coupling Coordination Model

Coupling coordination degree is used to measure the degree of coordination between various modules in the overall project [44,45]. The principle is described by Equations (9) to (13).

$$Y_{ij} = W_j X_{ij}^2 \tag{9}$$

$$M_j^+ = \sqrt{\sum_{j=1}^{m}\left(Y_j^+ - Y_{ij}\right)^2};\ M_j^- = \sqrt{\sum_{j=1}^{m}\left(Y_j^- - Y_{ij}\right)^2}; \tag{10}$$

$$U = \frac{M_j^-}{M_j^- + M_j^+} \tag{11}$$

$$C = \left[\frac{\prod_{i=1}^{K} CECC_K}{\left(\frac{\sum_{i=1}^{K} CECC_K}{K}\right)^K}\right]^{\frac{1}{K}} \tag{12}$$

$$D = \sqrt{C \times U} \tag{13}$$

where Y_{ij} represents the weighted standardized value of indicator j in year i, Y_j^+ and Y_j^- are the ideal values for positive and negative indicators, respectively, i.e., the maximum and minimum values of indicator j in year i. M_j^+ and M_j^- represent the Euclidean distances from Y_{ij} to the positive and negative ideal values, respectively. U reflects the proximity of the actual value to the ideal value, C represents the system coupling degree, and D denotes the coupling coordination degree. The interpretation of D values is as follows: $0.0 \leq D < 0.2$, low coordination; $0.2 \leq D < 0.4$, moderate coordination; $0.4 \leq D < 0.6$, good coordination; $0.6 \leq D < 0.8$, high coordination; $0.8 \leq D \leq 1.0$, perfect coordination.

2.3.6. Geographic Detector

A geographic detector is a statistical method used to detect spatial differentiation and reveal the driving forces behind it. It has promising applications in fields such as biomedicine, environmental protection, and social sciences [46]. By exploring the spatial similarity between independent and dependent variables, it can uncover and analyze the characteristics of spatial data changes and driving factors [47]. The implementation principle is described by Equations (14) to (15).

$$q = 1 - \frac{SSW}{SST} \quad (14)$$

$$SSW = \sum_{i=1}^{n} M_i\sigma_i^2; SST = M\sigma^2 \quad (15)$$

where q represents the explanatory power of the independent variable on the dependent variable, with a larger value indicating a greater influence of the independent variable on the spatial distribution of the dependent variable. $i = 1, 2, \cdots, n$, is the number of layers or categories of the independent or dependent variable, where M_i and M denote the number of units in $i - th$ layer and globally, respectively. σ_i^2 and σ^2 represent the variance within $i - th$ layer and across the entire region, respectively. SSW and SST, respectively, represent the total within-group variance and total global variance.

3. Result

3.1. Temporal and Spatial Variations of CECC and Sub-System Carrying Capacity

Based on the CECC evaluation system, the calculation results of the CECC and the carrying capacity of the four primary indicators (subsystems) in Shandong Province from 2010 to 2020 are presented in Figure 3, and the annual change rate of the carrying capacity is shown in Table 4.

The results indicated significant spatiotemporal differences in CECC. The overall CECC in Shandong Province (the mean CECC of each city in the corresponding year) showed a "U"-shaped development trend, fluctuating between 0.632 and 0.707 (Figure 3). During 2010–2020, the annual change rates of environmental carrying capacity (ECC), population carrying capacity (PCC), resource carrying capacity (CRCC), and economic–social carrying capacity (ESCC) were −2.60%, −4.86%, −0.49%, and 1.99%, respectively (Table 4). In terms of time difference, the decline of overall CECC in Shandong Province was the most significant from 2010 to 2013, and the annual change rate was as high as −6.29%. However, the upward trend of CECC in 2016–2019 and 2019–2020 is very remarkable, with annual rates of change of 6.73% and 5.74%, respectively. It is worth noting that from 2016 to 2020, in addition to the population carrying capacity showing a decline, the other subsystems showed an upward trend. At the municipal level, the CECC of most coastal cities in Shandong Province presents a "U"-shaped development trend. The CECC of each city reached its peak in 2016, and the CECC turned a corner this year. In terms of spatial differences, the average CECC of each municipality over a 10-year period showed the spatial pattern of "Weihai > Dongying > Yantai > Rizhao > Qingdao > Binzhou > Weifang", and from 2019 to 2020, except Dongying and Weifang, the CECC of other cities experienced an upward trend.

The four subsystems of carrying capacity also exhibited significant spatiotemporal differences. Apart from population carrying capacity, the primary indicators of carrying capacity all decreased from 2010 to 2013 and increased from 2016 to 2020. The changes in overall environmental carrying capacity (ECC), population carrying capacity (PCC), resource carrying capacity (CRCC), and economic–social carrying capacity (ESCC) in Shandong Province from 2010 to 2020 were −0.08, −0.19, −0.02, and 0.07, respectively. Additionally, from 2010 to 2020, Dongying and Rizhao were cities with relatively significant changes in ECC. The cities where ECC, PCC, and CRCC decreased the most over the 10-year period were Rizhao, Weifang, and Binzhou, while the areas where CRCC and ESCC increased significantly were Weihai.

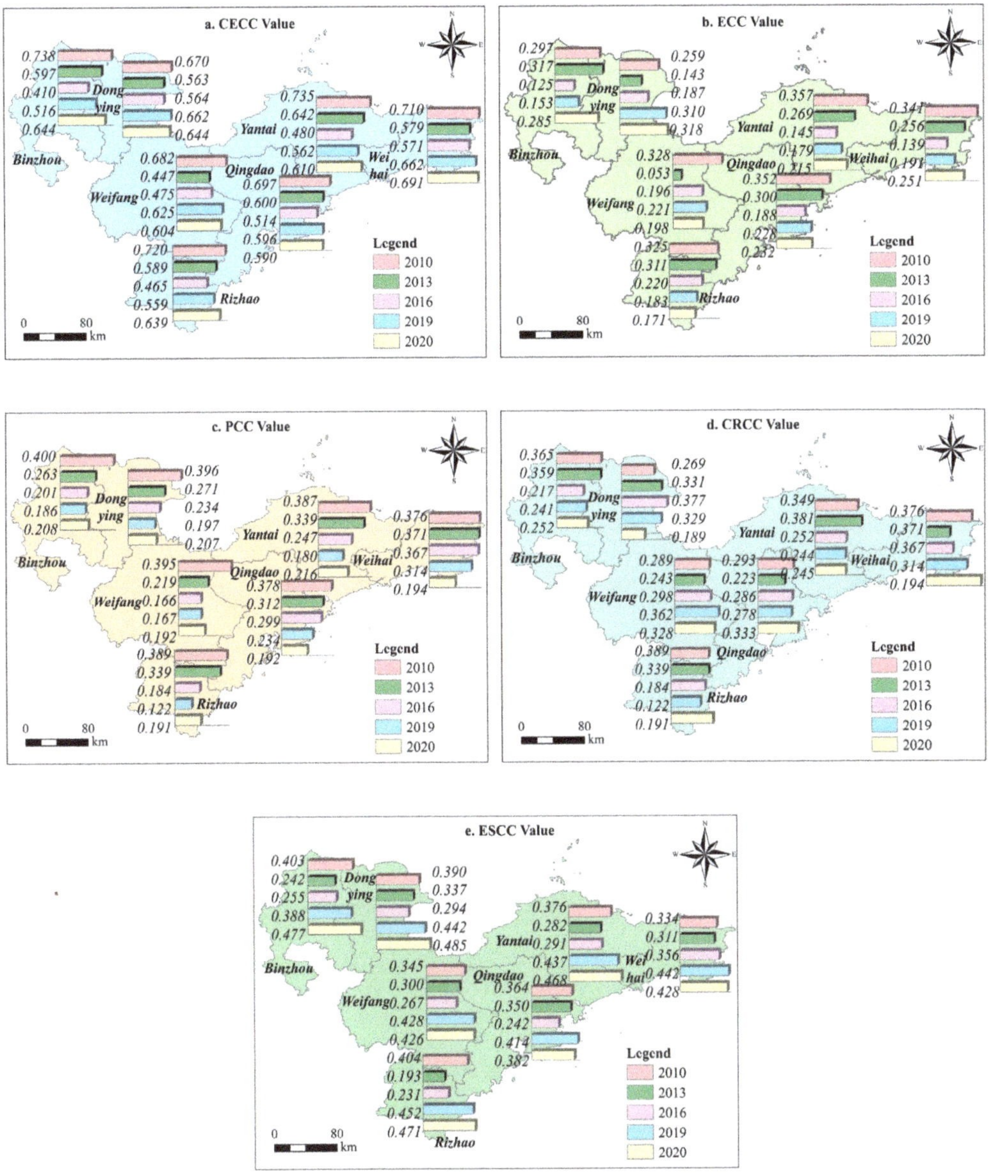

Figure 3. Spatial distribution of various carrying capacities in the Shandong coastal area from 2010 to 2020. (Note: (**a–e**) represents the spatial distribution of CECC, ECC, PCC, CRCC, and ESCC, respectively).

Table 4. Annual rate of change in carrying capacity in the Shandong coastal area from 2010 to 2020.

Region	Carrying Capacity	Period 2010–2013	2013–2016	2016–2019	2019–2020	2010–2020
Total	CECC	−6.29	−4.46	6.73	5.74	−1.07
	ECC	−9.00	−9.08	7.39	13.99	−2.60
	PCC	−7.46	−6.57	−5.81	−0.08	−4.86
	CRCC	−3.06	−1.85	1.70	5.42	−0.49
	ESCC	−7.67	−1.29	18.38	4.43	1.99
Qingdao	CECC	−4.64	−4.75	5.31	−1.10	−1.54
	ECC	−4.91	−12.48	7.20	1.80	−3.40
	PCC	−5.83	−1.47	−7.21	−17.95	−4.93
	CRCC	−7.99	9.38	−0.89	19.80	1.36
	ESCC	−1.29	−10.28	23.67	−7.70	0.49
Dongying	CECC	−5.32	0.07	5.77	−2.68	−0.38
	ECC	−14.87	10.19	21.77	2.69	2.28
	PCC	−10.56	−4.52	−5.23	4.97	−4.77
	CRCC	7.72	4.56	−4.26	−42.50	−2.98
	ESCC	−4.58	−4.21	16.76	9.74	2.43
Yantai	CECC	−4.24	−8.42	5.70	8.67	−1.70
	ECC	−8.25	−15.39	7.95	20.17	−3.97
	PCC	−4.14	−9.03	−9.01	19.76	−4.42
	CRCC	3.00	−11.26	−1.08	0.52	−2.98
	ESCC	−8.30	1.00	16.84	7.08	2.47
Weifang	CECC	−11.51	2.11	10.53	−3.37	−1.15
	ECC	−27.96	90.15	4.25	−10.20	−3.95
	PCC	−14.89	−8.06	0.36	14.75	−5.14
	CRCC	−5.28	7.51	7.17	−9.43	1.35
	ESCC	−4.34	−3.68	20.12	−0.50	2.35
Weihai	CECC	−6.15	−0.46	5.32	4.39	−0.26
	ECC	−8.32	−15.22	12.57	31.05	−2.64
	PCC	−0.47	−0.35	−4.84	−38.21	−4.85
	CRCC	−16.22	4.33	18.09	34.25	2.02
	ESCC	−2.31	4.83	8.10	−3.25	2.81
Rizhao	CECC	−6.06	−7.01	6.68	14.30	−1.13
	ECC	−1.44	−9.73	−5.63	−6.41	−4.73
	PCC	−4.31	−15.27	−11.22	56.58	−5.10
	CRCC	0.07	−3.14	−4.84	42.64	1.06
	ESCC	−17.44	6.69	31.85	4.11	1.65
Binzhou	CECC	−6.33	−10.45	8.64	24.75	−1.26
	ECC	2.24	−20.20	7.57	85.65	−0.42
	PCC	−11.43	−7.86	−2.40	11.64	−4.80
	CRCC	−0.57	−13.19	3.72	4.63	−3.09
	ESCC	−13.33	1.88	17.29	22.93	1.83

3.2. Change in Each Indicator

The standardized values of each indicator from 2010 to 2020 are shown in Figure 4.

In the environmental and population pressure systems, In the past ten years, the coastal areas of Shandong Province have implemented sewage discharge control and energy conservation and emission reduction policies, resulting in a downward trend in industrial wastewater discharge (X1) and million GDP energy consumption (X4) (note: since they are negative indicators, their standardized values show an upward trend), providing good environmental conditions for regional sustainable development. The values of land use intensity (X2) and passenger transport volume (X6) fluctuated around the years 2019 and 2013, respectively. From 2010 to 2020, the population surge and urban expansion have led to a significant increase in the development and utilization degree and population density of the coastline, which has brought great pressure on sustainable coastal development.

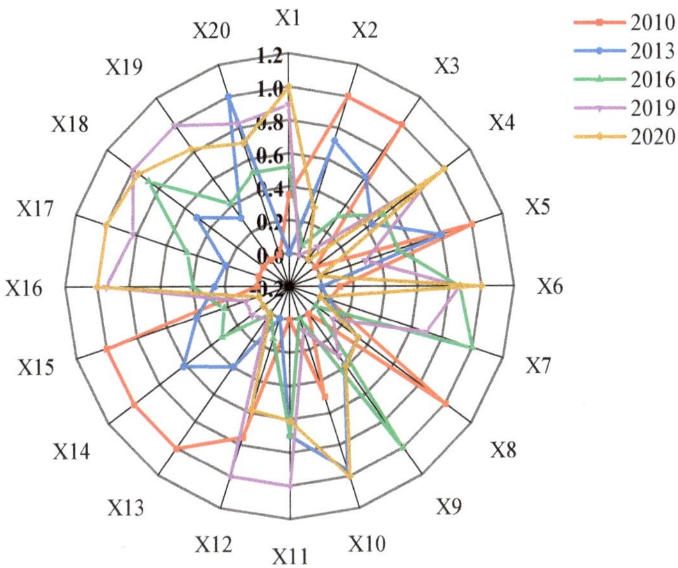

Figure 4. Changes in 20 indicators of CECC in Shandong Province from 2010 to 2020.

In the coastal resource system, the per capita wetland area (X7), net primary productivity of vegetation (X8), and fishery resources (X9) exhibited a trend of first decreasing and then increasing, with 2016 being an important turning point for changes in coastal resource carrying capacity. The per capita water resources (X10), vegetation coverage (X11), and nearshore water quality (X12) fluctuated from 2010 to 2020.

In the economic development and social support system, the decrease in port throughput (X14) and urbanization rate (X15) values has brought a significant burden to coastal areas. At the same time, the values of total marine economic output (X16), social security and employment input (X17), and education input (X18) have continued to increase over the decade, alleviating this pressure and promoting economic development in coastal areas. Per capita GDP (X13), scientific and technological input (X19), and environmental protection input (X20) have varied within a certain range, but their indicator values have all changed in a direction favorable to the economic development of coastal areas over the decade, to some extent alleviating multiple pressures and promoting the improvement of ecological carrying capacity.

3.3. *Analysis of Contribution Degree*

Figure 5 demonstrates the differences in the contribution degree (CD) of environmental pressure, population pressure, coastal resource economic development, and social support system to the CECC.

There were significant differences in the changes in CD for the four subsystems in coastal cities. From 2010 to 2020, except for Dongying and Binzhou, the CD of environmental pressure decreased in other cities, with the most significant changes observed in Rizhao (−13.17%) and Dongying (9.43%). The CD of coastal resources showed an upward trend in this decade, except for Dongying, Yantai, and Binzhou, with the most significant increases and decreases observed in Qingdao (14.19%) and Binzhou (−9.18%), respectively. The CD of population pressure in the seven coastal cities showed a fluctuating downward trend. However, the CD of economic development–social support in these cities was in the opposite trend to population pressure. The most significant changes were observed in Weifang (−23.4%) and Yantai (32.78%).

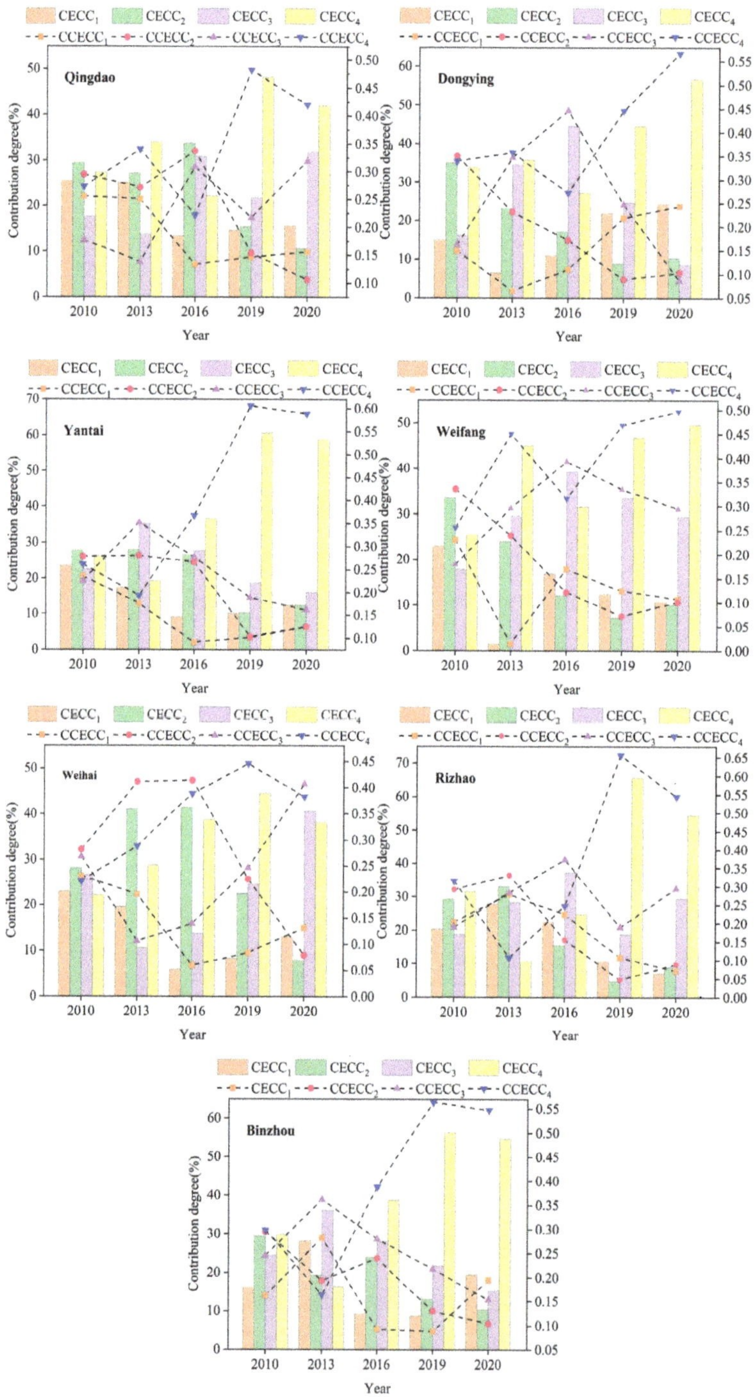

Figure 5. Contribution degree of factor layer indicators in coastal areas of Shandong Province from 2010 to 2020. (Note: $CECC_1$–$CECC_4$ represents the spatial distribution of ECC, PCC, CRCC, and ESCC, respectively. $CCECC_1$–$CCECC_4$ shows the contribution degree corresponding to the above parameters).

The trends in the changes in the CD of the four subsystems in each city exhibit significant differences. Regarding environmental pressure, Qingdao, Yantai, and Weihai showed a pattern of decreasing CD in environmental pressure followed by an increase, with peak and trough values observed in 2010 and 2016. In contrast, the CD of environmental pressure in Rizhao showed an opposite trend, while the CD in Weifang and Binzhou fluctuated within a certain range. In terms of population pressure, Dongying, Yantai, and Weifang exhibited a trend of decreasing CD in population pressure followed by an increase, with peak and trough values observed in 2010 and 2019. Conversely, Weihai demonstrated the opposite trend, and the population pressure CD in Qingdao, Rizhao, and Binzhou fluctuated. For coastal resources, Dongying, Yantai, Weifang, and Binzhou showed a trend of initially increasing and then decreasing CD in coastal resources. The changes in the CD of coastal resources in Qingdao and Rizhao were relatively complex. In the aspect of economic development and social support, except for Weihai showing a trend of initially increasing and then decreasing CD, the trends in other cities exhibited fluctuations, with the CD in 2020 being greater than in 2010.

4. Discussion and Suggestions

4.1. Analysis of Driving Factors of CECC

The analysis of the driving factors of CECC is an important foundation for supporting sustainable development in coastal areas [48,49]. By taking the change in each indicator as the independent variable and ecological carrying capacity as the dependent variable, we conducted a driving factor detection analysis using geographic detectors, with the results shown in Figure 6.

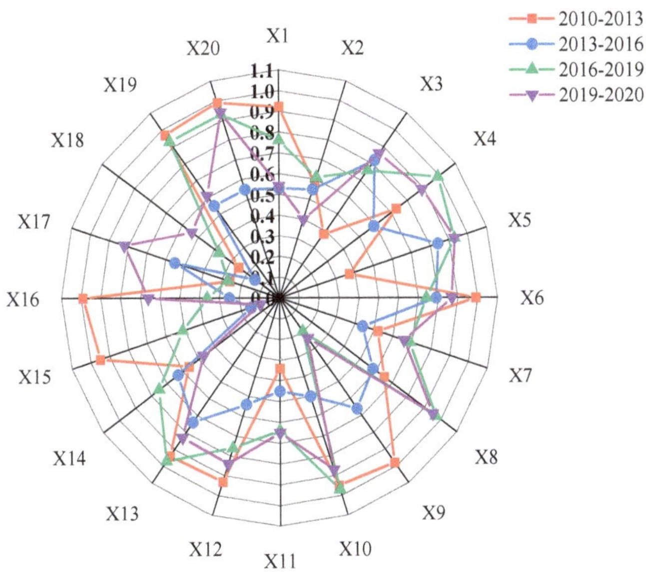

Figure 6. Detection results of driving factors of CECC in Shandong Province. (Note: X1–X20 represents the secondary index of the CECC evaluation system, and the results in the picture are the results of factor detection q value in the geographic detector).

As shown in Figure 6, from 2010 to 2013, the q value of factor detection of scientific and technological input (X19), environmental protection input (X20), passenger transport volume (X6), marine economy (X16), fishery resources (X9), and other indicators was at a high level, which indicated that they are the main factors driving the change in CECC. This indicated that although Shandong Province has increased its efforts in environmental protection and promoted the transformation of heavy industry to the green industry, for example, the discharge of industrial wastewater and million GPD energy consumption have decreased significantly, the rapid development of the marine economy and tourism industry, as well as the acceleration of urbanization in coastal areas [50–52], have exerted significant pressure on the coastal ecological environment, which was greater than the positive benefits brought by environmental protection input, resulting in a significant decline in CECC. In the following three years, the q values of population density (X5) and coastline utilization degree (X3) were much higher than other indicators, which was the main driving force for the change in CECC. During the period of 2016–2019, million GDP energy consumption (X4), net primary productivity of vegetation (X8), per capita water resources (X10), and per capita GDP (X13) emerged as the dominant factors influencing carrying capacity. Scientific and technological input and environmental protection input, to some extent, impacted carrying capacity. Shandong Province focused on the coordinated development of economic growth and environmental protection during this period, actively responding to energy conservation and emission reduction policies and optimizing and upgrading energy-intensive industries to green industries. This initiative facilitated economic development and the restoration of the coastal ecological environment, resulting in a significant increase in CECC. With the increase in environmental protection input and the optimization and upgrading of energy-intensive industries, the growth environment of vegetation in coastal areas and the nearshore water environment improved, leading to a further enhancement of CECC.

4.2. Coordination Between Economic Development and Environmental Protection

Urban economic carrying capacity is the basic condition for sustainable urban development [53–55], and in this study, the economic development–social support system is also the system with the largest weight ratio in ecological carrying capacity (about 37%). For example, from 2010 to 2013, the Qingdao municipal government increased investment in environmental protection to alleviate the huge pressure brought by resource depletion, population surge, and environmental pollution, resulting in a smaller decline in its ecological carrying capacity than other coastal cities in Shandong. From 2016 to 2019, the relevant governments increased investment in environmental protection, education, scientific and technological research and development, and social security and made use of the advantages of coastal areas to develop the economy rapidly, resulting in an upward trend in the contribution degree of the economic development–social support system of coastal cities in Shandong Province. This finding is consistent with other studies [56–58].

In addition, the premise of sustainable development is the coordinated development of economic growth and resource and environmental protection [59,60]. This study found that the problems of population surge, resource loss, and environmental pollution existed in the coastal areas of Shandong Province in the early stage of development, which destroyed the balance between economic development and resource and environmental protection and led to the continuous decline of CECC. However, after CECC drops to a certain threshold, with the compensation of resources and environmental carrying capacity by economic development and social support, the degree of coupling coordination between the subsystems of CECC is improved, and CECC recovers (Figure 7). This phenomenon fully proves that only by realizing the dynamic balance between economic growth and resources and environment can regional sustainable development be realized.

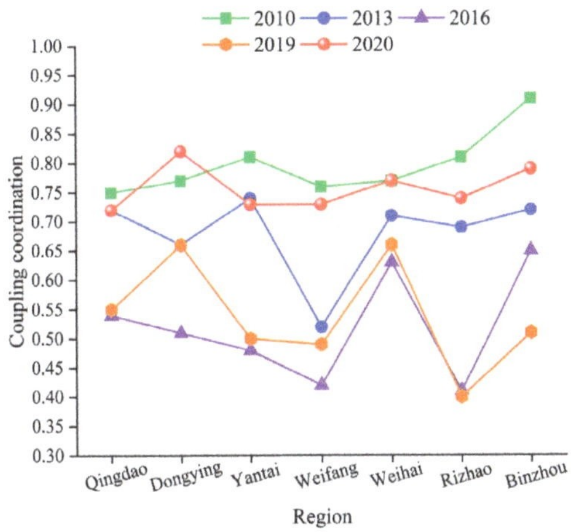

Figure 7. The coupling degree between the environmental system, population system, coastal resource system, and economic development—social development system in the coastal area of Shandong province.

4.3. Strengths and Weakness in This Study

By establishing the CECC monitoring and evaluation system in Shandong Province, this study dynamically monitored the change characteristics of CECC and analyzed the internal mechanism of the CECC subsystem (environment, population, coastal resources, economic development, and social support system) and the key driving factors of CECC change. On the one hand, it enhanced the research on the evaluation system of CECC in Shandong Province. On the other hand, it was beneficial for achieving dynamic monitoring of the coastal ecological environment and providing a practical basis for sustainable development in coastal regions. However, there are some shortcomings in this study: First, most of the index data comes from the statistical yearbook, and there are problems with the index coverage being not comprehensive, which may make the CECC quantification results inaccurate. Secondly, coastal ecosystems are often complex, nonlinear systems. Therefore, there may be some errors in the calculation of ecological carrying capacity using the state–space model. These defects affect the quantification of CECC in Shandong Province and the accuracy of related analysis results.

The research on the CECC is a comprehensive issue involving multiple interdisciplinary fields. In the future, by enhancing interdisciplinary collaboration, in-depth simulation of coupled dynamic models of internal subsystems of carrying capacity, such as economic development and environmental resources, will be conducted to formulate specific plans for regional green development.

4.4. Suggestions

Shandong coastal areas are rich in fishery resources and ports, and the rapid development of the marine economy is one of its advantageous industries [61,62]. Therefore, we suggest that the benefits of the marine economy should be fed back into environmental protection, social security and employment, education, and other work to improve socially sustainable development and promote the construction of the Shandong coastal zone into a dynamic area with sustainable development. Adhere to the green economic development model, namely, take sustainable development as the guide and achieve coordination between economic growth and environmental protection by improving resource utilization efficiency and reducing environmental pollution and ecological damage [63,64]. Specific measures are as follows:

(1) Delimit basic farmland protection areas, establish nature reserves, and restrict the development and utilization of land and coastlines. (2) Promote water conservation, develop agricultural and industrial water-saving technologies, strengthen the construction of urban water supply pipe networks, strictly control industrial wastewater discharge in coastal areas, and use coastal advantages to develop seawater desalination technology. (3) Promote the upgrading of marine fisheries, marine transportation, and other industries; strengthen regional cooperation; promote the complementary advantages of coastal areas and inland areas; and increase per capita income. (4) Rationally control population growth, promote the transformation and upgrading of energy-consuming industries to green industries, increase investment in environmental protection, improve education levels, and improve the social security system.

5. Conclusions

In this study, based on multi-source data, including remote sensing images, a dynamic monitoring model of CECC in Shandong Province was constructed by combining a multi-dimensional evaluation index system and a state space model. Secondly, a variety of ecological analysis models were used to evaluate the CECC driving mechanism and the balance between CECC subsystems in Shandong Province.

It was found that CECC in the study area showed a "U"-shaped development trend in the past 10 years, and the coastal ecosystems in Shandong were under great pressure, with significant spatio-temporal differences in carrying capacity of each dimension. Reasonable control of population growth, promoting the upgrading of energy-consuming industries to green industries, saving and protecting water resources, and strengthening the innovation of marine science and technology are important ways to promote the sustainable development of coastal areas in Shandong Province. The economic development–social support system has gradually become the dominant driving force for sustainable development in coastal zones. In addition, it was found that the degree of coordinated development between the economic development–social support system and the resource and environment system affects the sustainable development of the coastal zone. This coordination mechanism has enabled the relevant departments to pay more attention to the relationship between economic–social systems and natural ecosystems in the process of coastal management. Because of the difference between the goals and interests of economic development and ecological protection, decision-makers need to coordinate the interests of all parties and establish an effective coordination mechanism.

Author Contributions: H.L.: conceptualization, methodology, data processing, formal analysis, data curation, writing—original draft preparation, writing—review and editing, and visualization. Y.Z.: conceptualization, data processing, validation, methodology, review, and funding acquisition. X.W.: conceptualization and review. P.G.: methodology and review. K.L.: data processing. All authors have read and agreed to the published version of the manuscript.

Funding: This work was supported by the Director Fund of the International Research Center of Big Data for Sustainable Development Goals (Grant No. CBAS2022DF019) and the Shandong Provincial Natural Science Foundation (Grant No. ZR2022QD072).

Data Availability Statement: Data will be made available on request. The data are not publicly available due to privacy.

Acknowledgments: We would like to express thanks for the constructive comments from the editor and anonymous referees.

Conflicts of Interest: The authors declare that they have no known competing financial interests or personal relationships that could have appeared to influence the work reported in this paper.

References

1. Wei, B.; Li, Y.; Suo, A.; Zhang, Z.; Xu, Y.; Chen, Y. Spatial suitability evaluation of coastal zone, and zoning optimisation in ningbo, China. *Ocean Coast. Manag.* **2021**, *204*, 105507. [CrossRef]
2. Cabrera, J.S.; Lee, H.S. Coastal zone environment Integrity assessment for sustainable management: Part 1.Development of Adaptive Expert-Driven coastal zone Health index framework. *J. Mar. Sci. Eng.* **2022**, *10*, 1183. [CrossRef]
3. Fogarin, S.; Zanetti, M.; Dal Barco, M.K.; Zennaro, F.; Furlan, E.; Torresan, S.; Pham, H.V.; Critto, A. Combining remote sensing analysis with machine learning to evaluate short-term coastal evolution trend in the shoreline of Venice. *Sci. Total Environ.* **2023**, *859*, 160293. [CrossRef] [PubMed]
4. Yin, S.; Guo, J.; Han, Z. County-level environmental carrying capacity and spatial suitability of coastal resources:A case study of Zhuanghe, China. *Front. Mar. Sci.* **2022**, *9*, 1022382. [CrossRef]
5. Xu, N.; Gong, P. Significant coastline changes in China during 1991–2015 tracked by Landsat data. *Sci. Bull.* **2018**, *63*, 883–886. [CrossRef]
6. Huang, P.; Shi, H.; Wang, Z. Integrated Zoning and Spatial Heterogeneity of Coastal Watershed-Nearshore Waters. *Remote Sens.* **2023**, *15*, 3597. [CrossRef]
7. Liu, R.; Pu, L.; Huang, S. Simulation of coastal resource and environmental carrying capacity in the Yangtze River delta coastal zone based on shared socioeconomic pathways. *Front. Mar. Sci.* **2023**, *10*, 1008231. [CrossRef]
8. Zhang, Y.; Li, H.; Hou, X.; Guo, P.; Guo, J. Coastline protection and restoration: A comprehensive review of China's developmental trajectory. *Ocean Coast. Manag.* **2024**, *251*, 107094. [CrossRef]
9. Jiang, S.; Xu, N.; Li, Z.; Huang, C. Satellite derived coastal reclamation expansion in China since the 21st century. *Glob. Ecol. Conserv.* **2021**, *30*, e01797. [CrossRef]
10. Xu, N.; Wang, Y.; Huang, C.; Jiang, S.; Jia, M.; Ma, Y. Monitoring coastal reclamation changes across Jiangsu Province during 1984–2019 using landsat data. *Mar. Policy* **2022**, *136*, 104887. [CrossRef]
11. Naylor, L.A.; Venn, O.; Coombes, M.A.; Jackson, J. *Including Ecological Enhancements in the Planning, Design and Construction of Hard Coastal Structures: A process Guide*; Report to the Environment Agency (PID 110461); University of Exeter: Exeter, UK, 2011; 66p.
12. Chouhan, H.A.; Parthasarathy, D.; Pattanaik, S. Urban development, environmental vulnerability and CRZ violations in India: Impacts on fishing communities and sustainability implications in Mumbai coast. *Environ. Dev. Sustain.* **2017**, *19*, 971–985. [CrossRef]
13. Li, H.; Zhang, Y.; Fan, C.; Hou, X.; Zeng, L.; Guo, P. Characteristics and models of anthropogenic disturbances on islands from perspective of coastline: Extensive cases from Indian Ocean and mediterranean sea. *Ecol. Indic.* **2024**, *160*, 111835. [CrossRef]
14. Sowman, M.; Raemaekers, S. Socio-ecological vulnerability assessment in coastal communities in the BCLME region. *J. Mar. Syst.* **2018**, *188*, 160–171. [CrossRef]
15. Yan, H.; Du, W.; Feng, Z.; Yang, Y.; Xue, Z. Exploring adaptive approaches for social-ecological sustainability in the Belt and Road countries: From the perspective of ecological resource flow. *J. Environ. Manag.* **2022**, *311*, 114898. [CrossRef] [PubMed]
16. Seidl, I.; Tisdell, C.A. Carrying capacity reconsidered: From Malthus' population theory to cultural carrying capacity. *Ecol. Econ.* **1999**, *31*, 395–408. [CrossRef]
17. Shen, L.; Shu, T.; Liao, X.; Yang, N.; Ren, Y.; Zhu, M.; Cheng, G.; Wang, J. A new method to evaluate urban resources environment carrying capacity from the load-and-carrier perspective. *Resour. Conserv. Recycl.* **2020**, *154*, 104616. [CrossRef]
18. Gilland, B.; Higgins, G.M.; Kassam, A.; Naiken, L. *Potential Population Supporting Capacities of Lands in the Developing World*; Food and Agriculture Organization of the United Nations: Roma, Italy, 1982.
19. Lei, K.; Hu, D.; Wang, Z.; Yu, Y.; Zhao, Y. An analysis of ecological footprint trade and sustainable carrying capacity of the population in Macao. *Int. J. Sustain. Dev. World Ecol.* **2009**, *16*, 127–136. [CrossRef]
20. Diansheng, L. Comprehensive load-bearing capacity of resource and environment. *Res. Environ. Sci.* **1995**, *8*, 7–12.
21. Cui, G.; Zhang, X.; Zhang, Z.; Cao, Y.; Liu, X. Comprehensive land carrying capacities of the Cities in the Shandong Peninsula blue economic zone and their spatio-temporal variations. *Sustainability* **2019**, *11*, 439. [CrossRef]
22. Wu, X.; Hu, F. Analysis of ecological carrying capacity using a fuzzy comprehensive evaluation method. *Ecol. Indic.* **2020**, *113*, 106243. [CrossRef]
23. Wu, M.; Wu, J.; Zang, C. A comprehensive evaluation of the eco-carrying capacity and green economy in the Guangdong-Hong Kong-Macao Greater Bay Area, China. *J. Clean. Prod.* **2021**, *281*, 124945. [CrossRef]
24. Yang, J.; Ding, H. A quantitative assessment of sustainable development based on relative resource carrying capacity in Jiangsu Province of China. *Int. J. Environ. Res. Public Health* **2018**, *15*, 2786. [CrossRef]
25. Tang, M.; Wu, D.; Fu, X.; Cao, H. An assessment of ecological carrying capacity of Xilingol, Inner Mongolia. *Int. J. Sustain. Dev. World Ecol.* **2017**, *24*, 408–414. [CrossRef]
26. Xu, B.; Pan, J. Estimation of potential ecological carrying capacity in China. *Environ. Sci. Pollut. Res.* **2020**, *27*, 18044–18063. [CrossRef] [PubMed]
27. Li, S.; Liu, C.; Ge, C.; Yang, J.; Liang, Z.; Li, X.; Cao, X. Ecosystem health assessment using PSR model and obstacle factor diagnosis for Haizhou Bay, China. *Ocean Coast. Manag.* **2024**, *250*, 107024. [CrossRef]
28. Tang, Z. An integrated approach to evaluating the coupling coordination between tourism and the environment. *Tour. Manag.* **2015**, *46*, 11–19. [CrossRef]

29. Liu, R.; Pu, L.; Zhu, M.; Huang, S.; Jiang, Y. Coastal resource-environmental carrying capacity assessment: A comprehensive and trade-off analysis of the case study in Jiangsu coastal zone, eastern China. *Ocean Coast. Manag.* **2020**, *186*, 105092. [CrossRef]
30. Zhang, H.; Zhang, G. Evaluation and influencing factors of resources and environment carrying capacity of Guangdong–Hong Kong–Macao Greater Bay Area Economic Belt. *Environ. Dev. Sustain.* **2024**, 1–17. [CrossRef]
31. Liu, B.; Deng, M.; Yang, J.; Shi, Y.; Huang, J.; Li, C.; Qiu, B. Detecting anomalous spatial interaction patterns by maximizing urban population carrying capacity. *Comput. Environ. Urban Syst.* **2021**, *87*, 101616. [CrossRef]
32. Rockström, J.; Steffen, W.; Noone, K.; Persson, Å.; Chapin III, F.S.; Lambin, E.F.; Lenton, T.M.; Scheffer, M.; Folke, C.; Schellnhuber, H.J. A safe operating space for humanity. *Nature* **2009**, *461*, 472–475. [CrossRef]
33. Tian, Y.; Sun, C. Comprehensive carrying capacity, economic growth and the sustainable development of urban areas: A case study of the Yangtze River Economic Belt. *J. Clean. Prod.* **2018**, *195*, 486–496. [CrossRef]
34. Zou, H.; Duan, X.; Ye, L.; Wang, L. Locating sustainability issues: Identification of ecological vulnerability in Mainland China's mega-regions. *Sustainability* **2017**, *9*, 1179. [CrossRef]
35. Tan, F.; Lu, Z. Regional sustainability system as ecosystem: Case study of China's two leading economic circles from a keystone perspective. *Environ. Dev. Sustain.* **2019**, *21*, 961–983. [CrossRef]
36. Sun, L.; Miao, C.; Yang, L. Ecological-economic efficiency evaluation of green technology innovation in strategic emerging industries based on entropy weighted TOPSIS method. *Ecol. Indic.* **2017**, *73*, 554–558. [CrossRef]
37. Cheng, W.; Xi, H.; Sindikubwabo, C.; Si, J.; Zhao, C.; Yu, T.; Li, A.; Wu, T. Ecosystem health assessment of desert nature reserve with entropy weight and fuzzy mathematics methods: A case study of Badain Jaran Desert. *Ecol. Indic.* **2020**, *119*, 106843. [CrossRef]
38. Saaty, T.L. How to make a decision: The analytic hierarchy process. *Eur. J. Oper. Res.* **1990**, *48*, 9–26. [CrossRef]
39. Goepel, K.D. Implementation of an online software tool for the analytic hierarchy process (AHP-OS). *Int. J. Anal. Hierarchy Process* **2018**, *10*, 469–487. [CrossRef]
40. Li, Y.; Wu, L.; Han, Q.; Wang, X.; Zou, T.; Fan, C. Estimation of remote sensing based ecological index along the Grand Canal based on PCA-AHP-TOPSIS methodology. *Ecol. Indic.* **2021**, *122*, 107214. [CrossRef]
41. Mao, H.; Yu, D. A study on the quantitative research of regional carrying capacity. *Adv. Earth Sci.* **2001**, *16*, 549.
42. Tang, B.; Hu, Y.; Li, H.; Yang, D.; Liu, J. Research on comprehensive carrying capacity of Beijing–Tianjin–Hebei region based on state-space method. *Nat. Hazards* **2016**, *84*, 113–128. [CrossRef]
43. Zhan, Y.; Liu, X.; Li, Y.; Zhang, H.; Wang, D.; Fan, J.; Yang, J. Trends and contribution of different grassland types in restoring the Three River Headwater Region, China, 1988–2012. *Sci. Total Environ.* **2024**, *908*, 168161. [CrossRef] [PubMed]
44. Li, C.; Gao, X.; He, B.J.; Wu, J.; Wu, K. Coupling coordination relationships between urban-industrial land use efficiency and accessibility of highway networks: Evidence from Beijing-Tianjin-Hebei urban agglomeration, China. *Sustainability* **2019**, *11*, 1446. [CrossRef]
45. Kong, F.B.; Yang, W.C.; Xu, C.Y. Coordinated relationship and influencing factors of ecological environment and socio-economic coupling of urban agglomeration around Hangzhou Bay in China. *Acta Ecol. Sin.* **2023**, *43*, 2287–2297.
46. Wang, J.F.; Li, X.H.; Christakos, G.; Liao, Y.; Zhang, T.; Gu, X.; Zheng, X. Geographical detectors-based health risk assessment and its application in the neural tube defects study of the Heshun Region, China. *Int. J. Geogr. Inf. Sci.* **2010**, *24*, 107–127. [CrossRef]
47. Cao, F.; Ge, Y.; Wang, J.F. Optimal discretization for geographical detectors-based risk assessment. *GISci. Remote Sens.* **2013**, *50*, 78–92. [CrossRef]
48. Ou, M.; Lai, X.; Gong, J. Territorial pattern evolution and its comprehensive carrying capacity evaluation in the coastal area of Beibu Gulf, China. *Int. J. Environ. Res. Public Health* **2022**, *19*, 10469. [CrossRef]
49. Wang, C.; He, X.; Ma, M.; Xiong, L.; Zhang, W. Assessing coastal ecosystem carrying capacity by a comprehensive economy-resources-environment system: A case study of South Korea. *Ocean Coast. Manag.* **2022**, *227*, 106283. [CrossRef]
50. Sun, C.; Li, X.; Zou, W.; Wang, S.; Wang, Z. Chinese marine economy development: Dynamic evolution and spatial difference. *Chin. Geogr. Sci.* **2018**, *28*, 111–126. [CrossRef]
51. Zheng, Z.; Wu, Z.; Chen, Y.; Yang, Z.; Marinello, F. Exploration of eco-environment and urbanization changes in coastal zones: A case study in China over the past 20 years. *Ecol. Indic.* **2020**, *119*, 106847. [CrossRef]
52. Fang, X.; Shi, X.; Phillips, T.K.; Du, P.; Gao, W. The coupling coordinated development of urban environment towards sustainable urbanization: An empirical study of Shandong Peninsula, China. *Ecol. Indic.* **2021**, *129*, 107864. [CrossRef]
53. Ali-Toudert, F.; Ji, L. Modeling and measuring urban sustainability in multi-criteria based systems—A challenging issue. *Ecol. Indic.* **2017**, *73*, 597–611. [CrossRef]
54. Ren, Y.; Shen, L.; Wei, X.; Wang, J.; Cheng, G. A guiding index framework for examining urban carrying capacity. *Ecol. Indic.* **2021**, *133*, 108347. [CrossRef]
55. Yin, H.; Xiao, R.; Fei, X.; Zhang, H.; Gao, Z.; Wan, Y.; Tan, W.; Jiang, X.; Cao, W.; Guo, Y. Analyzing "economy-society-environment" sustainability from the perspective of urban spatial structure: A case study of the Yangtze River delta urban agglomeration. *Sustain. Cities Soc.* **2023**, *96*, 104691. [CrossRef]
56. Ma, P.; Ye, G.; Peng, X.; Liu, J.; Qi, J.; Jia, S. Development of an index system for evaluation of ecological carrying capacity of marine ecosystems. *Ocean Coast. Manag.* **2017**, *144*, 23–30. [CrossRef]
57. Shao, Q.; Liu, X.; Zhao, W. An alternative method for analyzing dimensional interactions of urban carrying capacity: Case study of Guangdong-Hong Kong-Macao Greater Bay Area. *J. Environ. Manag.* **2020**, *273*, 111064. [CrossRef]

58. Sun, J.; Miao, J.; Mu, H.; Xu, J.; Zhai, N. Sustainable development in marine economy: Assessing carrying capacity of Shandong province in China. *Ocean Coast. Manag.* **2022**, *216*, 105981. [CrossRef]
59. Georgeson, L.; Maslin, M.; Poessinouw, M. The global green economy: A review of concepts, definitions, measurement methodologies and their interactions. *GEO Geogr. Environ.* **2017**, *4*, e00036. [CrossRef]
60. Pournara, A.; Sakellariadou, F. Development of a Protocol for a Sustainable Blue Economy in the Coastal Zone: Case Study and Preliminary Results in a Coastal Industrial Area in the Eastern Mediterranean. *Sustainability* **2022**, *14*, 10323. [CrossRef]
61. Su, M.; Yang, Y. Evolution of district marine policies in China: The case of Shandong Province. *Mar. Policy* **2018**, *89*, 124–131. [CrossRef]
62. Fu, X.M.; Wang, N.; Jiang, S.S.; Tang, H.Y.; Xue, Z.K.; Li, J.M.; Wang, C.Y. Value evaluation of marine bioresources in Shandong offshore area in China. *Ocean Coast. Manag.* **2018**, *163*, 296–303. [CrossRef]
63. Loiseau, E.; Saikku, L.; Antikainen, R.; Droste, N.; Hansjürgens, B.; Pitkänen, K.; Leskinen, P.; Kuikman, P.; Thomsen, M. Green economy and related concepts: An overview. *J. Clean. Prod.* **2016**, *139*, 361–371. [CrossRef]
64. Mealy, P.; Teytelboym, A. Economic complexity and the green economy. *Res. Policy* **2022**, *51*, 103948. [CrossRef]

land

MDPI

Article

Assessing the Impact of Land-Based Anthropogenic Activities on the Macrobenthic Community in the Intertidal Zones of Anmyeon Island, South Korea

Jian Liang [1,†], Hai-Rui Huang [2,†], Meng-Yuan Shu [3] and Chae-Woo Ma [1,*]

1 Department of Life Science and Biotechnology, Soonchunhyang University, Asan 31538, Republic of Korea; 20217196@sch.ac.kr
2 Department of Business Administration Management, Sejong University, Seoul 05006, Republic of Korea; huang@sju.ac.kr
3 Department of Sports Medicine, Soonchunhyang University, Asan 31538, Republic of Korea; 20227611@sch.ac.kr
* Correspondence: cwooma@sch.ac.kr
† These authors contributed as co-first authors.

Abstract: Anthropogenic activities, particularly land reclamation and industrialization, have severely damaged South Korea's intertidal zones, resulting in a decline in biodiversity. In our study, we assessed the macrobenthic community in the intertidal zone of Anmyeon Island, South Korea, and used remote sensing to evaluate the impact of anthropogenic activities on the adjacent land areas. Spearman and Principal Coordinate Analysis (PCoA) indicated that the remote-sensing ecological index (RSEI) is a viable indicator for assessing the dissimilarity of macrobenthic communities in these zones. Moreover, biota–environment matching (BIO–ENV) and distance-based redundancy analysis (dbRDA) demonstrated that land cover types significantly influence the macrobenthic communities in nearby intertidal zones. Our study suggested that urbanization and agricultural activities have affected the terrestrial ecological environment and the adjacent intertidal communities. Consequently, the protection of these zones should extend beyond their borders to include the management of anthropogenic activities on adjacent lands. Our research contributes valuable insights to help inform conservation strategies and the policy-making necessary to safeguard South Korea's intertidal zones.

Keywords: remote sensing; ecological index; macrobenthos; biota–environment matching; conservation policies

1. Introduction

Intertidal zones, bridging land and sea, provide unique ecosystem services and socio-economic benefits [1,2]. However, as the global population grows rapidly, the impact of anthropogenic activities on these zones is escalating [3–5]. Specific activities, such as land reclamation, urbanization, aquaculture, tourism, and sewage discharge, are contributing to the decline in biodiversity within these zones [6,7].

Macrobenthos are a vital component of biodiversity in intertidal zones. They play a key role in the global carbon and nitrogen cycles and are a primary food source for predators [8]. Additionally, due to their sensitivity to human activities, macrobenthos are frequently used to assess the ecological quality of benthic environments [9,10]. Although the biodiversity of macrobenthos along the South Korean coast has been extensively studied [11], research focusing on the impact of anthropogenic activities on their diversity in intertidal zones remains limited.

Since the 1960s, South Korea's economy has experienced rapid growth [12]. Despite this remarkable economic success, the country's intertidal zones have been extensively impacted by human activities [13,14]. For instance, the Saemangeum Project, initiated by the South Korean government in 1991, is the world's largest land reclamation project. This project significantly reduced the area of intertidal zones and led to substantial habitat loss for numerous wild animals [15,16]. Although there has been an increased focus by the South Korean government on protecting intertidal zones [13], conservation policies remain largely confined to these areas. Measures include access restrictions, a halt to further land reclamation projects, and limitations on aquaculture activities [17].

With advancements in technology such as satellite imaging and remote sensing, it has become more feasible to assess environmental and ecological changes due to anthropogenic activities [18,19]. Remote sensing technology is extensively utilized to manage and monitor coastal environments [20]. Intertidal zones, due to their unique physical characteristics, require significant human and material resources for field surveys. Consequently, some studies have employed remote sensing technology to analyze the epifauna in intertidal zones [21,22]. However, the unique living habits of macrobenthos pose challenges to using remote sensing technology for assessing their diversity.

In this study, we employed remote sensing to evaluate anthropogenic activities in the land areas adjacent to intertidal zones and to ascertain their impact on the community of macrobenthic intertidal zones. This research will provide essential data to aid the South Korean government in developing policies to protect these zones.

2. Materials and Methods

2.1. Study Area

The study area is Anmyeon Island (36°25′–36°36′ E, 126°19′–126°22′ N), located in central South Korea, bordered by the Yellow Sea to the west and Cheonsu Bay to the east (Figure 1). Anmyeon Island stretches approximately 23 km from north to south and spans about 6 km from east to west, covering a total area of roughly 123 km^2. The annual average temperature on Anmyeon Island is 11.8 °C, with most rainfall occurring during the summer. The intertidal zone sediments on the west side of Anmyeon Island consist of sand, while those on the east side are composed of silt. The intertidal zone on the west side is primarily influenced by tourism and artisanal invertebrate fisheries, while the east side is affected mainly by shellfish culture and agricultural wastewater [23,24].

2.2. Sample Collection and Processing

Macrobenthic samples were collected using a quadrat. A 0.25 m side-length quadrat was utilized at western stations, with four replicates per sampling station, encompassing a total sampling area of 0.25 m^2. At eastern stations, a 0.5 m side-length quadrat was used with two replicates per station, resulting in a total sampling area of 0.5 m^2.

Three sampling stations were established in the upper, middle, and lower intertidal zones. In August 2018, samples were collected from stations E2 and E4. The subsequent year, station E3 was sampled in August 2019, followed by station E1 in August 2021. Most recently, in September 2023, samples were collected from stations W1, W2, W3, and W4, all during spring tides (Figure 1). Tidal regions were determined using a GPSMAP 64S (Garmin Ltd., Lenexa, KS, USA) and a LASER 600 rangefinder (Nikon Ltd., Tokyo, Japan). The western stations were characterized by sandy beaches, whereas eastern stations were located in shellfish aquaculture areas.

In the field, macrobenthic samples were sifted through a 0.5 mm sieve and preserved in a 5% formalin solution (Samchun Co., Ltd., Seoul, Republic of Korea). Furthermore, 300 g of sediment was collected at each station using a plastic spoon and stored at −25 °C in an icebox before transport to the laboratory for further analysis.

Macrobenthos were identified to the species level in the laboratory using an Olympus SZX-10 microscope (Olympus Co., Ltd., Tokyo, Japan). The mean grain size was measured utilizing an X-ray particle size analyzer (SediGraph 5100, Micromeritics Instrument Co., Ltd., Norcross, GA, USA). Acid volatile sulfide (AVS), chemical oxygen demand (COD), and ignition loss (IL) were analyzed in accordance with the Marine Environmental Process Test Method [25]. Sediment samples from western stations were analyzed for mean grain size and IL, whereas those from eastern stations were tested for AVS, COD, mean grain size, and IL.

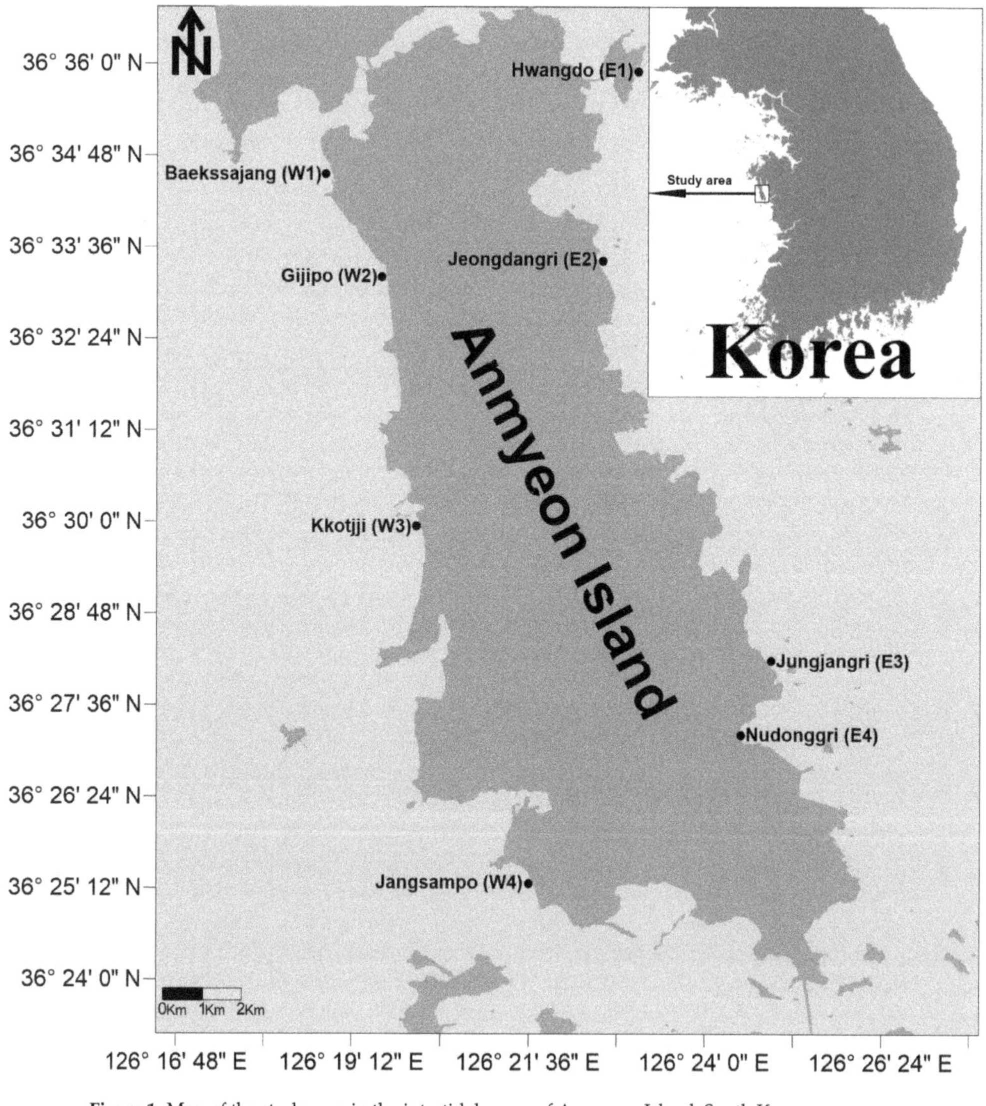

Figure 1. Map of the study area in the intertidal zones of Anmyeon Island, South Korea.

2.3. Diversity Indices

Four traditional diversity indices (i.e., species richness index, Pielou's evenness index, Simpson index, and Shannon–Wiener diversity index) and two taxonomic distinctness indices (i.e., average taxonomic distinctness and variation in taxonomic distinctness) were utilized to evaluate macrobenthos diversity. These indices were calculated using PRIMER version 7 (PRIMER-E Ltd., Albany, New Zealand) [26]. The taxonomic distinctness index considers species' evolutionary relatedness, overcoming many of the limitations of traditional species diversity indices. When the value falls below the 95% confidence interval, it indicates that anthropogenic activities may impact the environment [26]. The formulas for calculating the diversity indices are depicted in Table 1.

Table 1. The formulas for calculating the diversity indices.

Indices	Algorithm	Note	Reference
Species richness index (d)	$= (S-1)/\log(N)$	S: The total number of species; N: The number of individual organisms.	[27]
Pielou's evenness index (J')	$= H'/\log(S)$	H': Shannon–Wiener diversity index; S: Total number of species.	[27]
Simpson index (1-Lambda′)	$= 1 - \mathrm{SUM}(Ni^*(Ni-1)/(N^*(N-1))$	Ni: Number of individuals of the ith species; N: Total number of individual organisms.	[27]
Shannon–Wiener diversity index (H')	$= -\Sigma[(\frac{n_i}{N})\log_2(\frac{n_i}{N})]$	n_i :umber of individuals of the ith species; N: Total number of individuals.	[27]
Average taxonomic distinctness (Delta+)	$= \left[\Sigma\Sigma_{i<j}\omega_{ij}\right]/[S(S-1)/2]$	ω_{ij}: Weight assigned to the path length linking species i and species j in the hierarchical taxonomic tree; S: Total number of species.	[28]
Variation in taxonomic distinctness (Lambda+)	$= \left[\Sigma\Sigma_{i<j}\left(\omega_{ij}-\Delta^+\right)^2\right]/[S(S-1)/2]$	ω_{ij}: Weight assigned to the path length linking species i and species j in the hierarchical taxonomic tree; S: Total number of species.	[28]

2.4. Remote-Sensing Ecological Index

The Remote-Sensing Ecological Index (RSEI) utilizes remote-sensing data to assess the ecological condition of specific geographical areas [29]. The RSEI index was initially developed to evaluate the impact of urbanization on the environmental ecology of Fuzhou City. It is now used for assessing urban environmental ecological quality and other landscape ecosystems, such as islands and mining areas [30,31]. Most studies indicate that changes in RSEI values are closely related to anthropogenic activities. For example, the study by Zheng et al. (2020) suggested that changes in RSEI in China's coastal areas are closely associated with anthropogenic activities [32].

The RSEI value derives from the normalization and principal component analysis (PCA) of four remote-sensing indicators. RSEI is calculated using the formula:

$$\mathrm{RSEI} = f(\mathrm{Wet},\ \mathrm{NDVI},\ \mathrm{LST},\ \mathrm{NDBSI})$$

In the RSEI formula, 'Wet' represents land surface moisture; 'NDVI' (Normalized Difference Vegetation Index) relates to vegetation biomass, leaf area index, and coverage; 'LST' indicates land surface temperature; and 'NDBSI' (Normalized Differential Build-up and Bare Soil Index) reflects land surface desiccation. The methods for calculating these remote-sensing indicators are detailed in Table 2. The ecological quality of an environment can be categorized into five levels based on RSEI values (Table S1) [33].

Remote sensing images of the western stations were sourced from Landsat 9, while those of the eastern stations came from Landsat 8. The remote sensing imagery acquisition coincided with the period of field data collection. These images were obtained from the United States Geological Survey (www.usgs.gov) (accessed on 1 August 2024) with a spatial resolution of 30 m. Prior to calculating RSEI, surface reflectance data from the remote sensing images underwent radiometric, atmospheric, and geometric accuracy corrections [34]. The computation area for RSEI is defined as a 1 km^2 square area, with the midpoint of one side of the coastline of the square sampling area determined by the vertical line from the sampling station in the upper tidal regions to the coastline (Figure S1).

Table 2. The formulae for calculating the remote-sensing indicators.

Indices	Algorithm	Note	Reference
Wet	$Wet = 0.2626\rho_1 + 0.2141\rho_2 + 0.0926\rho_3 + 0.0656\rho_4 - 0.7629\rho_5 - 0.5388\rho_7$	p_i (i = 1, 2, 3, 4, 5, 6, 7): represents the different bands of Landsat image data	[29]
NDVI	$NDVI = (\rho_4 - \rho_3)/(\rho_4 + \rho_3)$	p_i (i = 1, 2, 3, 4, 5, 6, 7): represents the different bands of Landsat image data	[29]
LST	$\mathrm{LST} = T/[1 + (\lambda T/0.01438)\ln\varepsilon]$	T: brightness temperature at the sensor; λ: characteristic wavelength of the thermal infrared band; ε: land surface emissivity	[30]
NDBSI	$\mathrm{IBI} = \{2\rho_5/(\rho_5 + \rho_4) - [(\rho_4/(\rho_4 + \rho_3) + \rho_2/(\rho_2 + \rho_5)]\}/\{2\rho_5/(\rho_5 + \rho_4) + [(\rho_4/(\rho_4 + \rho_3) + \rho_2/(\rho_2 + \rho_5)]\}$ $\mathrm{SI} = [(\rho_5 + \rho_3) - (\rho_4 + \rho_1)]/[(\rho_5 + \rho_3) + (\rho_4 + \rho_1)]$ $\mathrm{NDBSI} = (\mathrm{IBI} + \mathrm{SI})/2$	p_i (i = 1, 2, 3, 4, 5, 6, 7): represents the different bands of Landsat image data	[30]

2.5. Data Analysis

The relationship between macrobenthic biodiversity, the Remote Sensing Ecological Index (RSEI), and various environmental factors was analyzed using Spearman correlation analysis. Principal Coordinate Analysis (PCoA) was used to explore the relationship between the macrobenthic community structure and ecological environment quality classification. Biota–Environment Matching (BIO-ENV) analysis was conducted to examine the association between benthic community structure, environmental factors, remote sensing indicators, and RSEI. Distance-based redundancy analysis (dbRDA) was used to assess the environmental variables influencing the spatial distribution of macrobenthic communities. Spearman correlation analysis utilized Origin version 2021 (OriginLab Corp., Northampton, MA, USA); Principal Coordinate Analysis (PCoA) was performed using OmicShare Tools (https://www.omicshare.com/tools) (accessed on 1 August 2024); BIO–ENV and dbRDA analysis were implemented using PRIMER version 7 (PRIMER-E, NZL). Due to variations in sampling areas and sediment types at the eastern and western stations, analyses for these locations were carried out separately.

3. Results

3.1. Environment Characteristics

IL and mean grain size ranged from 0.3–1.31 and 2.84–3.23, respectively, in the western stations. In contrast, in the eastern stations, IL and mean grain size ranged from 1.64–3.75 and 4.38–7.37, respectively. AVS and COD also varied from 1.64–3.75 and 4.38–7.37 in the eastern stations. The highest coefficient of variation was observed for IL in the eastern stations (Table 3). The values of environmental factors at each station are detailed in Table S2.

3.2. The Macrobenthic Composition

In our study, 55 macrobenthic taxa were identified in the western stations, including 23 arthropod taxa, 21 annelid taxa, 9 mollusk species, 1 echinoderm taxon, and 1 taxon classified nemertea. Conversely, 73 macrobenthic taxa were identified in the eastern stations, comprising 29 annelid taxa, 23 mollusk taxa, 17 arthropod taxa, 1 echinoderm taxa, 1 nemertea taxa, and 2 chordata taxa (Figure 2). In the western stations, the average species abundance was 2118 individuals per m^2, with the highest being 5208 individuals per m^2 at the W2M station (Figure 2). In the eastern stations, the average species abundance was 142 individuals per m^2, with the highest at 422 individuals per m^2 at the E4U station (Figure 2). In the shaded plots, species *Pygospio* sp. and *Urothoe convexa* exhibited high abundance at the western stations. Meanwhile, *Pirenella cingulata* and *Macrophthalmus japonicus* displayed high abundance at the eastern stations (Figure 3). Lists of macrobenthic species at each station are found in Tables S3 and S4.

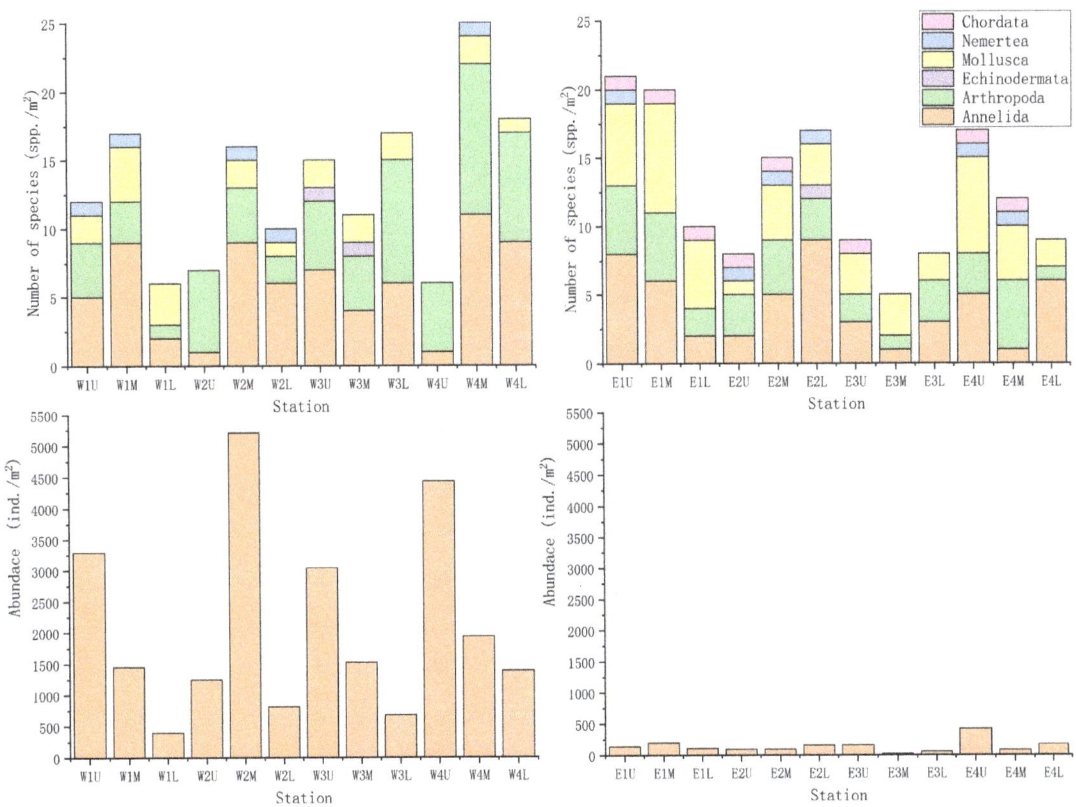

Figure 2. Number of species (spp./m^2) and species abundance (ind./m^2) at each station. Note: W: western station; E: eastern station; U: upper tidal region; M: middle tidal region; L: lower tidal region.

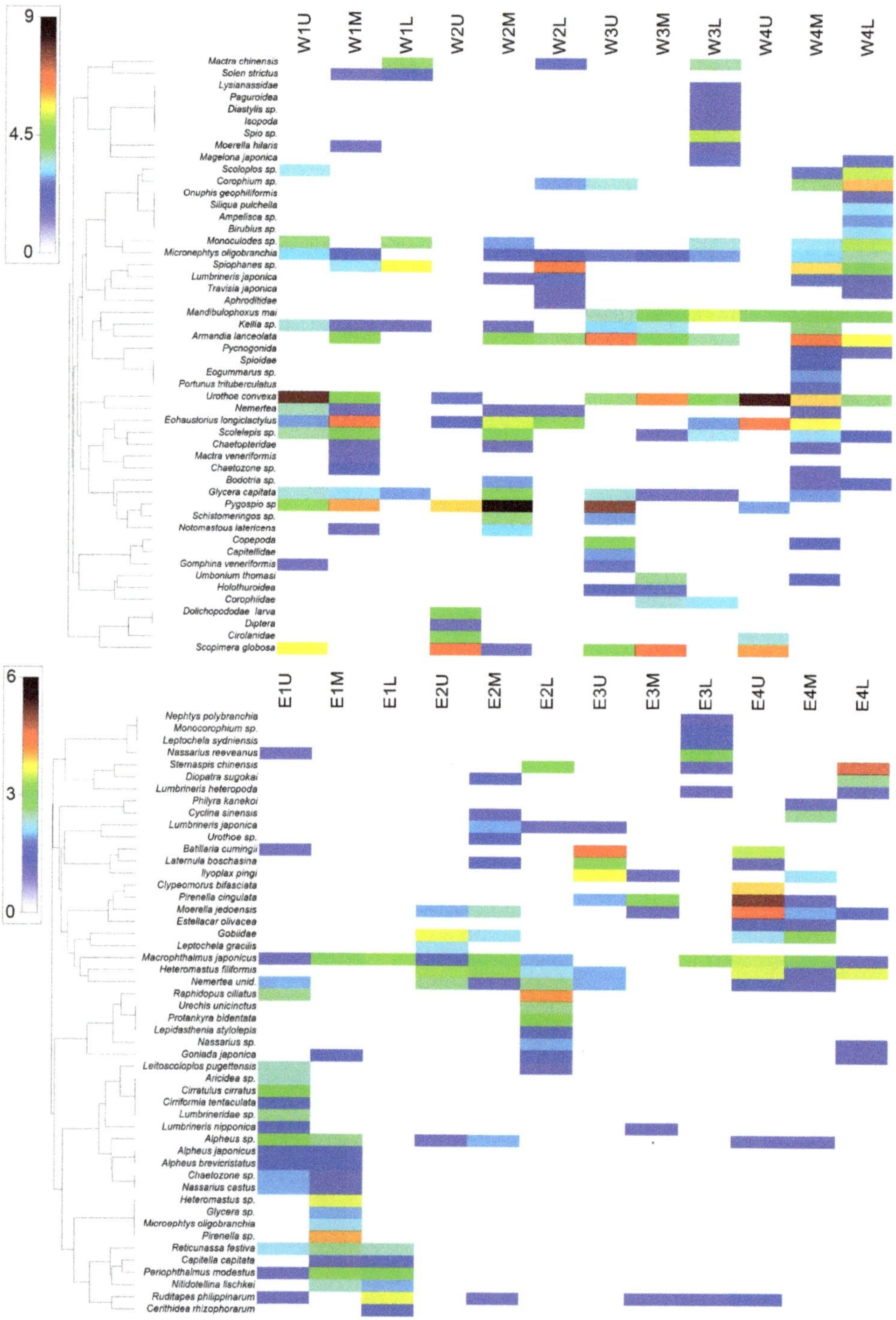

Figure 3. Shaded plots display the top fifty macrobenthic species in the intertidal zone of Anmyeon Island. Note: W: western station; E: eastern station; U: upper tidal region; M: middle tidal region; L: lower tidal region.

3.3. Values of Diversity Indices

At the western stations, the average value of d was 1.67 ± 0.77, peaking at 3.17 at the W4M station. The average value of J' was 0.53 ± 0.16, with a maximum of 3.17 at the W3L station. For H', the average was 1.91 ± 0.77, and the highest value was 3.17 at the W4L station. The average value of 1-Lambda′ was 0.58 ± 0.20, peaking at 0.84 at the W4L station. The average value of Delta+ was 86.35 ± 4.73, with the peak value of 91.21 at the W1L station. Lastly, the average value of Lambda+ was 311.07 ± 47.75, with the highest value of 381.64 at the W4L station (Figure S2).

At the eastern stations, the average value of d was 2.37 ± 0.90, with the highest value of 3.17 at the E1U station. The average value of J' was 0.76 ± 0.10, reaching a maximum of 0.90 at the E1U station. For H', the average was 2.70 ± 0.72, with a peak value of 3.96 at the E1U station. The average value of 1-Lambda′ was 0.77 ± 0.12, peaking at 0.93 at the E1U station. The average value of Delta+ was 91.07 ± 2.49, with the highest value of 91.21 at the E3U station. Lastly, the average Lambda+ value was 266.83 ± 92.30, with the highest value of 413.85 at the E1M station (Figure S3).

The average taxonomic distinctness index values for the W4U and W4L stations were outside the 95% probability distribution (Figure 4a). The taxonomic distinctness index values for the E1U, E1M, and E1L stations also fell outside the 95% probability distribution (Figure 4b).

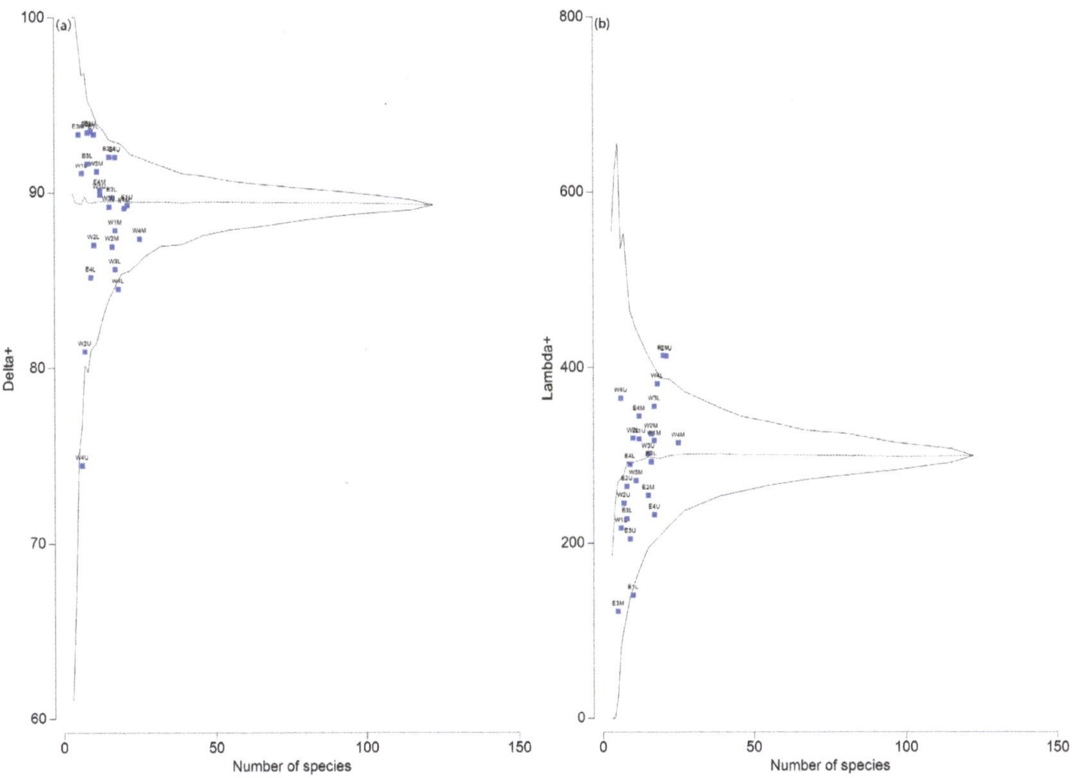

Figure 4. Funnel plots for the Average taxonomic distinctness index (**a**) and Variation in taxonomic distinctness index (**b**) for the number of species assemblages at each station. Note: W: western station; E: eastern station; U: upper tidal region; M: middle tidal region; L: lower tidal region; Dotted line: average value; Line: the 95% confidence limit.

3.4. Values of Remote Sensing Indicators and Remote Sensing Ecological Indices

The highest average NDVI value recorded was 0.46 at station E4 (Figure 5a). The Wet index peaked at an average of 0.00053 at E1 station, with the lowest average of 0.00045 recorded at E2 station (Figure 5b). The highest average NDBST value was −0.48 at W3 station, and the lowest was -0.67 at E4 station (Figure 5c). The LST registered its highest average at 46.33 at E3 station and its lowest at 33.30 at E1 station (Figure 5d). Lastly, the RSEI attained its highest average of 0.62 at W1 station and its lowest average of 0.26 at E4 station (Figure 6).

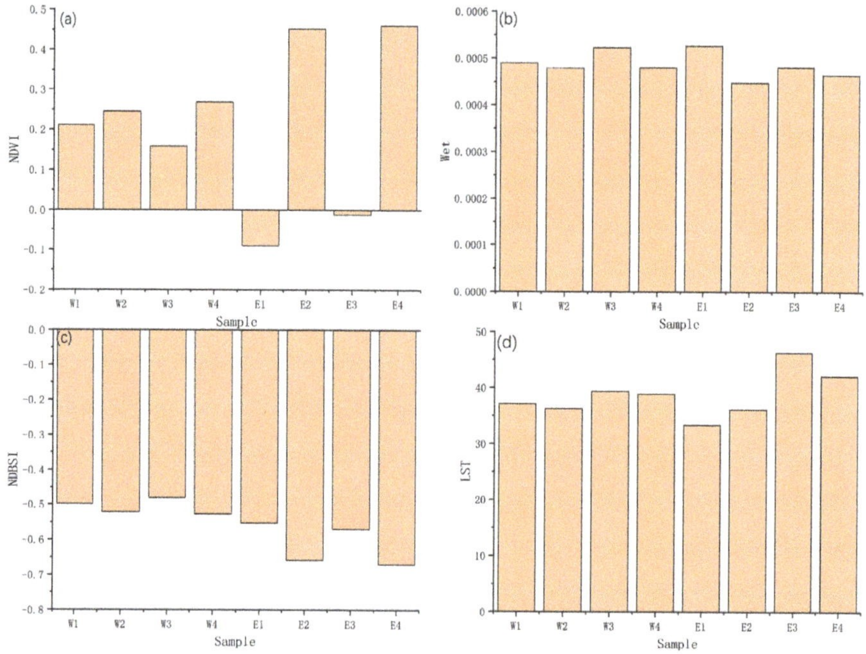

Figure 5. Average values for remote sensing indices at each station. Note: (**a**) NDVI, normalized difference vegetation index; (**b**) Wet, land surface moisture; (**c**) NDBSI, normalized differential build-up and bare soil index; (**d**) LST, land surface temperature; W, western station; E, eastern station.

The RSEI evaluation classified the W1 station's ecological environment quality as good. In contrast, the W3, E1, and E3 stations were classified as moderate, and the W2, W4, E2, and E4 stations were classified as fair. Overall, anthropogenic activities have affected both eastern and western stations, with the former exhibiting a higher level of influence (Figure 6).

Table 3. Environmental characteristics in the study area.

Sampling Stations	Environmental Factors	Range (Min–Max)	Mean ± CV
Western stations (W1U-W4L)	IL, %	0.3–1.31	0.57 ± 0.47
	Mean grain size, ∅	2.84–3.23	3 ± 0.05
Eastern stations (E1U-E4L)	AVS, mg/g	0–0.11	0.02 ± 1.46
	COD, mg/g	6.64–14.64	9.27 ± 0.27
	IL, %	1.64–3.75	2.42 ± 0.25
	Mean grain size, ∅	4.38–7.37	5.92 ± 0.12

Note: AVS, acid volatile sulfide; COD, chemical oxygen demand; CV, coefficient of variation; IL, ignition loss; U: upper tidal region; L: lower tidal region.

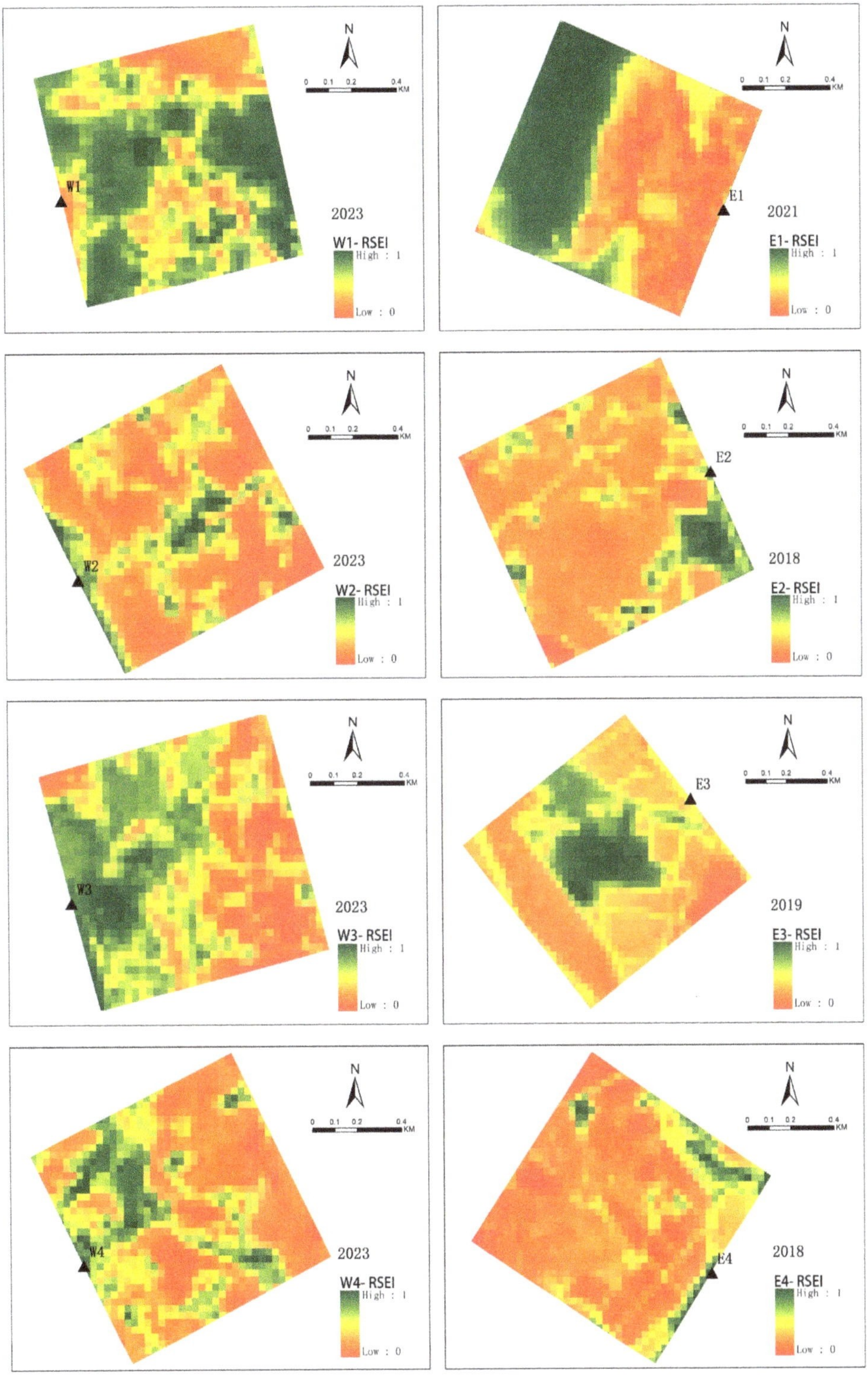

Figure 6. Spatial distribution of RSEI at each station. Note: W: Western station; E: Eastern station.

3.5. Results of Statistical Analysis

In Spearman correlation analysis at western stations, S correlated with d, H', and 1-Lambda'; N correlated with mean grain size; d correlated with H' and 1-Lambda'; J' correlated with H' and 1-Lambda'; H' correlated with 1-Lambda'; Delta+ correlated with Lambda+, RSEI, and IL. IL correlated with mean grain size (Figure 7).

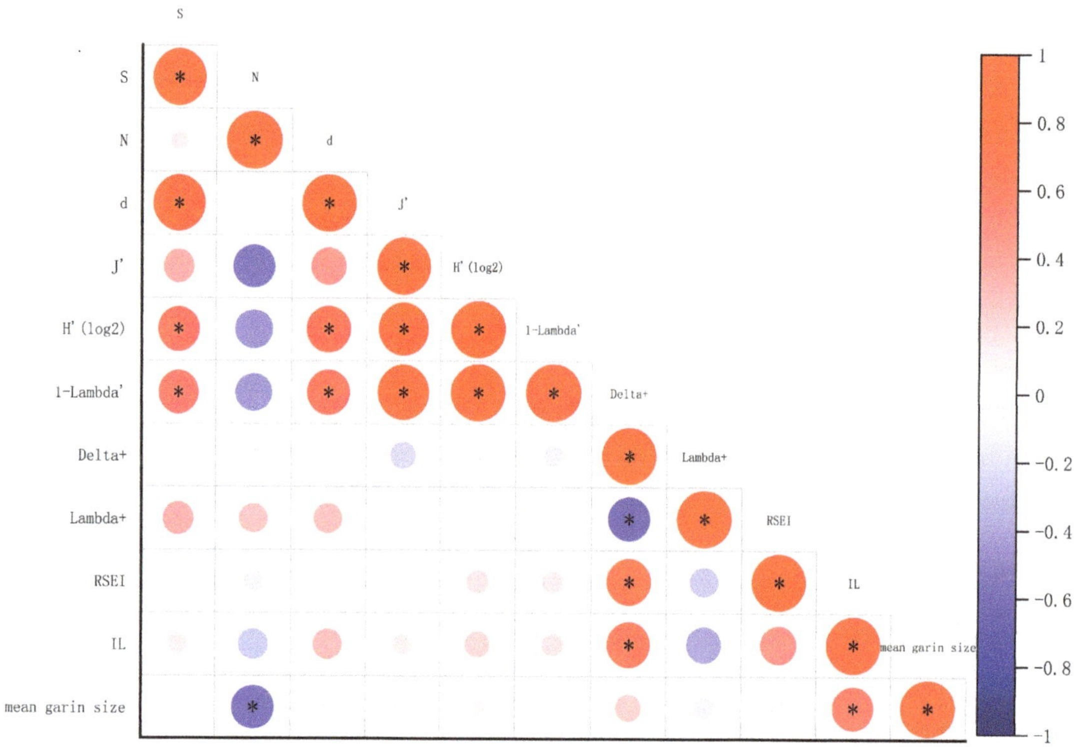

Figure 7. Results of Spearman correlation analysis at western stations. Note: S, number of species; N, number of individuals; d: species richness index; J': Pielou's evenness index; 1-Lambda': Simpson index; H': Shannon–Wiener diversity index; Delta+: average taxonomic distinctness; Lambda+: variation in taxonomic distinctness; RSEI, remote sensing ecological index; IL, ignition loss; *, $p < 0.05$; the larger the circle, the higher the correlation coefficient.

In Spearman correlation analysis at eastern stations, S correlated with N, d, J', H', 1-Lambda', and Lambda+; d correlated with J', H', 1-Lambda', and Lambda+; J' correlated with H', 1-Lambda', and Lambda+; H' correlated with 1-Lambda' and Lambda+; Delta+ correlated with Lambda+, RSEI was correlated with IL and COD; IL correlated with COD; Mean grain size was correlated with AVS (Figure 8).

In principal coordinate analysis, the station's ecological environment quality assigned as "Good" and "Medium" exhibited distinct separations at western stations (Figure 9a), and the eastern station's ecological environment quality assigned as "Medium" and "Fair" exhibited distinct separations at eastern stations (Figure 9b).

In the dbRDA analysis of the western stations, the dbRDA axes 1 and 2 together explained 71.2% of the variability in the fitted model, with 39% represented in Figure 10a. The dbRDA axis 1 was positively correlated with Wet and NDBSI and negatively correlated with other environmental factors. Similarly, the dbRDA axis 2 was positively correlated with NDBSI and negatively correlated with the remaining environmental factors (Table S5). Stations at W1 and W2 are located in the upper part of Figure 10a compared to W3 and W4.

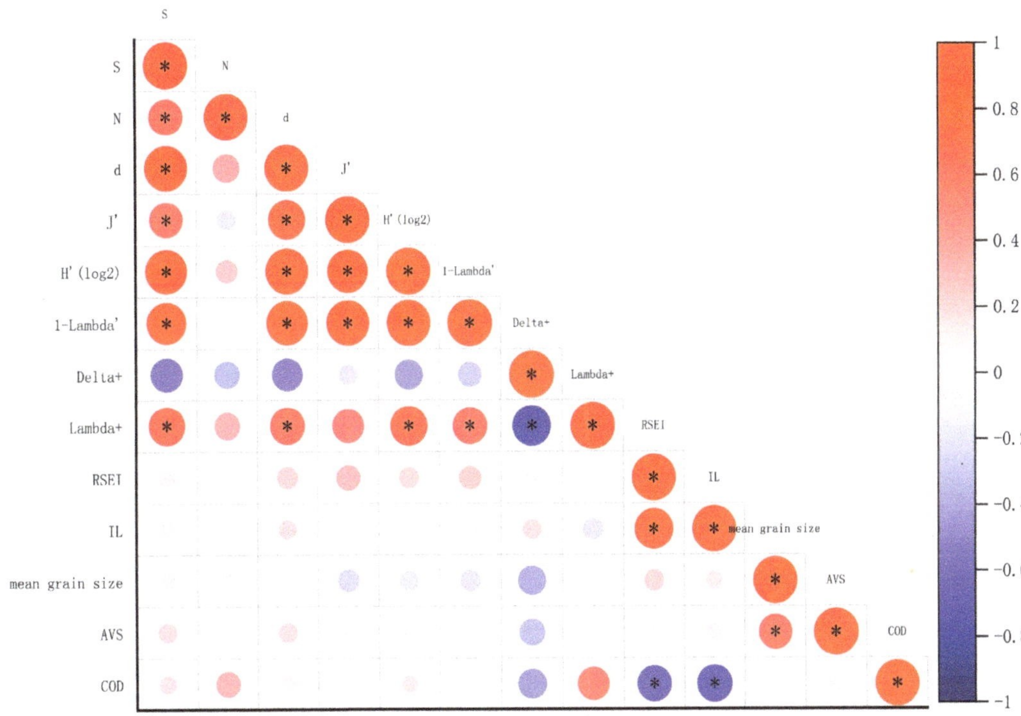

Figure 8. Results of Spearman correlation analysis at eastern stations. Note: *S*, number of species; *N*, number of individuals; *d*: species richness index; *J*′: Pielou's evenness index; 1-Lambda′: Simpson index; *H*′: Shannon–Wiener diversity index; Delta+: average taxonomic distinctness; Lambda+: variation in taxonomic distinctness; RSEI, remote sensing ecological index; AVS, acid volatile sulfide; COD, chemical oxygen demand; IL, ignition loss; *, $p < 0.05$; the larger the circle, the higher the correlation coefficient.

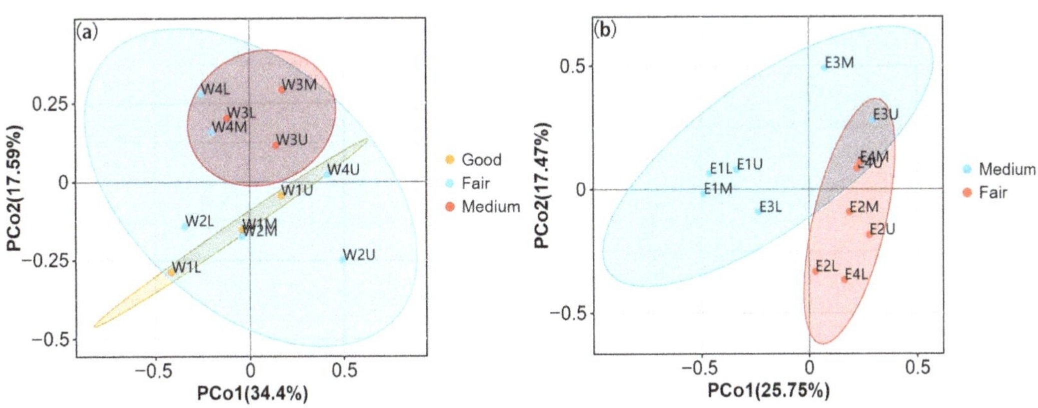

Figure 9. Principal Coordinate Analysis (PCoA) based on the macrobenthic community at western (**a**) and eastern stations (**b**). Note: W: western station; E: eastern station; U: upper tidal region; M: middle tidal region; L: lower tidal region.

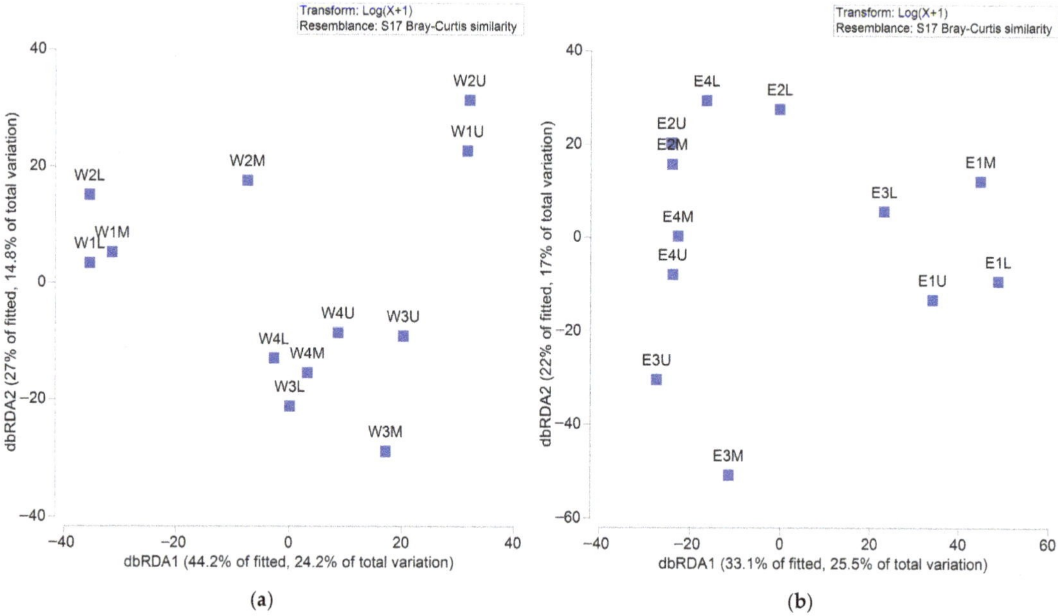

Figure 10. Distance-based redundancy analysis (dbRDA) diagram of the Bray–Curtis similarity index for the macrobenthic community at western (**a**) and eastern stations (**b**). Note: W: western station; E: eastern station; U: upper tidal region; M: middle tidal region; L: lower tidal region.

At the eastern stations, dbRDA axes 1 and 2 together explained 55.1% of the variability in the fitted model, with 42.5% represented in Figure 10b. dbRDA axis 1 was negatively correlated with NDVI, LST, COD, and IL, while it was positively correlated with other environmental factors. Similarly, dbRDA axis 2 was negatively correlated with Wet, LST, IL, and mean grain size, and positively correlated with the remaining environmental factors (Table S6). Stations at E1 are located in the right part of Figure 10b compared to E2, E3, and E4.

The BIO–ENV analysis indicated that LST and Mean grain size had the most significant impact on the macrobenthic community structure at western stations. At eastern stations, WET, LST, and COD were the most influential on the macrobenthic community structure (Table 4).

Table 4. Results from BIO–ENV analyses for the study area.

Sampling Station	Number of Environmental Variables	Correlation Coefficient	Environmental Variables
Western stations (W1U-W4L)	2	0.341	LST, Mean grain size
	3	0.289	RSEI, LST, Mean grain size
	3	0.232	NDBSI, LST, Mean grain size
Eastern stations (E1U-E4L)	3	0.628	Wet, LST, COD
	4	0.628	Wet, LST, AVS, COD
	4	0.614	NDVI, Wet, LST, COD

Note: NDVI, normalized difference vegetation index; Wet, the land surface moisture; NDBSI, normalized differential build-up and bare soil index; LST, land surface temperature; AVS, acid volatile sulfide; COD, chemical oxygen demand.

4. Discussion

4.1. Environment Characteristics and RSEI on Anmyeon Island

The findings of our study on the mean grain size at the eastern and western sites align with previous studies [35,36]. The average IL at the eastern stations was four times that of the western stations. The accumulation and preservation of organic matter in surface sediments are influenced by various environmental factors [37]. The eastern stations, situated in a semi-enclosed bay, experience significantly lower seawater flow rates compared to the western stations. Additionally, the discharge of large amounts of eutrophic freshwater from the artificial lake in the northern part of Cheonsu Bay further exacerbates the accumulation of organic matter in the surface sediments of the eastern stations [24].

RSEI has been extensively employed to evaluate the ecological environmental quality across various landscapes [38]. In this study, the average RSEI values ranged from 0.26 to 0.62, which are comparable to those observed in urban areas [39]. Various factors, including resource utilization and climate change, influence the RSEI values [40,41]. According to Liu et al.'s 2020 study, agricultural activities can alter RSEI values [42]. Anthropogenic activities significantly contribute to COD enhancement in marine sediments [43,44]. The correlation analysis revealed a negative association between RSEI and COD (Figure 9), suggesting that anthropogenic activities, specifically agricultural actions along the eastern land stations, affect the COD measurements in the intertidal zone.

4.2. Biodiversity of Macrobenthic on Anmyeon Island

At the western stations, arthropods were the most abundant taxa, consistent with Jung's (2006) findings [45]. At the eastern stations, annelids prevailed as the most abundant taxa, corroborating a previous study [46]. Despite the western site having a smaller sampling area than the eastern stations, species abundance at the western stations was significantly higher (Figure 2). This discrepancy is attributed to sediment type, a major environmental factor influencing macrobenthic communities [47–51]. The sediment type at the western stations comprised sand, whereas it consisted of silt at the eastern stations. The distinct physical characteristics of sandy beaches likely contribute to the notable increase in the abundance of specific species [51]. For instance, individuals of *Pygospio* sp. and *Urothoe convexa* constituted 29.4% and 27.5% of the total individual count, respectively.

Although traditional diversity indices (e.g., Pielou's evenness index and Simpson index) are widely used to assess the community structure of macrobenthic organisms, compared to taxonomic distinctness indices (e.g., the average taxonomic distinctness index and the variation in taxonomic distinctness index), they only consider diversity information at the species level [52]. Studies have shown that taxonomic distinctness indices are more effective than traditional indices in assessing human disturbances [53,54]. For instance, while the species number and abundance in W4U and W4L were not the lowest, their Delta+ values fell below the 95% confidence interval, indicating that human disturbances might have affected these areas. However, it is important to note that the Lambda+ values for W4U and W4L did not drop below this confidence interval, suggesting a nonlinear response of Lambda+ to human disturbances [55]. Caution is advised when using Lambda+ values, particularly when the species count is low but Lambda+ values are high [56]. No single biodiversity index perfectly evaluates human disturbances. The effectiveness of taxonomic distinctness indices varies depending on the specific taxonomic groups and habitats analyzed [57,58]. Moreover, some studies have highlighted that taxonomic distinctness indices are more responsive to natural environmental stress than to human disturbances [59,60]. In conclusion, we advise employing a combination of multiple biodiversity indices to assess the impact of anthropogenic disturbances on biodiversity.

4.3. Relationship Between Macroinvertebrate Communities and Anthropogenic Activities

In BIO–ENV analyses, LST has been identified as a primary factor influencing the macrobenthic community at both the western and eastern stations. The dbRDA further confirmed that LST is the primary environmental variable driving spatial distribution differences in these communities (Tables S5 and S6). Several studies have demonstrated that different land cover types significantly influence LST [61–63]. Specifically, at the western stations, W1 and W2 are predominantly influenced by agricultural activities due to the extensive farmland in these areas, while W3 and W4 are more significantly impacted by urbanization and tourism, given the considerable number of buildings near the coastline. At the eastern stations, E1 is situated near a town, primarily influenced by urbanization. At the same time, E2, E3, and E4 are located in predominantly agricultural areas, where the main impact comes from agricultural activities.

RSEI has been established as an effective metric for assessing the quality of terrestrial ecological environments [64]. A positive correlation exists between RSEI and Delta+ at the western stations. Anthropogenic activities can adversely affect the RSEI [65,66]. Similarly, Delta+ has been shown to be effective in evaluating anthropogenic activities [67]. The western stations are situated on sandy beaches, which possess tourist attributes. Land developments on the upper parts of these beaches, including the construction of parking lots, hotels, and recreational facilities, have intensified the impacts of anthropogenic activities on these ecosystems. Some studies have indicated that tourism adversely affects sandy beach ecosystems [68,69]. However, RSEI did not correlate with Delta+ at the eastern station, possibly because the land use in the upper part of the intertidal zone at the eastern station mainly comprises farmland. Diverse developmental activities in the upper intertidal zones can have varying impacts on the intertidal macrobenthic communities.

Agricultural activities often release significant amounts of wastewater containing pesticides and fertilizers into the surrounding environment [70]. When this untreated wastewater reaches intertidal zones, it can harm macrobenthic communities [71]. Urbanization refers to the physical transformation of the natural environment by human activities. For sandy beaches, urbanization-related infrastructure improvements lead to an increase in tourists, which can affect the macrobenthic diversity of these beaches [72]. For mudflats, the discharge of domestic or industrial wastewater due to urbanization can negatively impact macrobenthic communities [73].

Additionally, results from the principal coordinate analysis (Figure 9) suggest that the quality of the ecological environment on land can affect intertidal macrobenthic communities. Urbanization and agricultural activities are closely related to the quality of the terrestrial ecological environment [74]. As mentioned earlier, agricultural activities and urbanization also affect the macrobenthic communities in the intertidal zone. Our study indicates that RSEI also holds potential as an indirect indicator for evaluating the dissimilarity of macrobenthic communities in intertidal zones. Nevertheless, it is important to acknowledge some limitations of RSEI, such as possible instabilities in its time-series assessments [75].

This highlights the importance of extending biodiversity conservation strategies in intertidal zones beyond their current scopes. Addressing changes in land cover types caused by nearby anthropogenic activities is also critical. Effective management of intertidal areas and their surrounding environments is vital for the practical preservation of biodiversity.

4.4. Shortcomings of the Study

Due to limitations in funding and personnel resources, our study was conducted exclusively in the intertidal zone of Anmyeon Island during the summer. In this season, macrobenthic communities typically exhibit greater species richness, higher population densities, and a distinct taxonomic composition [73], making them more representative of the region's community structure than other seasons. The intertidal zone is also particularly vulnerable to anthropogenic activities during the summer, especially in South Korea, where many tourists frequent these areas and agricultural activities intensify.

Additionally, due to inconsistencies in sampling areas and the fact that surveys were conducted in different years, we separately assessed the impact of anthropogenic activities on macrobenthic communities for the eastern and western intertidal zones. However, despite the different survey years, the main anthropogenic activities on Anmyeon Island remained unchanged during the study period.

In future studies, it will be essential to conduct surveys across all seasons and ensure they are carried out within the same year. Standardizing both the sampling areas and the timing of surveys will allow for a more comprehensive assessment of the impact of anthropogenic activities on the macrobenthic communities in the intertidal zone of Anmyeon Island.

5. Conclusions

Anthropogenic activities have significantly impacted the ecosystems of South Korea's intertidal zones. Our study found the ecological environment quality of the land adjacent to the western station of Anmyeon Island, South Korea, to be superior compared to that of the eastern station. Combining field data and remote sensing revealed that differing levels of human use of the land adjacent to intertidal zones elicited varying negative responses in the macrobenthic communities, despite the South Korean government implementing numerous legislations since 1999 aimed at the sustainable use of coastal resources. However, protection policies for the intertidal zones are still limited to those areas. We urge the South Korean government to consider various aspects, particularly terrestrial anthropogenic activities adjacent to the intertidal zones, when drafting future protection policies. Additionally, the South Korean government must establish specialized legislation to protect these zones and address current regulatory gaps.

Supplementary Materials: The following supporting information can be downloaded at: https://www.mdpi.com/article/10.3390/land14010062/s1, Table S1: Five levels of ecological environmental quality are classified based on RSEI; Table S2: Environmental factors at each sampling station; Table S3: Macrobenthic species lists (ind./m2) at western stations; Table S4: Macrobenthic species lists (ind./m2) at eastern stations; Table S5: Eigenvectors of environmental factors to the dbRDA axes 1 and 2 at western stations; Table S6: Eigenvectors of environmental factors to the dbRDA axes 1 and 2 at eastern stations; Figure S1: Example of sampling for RSEI calculation; Figure S2: Diversity indices values at the western stations; and Figure S3: Diversity indices values at the eastern stations.

Author Contributions: Conceptualization, J.L. and H.-R.H.; methodology, J.L.; software, J.L.; validation, J.L.; formal analysis, J.L.; investigation, H.-R.H. and M.-Y.S.; resources, J.L.; data curation, J.L.; writing—original draft preparation, J.L.; writing—review and editing, J.L. and C.-W.M.; visualization, C.-W.M.; supervision, C.-W.M.; project administration, C.-W.M.; funding acquisition, C.-W.M. All authors have read and agreed to the published version of the manuscript.

Funding: This work was supported by the Soonchunhyang University Research Fund.

Data Availability Statement: Data are available in the Supplementary Materials.

Conflicts of Interest: The authors declare no conflicts of interest.

References

1. Mehvar, S.; Filatova, T.; Dastgheib, A.; De Ruyter Van Steveninck, E.; Ranasinghe, R. Quantifying Economic Value of Coastal Ecosystem Services: A Review. *J. Mar. Sci. Eng.* **2018**, *6*, 5. [CrossRef]
2. Liang, J.; Huang, H.-R.; Ma, C.-W.; Son, D.-S.; Kim, S.-K. Using the Heavy Metal Indices and Benthic Indices to Assess the Ecological Quality in the Tidal Flats of Garolim Bay, South Korea. *Water* **2024**, *16*, 736. [CrossRef]
3. Thompson, R.C.; Crowe, T.P.; Hawkins, S.J. Rocky Intertidal Communities: Past Environmental Changes, Present Status and Predictions for the next 25 Years. *Environ. Conserv.* **2002**, *29*, 168–191. [CrossRef]
4. Mead, A.; Griffiths, C.L.; Branch, G.M.; McQuaid, C.D.; Blamey, L.K.; Bolton, J.J.; Anderson, R.J.; Dufois, F.; Rouault, M.; Froneman, P.W.; et al. Human-Mediated Drivers of Change—Impacts on Coastal Ecosystems and Marine Biota of South Africa. *Afr. J. Mar. Sci.* **2013**, *35*, 403–425. [CrossRef]
5. Newton, A.; Icely, J.; Cristina, S.; Perillo, G.M.E.; Turner, R.E.; Ashan, D.; Cragg, S.; Luo, Y.; Tu, C.; Li, Y.; et al. Anthropogenic, Direct Pressures on Coastal Wetlands. *Front. Ecol. Evol.* **2020**, *8*, 144. [CrossRef]
6. Zhai, T.; Wang, J.; Fang, Y.; Qin, Y.; Huang, L.; Chen, Y. Assessing Ecological Risks Caused by Human Activities in Rapid Urbanization Coastal Areas: Towards an Integrated Approach to Determining Key Areas of Terrestrial-Oceanic Ecosystems Preservation and Restoration. *Sci. Total Environ.* **2020**, *708*, 135153. [CrossRef] [PubMed]
7. Wu, W.; Yang, Z.; Tian, B.; Huang, Y.; Zhou, Y.; Zhang, T. Impacts of Coastal Reclamation on Wetlands: Loss, Resilience, and Sustainable Management. *Estuar. Coast. Shelf Sci.* **2018**, *210*, 153–161. [CrossRef]
8. Sukumaran, S.; Vijapure, T.; Mulik, J.; Ridha, H. Marine Macrobenthos of NorthWest India-Reviewing the Known and Unknown. *Front. Mar. Sci.* **2021**, *8*, 671245. [CrossRef]
9. Liang, J.; Ma, C.-W.; Kim, K.-B.; Son, D.-S. Can the Ecological Quality of Several Bays in South Korea Be Accurately Assessed Using Multiple Benthic Biotic Indices? *J. Mar. Sci. Eng.* **2024**, *12*, 1179. [CrossRef]
10. Liang, J.; Ma, C.-W.; Son, D.-S. Using the Heavy Metal and Biotic Indices to Assess Ecological Quality in the Central Area of the East Sea, South Korea. *Water* **2024**, *16*, 1230. [CrossRef]
11. Park, J.; Song, S.J.; Ryu, J.; Kwon, B.-O.; Hong, S.; Bae, H.; Choi, J.-W.; Khim, J.S. Macrozoobenthos of Korean Tidal Flats: A Review on Species Assemblages and Distribution. *Ocean Coast. Manag.* **2014**, *102*, 483–492. [CrossRef]
12. Koo, H. Rising Inequality and Shifting Class Boundaries in South Korea in the Neo-Liberal Era. *J. Contemp. Asia* **2021**, *51*, 1–19. [CrossRef]
13. Yim, J.; Kwon, B.-O.; Nam, J.; Hwang, J.H.; Choi, K.; Khim, J.S. Analysis of Forty Years Long Changes in Coastal Land Use and Land Cover of the Yellow Sea: The Gains or Losses in Ecosystem Services. *Environ. Pollut.* **2018**, *241*, 74–84. [CrossRef] [PubMed]
14. Hong, S.-K.; Koh, C.-H.; Harris, R.R.; Kim, J.-E.; Lee, J.-S.; Ihm, B.-S. Land Use in Korean Tidal Wetlands: Impacts and Management Strategies. *Enviro. Manag.* **2010**, *45*, 1014–1026. [CrossRef] [PubMed]
15. Moores, N.; Rogers, D.I.; Rogers, K.; Hansbro, P.M. Reclamation of Tidal Flats and Shorebird Declines in Saemangeum and Elsewhere in the Republic of Korea. *Emu* **2016**, *116*, 136–146. [CrossRef]
16. Ryu, J.; Nam, J.; Park, J.; Kwon, B.-O.; Lee, J.-H.; Song, S.J.; Hong, S.; Chang, W.K.; Khim, J.S. The Saemangeum Tidal Flat: Long-Term Environmental and Ecological Changes in Marine Benthic Flora and Fauna in Relation to the Embankment. *Ocean. Coast. Manag.* **2014**, *102*, 559–571. [CrossRef]
17. Koh, C.-H.; De Jonge, V.N. Stopping the Disastrous Embankments of Coastal Wetlands by Implementing Effective Management Principles: Yellow Sea and Korea Compared to the European Wadden Sea. *Ocean Coast. Manag.* **2014**, *102*, 604–621. [CrossRef]
18. Li, J.; Pei, Y.; Zhao, S.; Xiao, R.; Sang, X.; Zhang, C. A Review of Remote Sensing for Environmental Monitoring in China. *Remote Sens.* **2020**, *12*, 1130. [CrossRef]
19. El-Naggar, H.A.; Salem, E.S.; El-Kafrawy, S.B.; Bashar, M.A.; Shaban, W.M.; El-Gayar, E.E.; Ahmed, H.O.; Ashour, M.; Abou-Mahmoud, M.E. An Integrated Field Data and Remote Sensing Approach for Impact Assessment of Human Activities on Epifauna Macrobenthos Biodiversity along the Western Coast of Aqaba Gulf. *Ecohydrology* **2022**, *15*, e2400. [CrossRef]
20. El Mahrad, B.; Newton, A.; Icely, J.; Kacimi, I.; Abalansa, S.; Snoussi, M. Contribution of Remote Sensing Technologies to a Holistic Coastal and Marine Environmental Management Framework: A Review. *Remote Sens.* **2020**, *12*, 2313. [CrossRef]
21. Davies, B.F.R.; Gernez, P.; Geraud, A.; Oiry, S.; Rosa, P.; Zoffoli, M.L.; Barillé, L. Multi- and Hyperspectral Classification of Soft-Bottom Intertidal Vegetation Using a Spectral Library for Coastal Biodiversity Remote Sensing. *Remote Sens. Environ.* **2023**, *290*, 113554. [CrossRef]
22. Thorner, J.; Kumar, L.; Smith, S.D.A. Fine-Scale Three-Dimensional Habitat Mapping as a Biodiversity Conservation Tool for Intertidal Rocky Reefs. *J. Coast. Res.* **2013**, *290*, 1184–1190. [CrossRef]
23. Liang, J.; Shu, M.-Y.; Huang, H.-R.; Ma, C.-W.; Kim, S.-K. Using Benthic Indices to Assess the Ecological Quality of Sandy Beaches and the Impact of Urbanisation on Sandy Beach Ecosystems. *J. Mar. Sci. Eng.* **2024**, *12*, 487. [CrossRef]
24. Liang, J.; Ma, C.-W.; Kim, S.-K.; Park, S.-H. Assessing the Benthic Ecological Quality in the Intertidal Zone of Cheonsu Bay, Korea, Using Multiple Biotic Indices. *Water* **2024**, *16*, 272. [CrossRef]

25. National Institute of Fisheries Science. *National Institute of Fisheries Science Notification of Marine Environmental Process Test Standards 2010;* National Institute of Fisheries Science: Busan, Republic of Korea, 2010.
26. Clarke, K.; Gorley, R.; Somerfield, P.; Warwick, R. *Changes in Marine Communities: An Approach to Statistical Analysis and Interpretation*, 3rd ed.; Plymouth Marine Laboratory, PRIMER-E Ltd.: Wellington, New Zealand, 2014.
27. Hill, M.O. Diversity and Evenness: A Unifying Notation and Its Consequences. *Ecology* **1973**, *54*, 427–432. [CrossRef]
28. Warwick, R.M.; Clarke, K.R. New "biodiversity" Measures Reveal a Decrease in Taxonomic Distinctness with Increasing Stress. *Mar. Ecol. Prog. Ser.* **1995**, *129*, 301–305. [CrossRef]
29. Hu, X.; Xu, H. A New Remote Sensing Index for Assessing the Spatial Heterogeneity in Urban Ecological Quality: A Case from Fuzhou City, China. *Ecol. Indic.* **2018**, *89*, 11–21. [CrossRef]
30. Liu, H.; Jiang, Y.; Misa, R.; Gao, J.; Xia, M.; Preusse, A.; Sroka, A.; Jiang, Y. Ecological Environment Changes of Mining Areas around Nansi Lake with Remote Sensing Monitoring. *Environ. Sci. Pollut. Res.* **2021**, *28*, 44152–44164. [CrossRef] [PubMed]
31. Wen, X.; Ming, Y.; Gao, Y.; Hu, X. Dynamic Monitoring and Analysis of Ecological Quality of Pingtan Comprehensive Experimental Zone, a New Type of Sea Island City, Based on RSEI. *Sustainability* **2019**, *12*, 21. [CrossRef]
32. Zheng, Z.; Wu, Z.; Chen, Y.; Yang, Z.; Marinello, F. Exploration of Eco-Environment and Urbanization Changes in Coastal Zones: A Case Study in China over the Past 20 Years. *Ecol. Indic.* **2020**, *119*, 106847. [CrossRef]
33. Zhang, T.; Yang, R.; Yang, Y.; Li, L.; Chen, L. Assessing the Urban Eco-Environmental Quality by the Remote-Sensing Ecological Index: Application to Tianjin, North China. *ISPRS. Int. J. Geo-Inf.* **2021**, *10*, 475. [CrossRef]
34. Geng, J.; Yu, K.; Xie, Z.; Zhao, G.; Ai, J.; Yang, L.; Yang, H.; Liu, J. Analysis of Spatiotemporal Variation and Drivers of Ecological Quality in Fuzhou Based on RSEI. *Remote Sens.* **2022**, *14*, 4900. [CrossRef]
35. Oh, J.-K.; Han, C.-H. Depositional Environment of Sandy Tidal Flat in Anmyeondo, Western Coast of Korea. *J. Kor. Earth Sci. Soc.* **2010**, *31*, 139–150. [CrossRef]
36. Woo, H.J.; Choi, J.U.; Ryu, J.H.; Choi, S.H.; Kim, S.R. Sedimentary Environments in the Hwangdo Tidal Flat, Cheonsu Bay. *J. Wetlands Res.* **2005**, *7*, 63–67.
37. Lu, X.; Zhou, F.; Chen, F.; Lao, Q.; Zhu, Q.; Meng, Y.; Chen, C. Spatial and Seasonal Variations of Sedimentary Organic Matter in a Subtropical Bay: Implication for Human Interventions. *Int. J. Environ. Res. Public Health* **2020**, *17*, 1362. [CrossRef]
38. Yuan, B.; Fu, L.; Zou, Y.; Zhang, S.; Chen, X.; Li, F.; Deng, Z.; Xie, Y. Spatiotemporal Change Detection of Ecological Quality and the Associated Affecting Factors in Dongting Lake Basin, Based on RSEI. *J. Clean. Prod.* **2021**, *302*, 126995. [CrossRef]
39. Yue, H.; Liu, Y.; Li, Y.; Lu, Y. Eco-Environmental Quality Assessment in China's 35 Major Cities Based on Remote Sensing Ecological Index. *IEEE Access* **2019**, *7*, 51295–51311. [CrossRef]
40. Xiong, Y.; Xu, W.; Lu, N.; Huang, S.; Wu, C.; Wang, L.; Dai, F.; Kou, W. Assessment of Spatial–Temporal Changes of Ecological Environment Quality Based on RSEI and GEE: A Case Study in Erhai Lake Basin, Yunnan Province, China. *Ecol. Indic.* **2021**, *125*, 107518. [CrossRef]
41. An, M.; Xie, P.; He, W.; Wang, B.; Huang, J.; Khanal, R. Spatiotemporal Change of Ecologic Environment Quality and Human Interaction Factors in Three Gorges Ecologic Economic Corridor, Based on RSEI. *Ecol. Indic.* **2022**, *141*, 109090. [CrossRef]
42. Liu, X.Y.; Zhang, X.X.; He, Y.R.; Luan, H.J. Monitoring and assessment of ecological change in coastal cities based on RSEI. *Int. Arch. Photogramm. Remote Sens. Spatial Inf. Sci.* **2020**, *XLII-3/W10*, 461–470. [CrossRef]
43. Yuan, Y.; Song, D.; Wu, W.; Liang, S.; Wang, Y.; Ren, Z. The Impact of Anthropogenic Activities on Marine Environment in Jiaozhou Bay, Qingdao, China: A Review and a Case Study. *Reg. Stud. Mar. Sci.* **2016**, *8*, 287–296. [CrossRef]
44. Irizuki, T.; Hirose, K.; Ueda, Y.; Fujihara, Y.; Ishiga, H.; Seto, K. Ecological Shifts Due to Anthropogenic Activities in the Coastal Seas of the Seto Inland Sea, Japan, since the 20th Century. *Mar. Pollut. Bull.* **2018**, *127*, 637–653. [CrossRef]
45. Jung, Y.-H. Ecological Studies and Spatial Patterns of Macrobenthos on Tidal Flat around Anmyon-Do. Master's Thesis, Soonchunhyang University, Asan, Republic of Korea, 2006.
46. Jung, R.-H.; Seo, I.-S.; Lee, W.-C.; Kim, H.-C.; Park, S.-R.; Kim, J.-B.; Oh, C.-W.; Choi, B.-M. Community Structure and Health Assessment of Macrobenthic Assemblages at Spring and Summer in Cheonsu Bay, West Coast of Korea. *J. Korean Soc. Oceanogr.* **2014**, *19*, 272–286.
47. Dong, J.-Y.; Zhao, L.; Sun, X.; Hu, C.; Wang, Y.; Li, W.-T.; Zhang, P.-D.; Zhang, X. Response of Macrobenthic Communities to Heavy Metal Pollution in Laoshan Bay, China: A Trait-Based Method. *Mar. Pollut. Bull.* **2021**, *167*, 112292. [CrossRef] [PubMed]
48. Mucha, A.P.; Vasconcelos, M.T.S.D.; Bordalo, A.A. Macrobenthic Community in the Douro Estuary: Relations with Trace Metals and Natural Sediment Characteristics. *Environ. Pollut.* **2003**, *121*, 169–180. [CrossRef] [PubMed]
49. Liang, J.; Ma, C.-W.; Kim, K.-B. Ecological Risk Assessment of Heavy Metals in Surface Sediments and Their Impact on Macrobenthos in Asan Bay, South Korea. *Front. Mar. Sci.* **2024**, *11*, 1450396. [CrossRef]
50. Liang, J.; Ma, C.-W.; Kim, K.-B. Comparing the environmental impacts of pollution from two types of industrial zones on the coast. *Front. Mar. Sci.* **2024**, 11. [CrossRef]
51. Defeo, O.; McLachlan, A.; Schoeman, D.S.; Schlacher, T.A.; Dugan, J.; Jones, A.; Lastra, M.; Scapini, F. Threats to Sandy Beach Ecosystems: A Review. *Estuar. Coast. Shelf Sci.* **2009**, *81*, 1–12. [CrossRef]

52. Li, S.; Zhang, X.; Su, H.; Chen, S.; Chen, W.; Li, F.; Wang, T.; Sun, Z.; Weng, X. Taxonomic Diversity of the Macrobenthos and Its Relationship with Environmental Factors in the Huanghe River (Yellow River) Estuary and Adjacent Seas, China. *J. Sea Res.* **2023**, *194*, 102402. [CrossRef]
53. Tweedley, J.R.; Warwick, R.M.; Potter, I.C. Can Biotic Indicators Distinguish between Natural and Anthropogenic Environmental Stress in Estuaries? *J. Sea Res.* **2015**, *102*, 10–21. [CrossRef]
54. Leonard, D.R.P.; Robert Clarke, K.; Somerfield, P.J.; Warwick, R.M. The Application of an Indicator Based on Taxonomic Distinctness for UK Marine Biodiversity Assessments. *J. Environ. Manag.* **2006**, *78*, 52–62. [CrossRef] [PubMed]
55. Mouillot, D.; Laune, J.; Tomasini, J.-A.; Aliaume, C.; Brehmer, P.; Dutrieux, E.; Do Chi, T. Assessment of Coastal Lagoon Quality with Taxonomic Diversity Indices of Fish, Zoobenthos and Macrophyte Communities. *Hydrobiologia* **2005**, *550*, 121–130. [CrossRef]
56. Hu, G.; Zhang, Q. Seasonal Variations in Macrobenthic Taxonomic Diversity and the Application of Taxonomic Distinctness Indices in Bohai Bay, Northern China. *Ecol. Indic.* **2016**, *71*, 181–190. [CrossRef]
57. Jiang, X.; Song, Z.; Xiong, J.; Xie, Z. Can Excluding Non-Insect Taxa from Stream Macroinvertebrate Surveys Enhance the Sensitivity of Taxonomic Distinctness Indices to Human Disturbance? *Ecol. Indic.* **2014**, *41*, 175–182. [CrossRef]
58. Tolimieri, N.; Anderson, M.J. Taxonomic Distinctness of Demersal Fishes of the California Current: Moving Beyond Simple Measures of Diversity for Marine Ecosystem-Based Management. *PLoS ONE* **2010**, *5*, e10653. [CrossRef]
59. Bevilacqua, S.; Sandulli, R.; Plicanti, A.; Terlizzi, A. Taxonomic Distinctness in Mediterranean Marine Nematodes and Its Relevance for Environmental Impact Assessment. *Mar. Pollut. Bull.* **2012**, *64*, 1409–1416. [CrossRef]
60. Salas, F.; Patrício, J.; Marcos, C.; Pardal, M.A.; Pérez-Ruzafa, A.; Marques, J.C. Are Taxonomic Distinctness Measures Compliant to Other Ecological Indicators in Assessing Ecological Status? *Mar. Pollut. Bull.* **2006**, *52*, 162–174. [CrossRef]
61. Chu, L.; Oloo, F.; Bergstedt, H.; Blaschke, T. Assessing the Link between Human Modification and Changes in Land Surface Temperature in Hainan, China Using Image Archives from Google Earth Engine. *Remote Sens.* **2020**, *12*, 888. [CrossRef]
62. Ren, Y.; Deng, L.-Y.; Zuo, S.-D.; Song, X.-D.; Liao, Y.-L.; Xu, C.-D.; Chen, Q.; Hua, L.-Z.; Li, Z.-W. Quantifying the Influences of Various Ecological Factors on Land Surface Temperature of Urban Forests. *Environ. Pollut.* **2016**, *216*, 519–529. [CrossRef] [PubMed]
63. Wang, X.; Guo, W.; Qiu, B.; Liu, Y.; Sun, J.; Ding, A. Quantifying the Contribution of Land Use Change to Surface Temperature in the Lower Reaches of the Yangtze River. *Atmos. Chem. Phys.* **2017**, *17*, 4989–4996. [CrossRef]
64. Gou, R.; Zhao, J. Eco-Environmental Quality Monitoring in Beijing, China, Using an RSEI-Based Approach Combined with Random Forest Algorithms. *IEEE Access* **2020**, *8*, 196657–196666. [CrossRef]
65. Tang, H.; Fang, J.; Xie, R.; Ji, X.; Li, D.; Yuan, J. Impact of Land Cover Change on a Typical Mining Region and Its Ecological Environment Quality Evaluation Using Remote Sensing Based Ecological Index (RSEI). *Sustainability* **2022**, *14*, 12694. [CrossRef]
66. Su, X.; Wu, J.; Li, P.; Li, R.; Cheng, P. RSEI-Based Modeling of Ecological Security and Its Spatial Impacts on Soil Quality: A Case Study of Dayu, China. *Sustainability* **2022**, *14*, 4428. [CrossRef]
67. Maghsoudlou, A.; Momtazi, F.; Hashtroudi, M.S. Ecological Quality Status (EcoQs) of Chabahar Sub-Tropical Bay Based on Multimetric Macrobenthos-Indexes Approach: Response of Bio-Indexes to Sediment Structural/Pollutant Variables. *Reg. Stud. Mar. Sci.* **2020**, *40*, 101524. [CrossRef]
68. Machado, P.M.; Suciu, M.C.; Costa, L.L.; Tavares, D.C.; Zalmon, I.R. Tourism Impacts on Benthic Communities of Sandy Beaches. *Mar. Ecol.* **2017**, *38*, e12440. [CrossRef]
69. Fanini, L.; Defeo, O.; Elliott, M. Advances in Sandy Beach Research–Local and Global Perspectives. *Estuar. Coast. Shelf Sci.* **2020**, *234*, 106646. [CrossRef]
70. Nagendran, R. Agricultural Waste and Pollution. In *Waste*; Elsevier: Amsterdam, The Netherlands, 2011; pp. 341–355, ISBN 978-0-12-381475-3.
71. Llanos, E.N.; Jaubet, M.L.; Elías, R. The Intertidal Benthic Community of SW Atlantic as an Environmental Indicator of 50 Years of Human-Mediated Changes. *Mar. Biol. Res.* **2019**, *15*, 458–475. [CrossRef]
72. Augusto, M.; Abude, R.R.S.; Cardoso, R.S.; Cabrini, T.M.B. Local Urbanization Impacts Sandy Beach Macrofauna Communities over Time. *Front. Mar. Sci.* **2023**, *10*, 1158413. [CrossRef]
73. Dissanayake, N.G.; Frid, C.L.J.; Caswell, B.A. Biodiversity, Trait Composition and Ecological Functioning: Impacts of Coastal Urbanisation on Subtropical Mudflats. *Mar. Freshwater Res.* **2020**, *71*, 1043. [CrossRef]
74. Shi, F.; Li, M. Assessing Land Cover and Ecological Quality Changes under the New-Type Urbanization from Multi-Source Remote Sensing. *Sustainability* **2021**, *13*, 11979. [CrossRef]
75. Zheng, Z.; Wu, Z.; Chen, Y.; Guo, C.; Marinello, F. Instability of Remote Sensing Based Ecological Index (RSEI) and Its Improvement for Time Series Analysis. *Sci. Total Environ.* **2022**, *814*, 152595. [CrossRef] [PubMed]

MDPI AG
Grosspeteranlage 5
4052 Basel
Switzerland
Tel.: +41 61 683 77 34

Land Editorial Office
E-mail: land@mdpi.com
www.mdpi.com/journal/land

www.ingramcontent.com/pod-product-compliance
Lightning Source LLC
Chambersburg PA
CBHW040746250926
51850CB00007BB/1420
* 9 7 8 3 7 2 5 8 6 0 3 9 5 *